Wind-Waves in Oceans

Springer
Berlin
Heidelberg
New York
Hong Kong
London
Milan
Paris
Tokyo

Physics and Astronomy ONLINE LIBRARY

http://www.springer.de/phys/

Physics of Earth and Space Environments

http://www.springer.de/phys/books/ese/

The series *Physics of Earth and Space Environments* is devoted to monograph texts dealing with all aspects of atmospheric, hydrospheric and space science research and advanced teaching. The presentations will be both qualitative as well as quantitative, with strong emphasis on the underlying (geo)physical sciences. Of particular interest are

- contributions which relate fundamental research in the aforementioned fields to present and developing environmental issues viewed broadly

- concise accounts of newly emerging important topics that are embedded in a broader framework in order to provide quick but readable access of new material to a larger audience

The books forming this collection will be of importance for graduate students and active researchers alike.

Series Editors:

Professor Dr. Rodolfo Guzzi
Responsabile di Scienze della Terra
Head of Earth Sciences
Via di Villa Grazioli, 23
00198 Roma, Italy

Professor Dr. Dieter Imboden
ETH Zürich
Hönggerberg
8093 Zürich, Switzerland

Dr. Louis J. Lanzerotti
Bell Laboratories, Lucent Technologies
700 Mountain Avenue
Murray Hill, NJ 07974, USA

Professor Dr. Ulrich Platt
Ruprecht-Karls-Universität Heidelberg
Institut für Umweltphysik
Im Neuenheimer Feld 366
69120 Heidelberg, Germany

Igor V. Lavrenov

Wind-Waves in Oceans

Dynamics and Numerical Simulations

With 121 Figures

 Springer

Professor Dr. Igor V. Lavrenov
Arctic and Antartic Research Institute
38 Bering Str.
199397 St. Petersburg
Russia

Library of Congress Cataloging-in-Publication Data:
Lavrenov, I.V. (Igor' Viktorovich)
Wind-waves in oceans: dynamics and numerical simulations/I. V. Lavrenov. p.cm.– (Physics of earth and space environments, ISSN 1610-1677)
Includes bibliographical references and index.
ISBN 3-540-44051-1 (alk. paper)
1. Wind waves–Mathematical models. I. Title. II. Series.
QC880.4.W3 L385 2003   551.47'022'015118–dc21   2002191127

ISBN 3-540-44015-1 Springer-Verlag Berlin Heidelberg New York

This work is subject to copyright. All rights are reserved, whether the whole or part of the material is concerned, specifically the rights of translation, reprinting, reuse of illustrations, recitation, broadcasting, reproduction on microfilm or in any other way, and storage in data banks. Duplication of this publication or parts thereof is permitted only under the provisions of the German Copyright Law of September 9, 1965, in its current version, and permission for use must always be obtained from Springer-Verlag. Violations are liable for prosecution under the German Copyright Law.

Springer-Verlag Berlin Heidelberg New York
a member of BertelsmannSpringer Science+Business Media GmbH

http://www.springer.de

© Springer-Verlag Berlin Heidelberg 2003
Printed in Germany

The use of general descriptive names, registered names, trademarks, etc. in this publication does not imply, even in the absence of a specific statement, that such names are exempt from the relevant protective laws and regulations and therefore free for general use.

Typesetting by the author
Data conversion by LE-TEX, Leipzig
Cover design: Erich Kirchner, Heidelberg

Printed on acid-free paper   SPIN: 10857378   54/3141/tr - 5 4 3 2 1 0

To my teachers

Preface

The study of sea waves has always been in the focus of mankind's attention. This is attributed not only to a desire to understand the behaviour in seas and oceans, but also, it has some practical necessity. Developing up-to-date wind wave numerical methods requires detailed mathematical modelling, starting with wave generation, development, propagation and transformation on the surface in different water areas under quasi-stationary conditions, up to a synthesis of climatic features observed under different wave generation conditions in oceans, sea or coastal areas.

The present monograph considers wind waves in terms of the most general formulation of the problem as a probable hydrodynamic process with wide spatial variability. It ranges between the global scale of the oceans, whose typical size is comparable with the Earth's radius, to the regional and local scales of the seas, including water areas limited in space with significant current or depth gradients in coastal zones, where waves cease their existence having propagated tens of thousand miles.

The importance of mathematical wind wave modelling as a random hydrodynamic process with a wide range of spatial–temporal variability at different scales is due to the stricter requirements for detailing, completeness and reliability of wave characteristic data. This is necessary for improving wave forecasting, elaborating principally new ocean diagnostic methods and means, and expanding areas of resource exploration of the world's oceans and their shelves. It is also important for an increasing number of tasks concerning special hydrometeorological (weather) services for marine and oceanic operations.

The essence of mathematical wind wave modelling in the ocean in conditions of spatial non-uniformity, taking into account currents, uneven bottoms and ice influencing the generation, evolution and propagation of wind waves, is presented in this monograph. Considerable attention has been given to this problem during the last few years, but the methodical presentation of the modern theory is undertaken here for the first time.

The monograph begins with the most general problem formulation of the mathematical wind wave modeling of the world's oceans. Some new approaches and optimal numerical methods of the solution of the problem are presented. The Earth's spherical effect on wave propagation is considered for global distances in the ocean. The main attention is given to the problem of wave interaction with the bottom, non-uniform currents and ice cover. The effects of spectrum evolution, wave reflection and blocking are considered at a current. The monograph gives a spectral model description of "abnormal waves", which are very dangerous to ship navigation. Practical recommendations are given, for the first time here, in order to estimate the wave elements in a non-uniform current.

The monograph can be used by specialists from many disciplines: oceanographers, meteorologists, hydraulic engineers and others, whose activity is connected with the ocean. It may also be used as an additional textbook for students in higher educational establishments for its methodical presentation of problems concerning physics and mathematical wind wave modelling.

Although the author's results serve as a basis for the monograph, I am sincerely grateful to persons who participated in scientific discussions, assisted in solving theoretical problems and carrying out field measurements at sea. They are V. Dymov, T. Pasechnik and others (Arctic and Antarctic Research Institute, St. Petersburg); I. Davidan, Ye. Gutshabash, V. Ryvkin, V. Rozhkov, L. Lopatukhin, B. Shatov and others (Branch of the State Oceanographic Institute in St. Petersburg); Yu. Abuzyarov and V. Ryabinin (Hydrometeorological Centre, Moscow); V. Zakharov, M. Zaslavskii and V. Krasitskii (Institute of Oceanology, the Russian Academy of Sciences); G. Matushevskiy and V. Polnikov (State Oceanographic Institute, Moscow); I. Kantarzhi (STANKIN); V. Korobov (NORDECO); S. Strekalov (Hydroproject, Moscow), E. Pelinovskiy (Institute of Applied Physics, Nizniy Novgorod); V. Kudriavtcev (Marine Hydrophysical Institute, Ukraine); V. Makin, G. Komen, and J. Onvlee (KNMI, Netherlands); F.J. Ocampo-Torres (CICESE, Mexico); D. Resio (ERDC, US); L. Cavaleri (ISDGM, Italy); H. Krogstad (NTNU, Norway); and G. Athanassoulis (NTUA, Greece).

The author is also grateful to N. Yakovleva for her assistance in the design of the monograph; and S. Kozhevnikov and M. Arsenchuk for helping in translation.

Acknowledgements. The monograph includes the results of some scientific programmes concerning fundamental studies and data retrieval undertaken at the Arctic and Antarctic Research Institute in 1994–2001 with support of the Russian Foundation for Basic Research (RFBR) – 01-05-64846 and by the INTAS Projects:99-666; 01-0025; 01-0234; 01-2156.

St. Petersburg *Igor V. Lavrenov*
November 2002

Contents

Introduction 1

1 **General Problem Formulation of Wind Wave Modelling in a Non-Uniform Ocean** 11
  1.1 Hydrodynamic Problem of Surface Wave Generation by Air Flow 11
  1.2 Geometric Optics Approximation 14
  1.3 Wave Action Conservation Law 18
  1.4 Statistical Wind Wave Description 23
  1.5 Kinetic Equation of the Evolution of the Wind Wave Spectrum 25
  1.6 General Problem Formulation for Determining the Wave Action Spectral Density in the Ocean 28
  1.7 Spatial–Temporal Scale Considerations in the Analysis of Problem Solution 32

2 **Mathematical Simulation of Wave Propagation at Global Distances** 35
  2.1 Problem Formulation of Ocean Wind Wave Modelling Using Spherical Variables 35
  2.2 Correspondence of Local and Global Coordinate Systems 38
  2.3 Numerical Simulation of Swell Propagation Using the Method of Characteristics 39
  2.4 The Influence of Current on the Evolution of Waves at the Global Scale 44

3 **Numerical Implementation of the Wave Energy Balance Equation** 49
  3.1 Statement of the Problem 49
  3.2 Formulation of the Wave Propagation Problem 51
  3.3 Standard Numerical Propagation Schemes 53
  3.4 Elimination of the "Garden Sprinkler Effect" 54
  3.5 Interpolation-Ray (INTERPOL) Method 59
  3.6 Comparison of Results of Numerical Wave Propagation Schemes 60

	3.7 Numerical Integration of the Source Function	65
	3.8 Conclusions	78

4 Study of Physical Mechanisms Forming the Wind Wave Energy Spectrum in Deep Water 81

4.1 Non-Linear Energy Transfer in Wind Wave Spectrum 81
    4.1.1 Problem Review 81
    4.1.2 Optimal Algorithm of Computation of Non-linear Energy Transfer 85
    4.1.3 Calculation Results of Non-linear Energy Transfer in Wind Wave Spectrum 92
    4.1.4 Numerical Study of Non-Stationary Solution of the Hasselmann Equation 105
4.2 Wind Wave Energy Input 123
4.3 Wave Energy Dissipation in Deep Water 127
4.4 Influence of Mesoscale Effects on Wind Wave Evolution 136

5 Wave Evolution in Non-uniform Currents in Deep Water . 153
5.1 Formulation of the Problem in the Local Coordinate System . 153
5.2 Frequency-Angular Spectrum Evolution in a Current 155
5.3 Spectral Model of Rips 164
5.4 Estimation of Non-Linear Wave Interaction in the Rip Spectrum 175
5.5 Wave Element Transformations in Current, Varying Along its Direction 183
5.6 Wave Diffraction Around a Caustic 190
5.7 Wind Wave Transformation in Cross-Velocity Shear Current . 198
5.8 Freak Wave Problem 217
5.9 Influence of Vertical Non-Uniform Current on Wind Wave Transformation 225
5.10 Wind Wave Generation in a Current 237
5.11 Some Practical Recommendations for Estimation of Current Influence on Waves 247

6 Wave Transformation in Shallow Water 253
6.1 Spectrum Transformation Due to Wave Refraction in Shallow Water 253
6.2 Weak Non-Linear Wave Interaction in Shallow Water 261
6.3 Joint Influence of Bottom and a Non-Uniform Current on Wave Transformation 268
6.4 Wave Energy Dissipation in Shallow Water 285
6.5 Numerical Model of Wind Wave Transformation in a Coastal Area 293

	6.6	Wind Wave Spectrum Transformation According to the Black Sea International Experiment 302
7	**Wave Transformation in Ice-Covered Water** 315	
	7.1	Wave Problem Formulation in Water with Ice Cakes 315
	7.2	Wave Spectrum Evolution in Water with Ice Cakes 317
	7.3	Kinetic Equation for Non-linear Waves in Sea with Ice Cakes 324
	7.4	Estimation of Non-Linear Energy Transfer in the Wave Spectrum in Water with Ice Cakes............. 332
	7.5	Numerical Simulation of Surface Gravity Wave Interaction with Elastic Ice Floes 335
8	**Conclusion** 347	

Bibliography 351

Index 371

Introduction

Surface gravity waves have always caused great interest, as they are an example of a well-known, but still very complicated phenomenon. They can easily be observed, but it is rather difficult to describe them mathematically. Sir Horace Lamb noted in his letter to the London Mathematical Society in 1904 that waves were almost the first hydrodynamic problem studied systematically on the basis of general equations. It gave an example of activity that provided efficient evidence of new and unusual analytical methods. Wave observations on the sea surface and their connection with wind have been performed from time immemorial. It was quite natural that the pioneers of theoretical hydrodynamics (J. Lagrange, G. Eiry, G. Stokes and J. Rayleigh) tried to find an explanation of the elementary properties of surface waves using the assumption of ideal fluid motion.

Many papers, some of which later became classics, were devoted to solving the problem of surface gravity waves. The well-known Cauchy–Poisson problem of wave generation caused by an irregular external force in the free surface of the fluid should be mentioned (Sretenskiy, 1977). Many outstanding researchers (J. Bussinesk, T. Levi-Civita, G. Stoker, M.S. Longuet-Higgins, A.I. Nekrasov, S.L. Chaplygin, L.N. Sretensky, N.Ye. Kochin, S.S. Voight, Ya.N. Sekerzh-Zenkovich, M.A. Lavrentyev, Yu.Z. Aleshkov, V. Zakharov, E. Pelinovskiy and others) should be mentioned in this connection. Their studies made a valuable contribution to understanding fluid wave motion and developed the analytical methods of theoretical studies.

However, real waves on the ocean surface are sufficiently complicated that theory is never adequate to explain actual observations. J. Rayleigh (1876) pointed out that the main law of sea waves was the absence of any law. The irregularity of wind waves made their description difficult.

In 1805, Beaufort was among the first to try to establish a relation between the wind force and sea surface state. Based on many sea surface observations during storms he discovered that typical wave heights and periods appeared at a specific wind speed, i.e. wind of a definite force induced sea waves with typical elements. As a result his famous scale of waves appeared and it is still used in modified form (Abuzyarov, 1981).

The problem of wave height correlation with wind speed was formulated by W. Kelvin (1871), although no real progress was achieved at that time. By 1850, T. Stevenson carried out surface wave observations in lakes and obtained some empirical ratios between the largest wave height and fetch. 75 years later H. Jeffreys (1971) tried to model wave generation by wind under laboratory conditions. And still in 1956 F. Ursell reasonably mentioned that wind blowing over the surface of water generated waves by means of physical processes that could not be considered to be well studied.

At the same time, applied wave prediction problems required the construction of mathematical methods for wind wave estimation. The energy wave balance equation, proposed by V.M. Makkaveyev in 1937, was one of the first attempts to apply a mathematical approximation of the evolution of wind waves in practice. This equation described wind wave energy evolution in its simplest form, depending on the energy supply due to wind and dissipation. The "energy approach" resulted in the creation of elementary practical methods for wave calculation and predication. It gave the possibility of estimating wave elements in deep and shallow water depending on wind speed and direction, duration, fetch and sea depth. Widespread practical methods and wave prediction were proposed by the Russian scientists V.V. Shuleykin (1959), Yu.M. Krylov (1956) for deep sea and A.P. Braslavskiy (1952) for shallow water conditions. The methods of H. Sverdrup and V. Munk (1961), modified by C. Bretschneider (1958), were worked out in Western countries.

Simultaneously with the study of the physical nature of sea waves, their statistical properties were investigated. M.S. Longuet-Higgins (1962), Yu.M. Krylov (1966) and I.S. Brovikov (1954) conducted theoretical studies and I.N. Davidan (1969), Ya.G. Vilenskiy and B.H. Glukhovskiy (1961, 1955) carried out empirical studies on the basis of processing wave records and visual wave observations. It was established that in spite of the variety of individual waves, their statistical characteristics (such as the mean height, mean period, mean length, dispersion, distribution function, correlation function and two-dimensional spectrum) were stable in the quasi-stationary and quasi-uniform process range.

A synthesis of energetic and statistical regularities of wind waves made it possible to extend the calculation of sea wave elements. This was used by G.V. Rzheplinskiy, Yu.M. Krylov and G.V. Matyshevskiy (1969) to create methods of wind wave calculation for complicated wave formation conditions at an uneven sea coastline.

A new important achievement in the theoretical description of wind sea was made in the 1950s. The investigators approached the study of wind waves in terms of spectra using the modern theory of random functions. Wind waves were considered to be a probable random process and were described on the basis of combined application of statistical and hydrodynamic methods. The Fourier representation of a random wave process allowed one to separate it into harmonic elements, whose conduct might be considered in terms of the classical theory of wave motion.

The results of M.S. Longuet-Higgins and co-workers (1957, 1960–1962, 1964), O.M. Phillips (1957, 1958), J. Miles (1957, 1960) and K. Hasselmann (1960, 1962, 1963, 1965, 1966) published after 1956, were the basis of the physics of wind wave development. O.M. Phillips proposed a linear spectral theory of the mechanism of resonance in wave generation due to normal pressure fluctuations of turbulent wind. A model of the influence of a pressure harmonic wave on an ideal incompressible fluid surface was further described by G. Lamb (1947), L.N. Sretenskiy (1977), et al.

The mechanism of wave generation, suggested by J. Miles, was based on the theory of the instability of the air–water interface in the presence of a flow with a velocity gradient in the boundary layer. The classical studies by W. Kelvin (1871) and G. Helmholtz (Le Blond & Maysek, 1981) giving the solution of the unstable interface problem between two fluids with different, but constant densities and motion velocities served as a basis of the theory of J. Miles.

The important part in forming the structure of the wind wave spectrum is explained with the help of wave–wave interactions. Theoretical studies of this problem were initiated by M.S. Longuet-Higgins (Ocean Wave Modeling, 1985) and continued by O.M. Phillips (1960). The non-linear wave interaction for the continuous spectrum case was investigated by K. Hasselmann (1960, 1962, 1963, 1965) and independently by V. Zakharov (1968). As a result of the four-wave resonance interaction, energy redistribution occurred in the wind wave spectrum. The complete wave action, energy and wave momentum were preserved. For a long time, the complicated form of the collision integral did not allow researchers to obtain correct numerical estimations of non-linear energy transfer in the wind wave spectrum.

At that time, O.M. Phillips (1958) showed the existence of an equilibrium or saturating interval in the wind wave spectrum in the high-frequency spectrum range. Governed by intense energy dissipation it reached its upper limit due to wave crest collapse. As it was shown later (Kitaigorodskii et al., 1975) the equilibrium interval of the spatial spectrum was an invariant irrespective of the basin depth or the current.

The results of M.S. Longuet-Higgins also served as a basis for the modern understanding of wave evolution in non-uniform currents and shallow water. In particular he was able to show (Longuet–Higgins, 1957) that the spatial spectrum preserved its value along the trajectory of a wave packet propagating with wave refraction. His studies, performed with R. Stewart, considered the effects of wave interaction with non-uniform currents. This interaction was described by the so-called radiation stresses (Longuet–Higgins and Stewart, 1960–1962, 1964). These ideas were further developed by F.P. Bretherton and C.J.R. Garret (1968) and F.P. Bretherton (1971) showing that the adiabatic invariant-wave action preserved the same value in non-uniform moving media.

In the 1960s, when powerful computers appeared, wind wave numerical modelling began to be developed as a result of integrating the spectral energy balance equation. Elaboration of such models began in Russia (Davidan, 1969; Pasechnik, 1975; Davidan et al., 1975, 1977, 1978; Dzhenyuk, 1976), in France (Gelci et al., 1963; Fons, 1966; Gelci & Devillaz, 1969, 1975), in the

USA (Pierson et al., 1955; Barnett, 1968; Bunting, 1970), in the UK (Pierson et. al., 1966; Darbyshire & Simpson, 1967; Ewing, 1971), in Japan (Isozaki & Uji, 1973, 1974; Uji, 1975) and in certain other countries. Some of these models are used nowadays for wave prediction in practical work.

The mechanisms of wave and wind interaction, dissipation and non-linear energy transfer were not fully investigated. The source functions in the wave energy balance equation in discrete-spectral models were usually represented using certain terms, the other terms being taken into account indirectly by fitting coefficients. This was, for example, a well-known method used by W. Pierson, L. Tick and L. Baer (1966) and its modifications were also used (Inoue, 1966; Darbyshire & Simpson, 1967; Bunting, 1970). The models of T. Barnett (1968), T. Barnett et al. (1969) and J. Ewing (1971) took into account the non-linear mechanism of energy transfer in the wind wave spectrum in the source function. It was parameterised by D. Cartwright and represented as a sum of a Fourier–Chebyshev series. The method of T. Barnett was modified by T. Pasechnik (1975) and used for the first time in Russia for wave calculation in the Atlantic Ocean and the Gulf of Finland.

Alongside with the development of wind wave mathematical models, theoretical and experimental studies continued as a basis for their verification and improvement. Full-scale studies at the St. Petersburg Branch of the State Oceanographic Institute (Davidan et al., 1978), the "SoyuzmorNIIProject" (Krylov et al., 1976), and the Marine Hydrophysical Institute of the Ukrainian Academy of Sciences (Yefimov & Solovyev, 1979) should be mentioned among these experimental investigations.

The JONSWAP international experiment (K. Hasselmann et al., 1973) carried out in 1973 in the North Sea is still attracting great attention. The obtained data allowed researchers to estimate a number of theoretical results. The peculiarities of wave spectrum development were revealed. The contribution of separate constituents to the source function describing different physical mechanisms forming a wind wave spectrum were determined. The use of experimental data and probabilistic methods allowed workers to estimate more reliably the frequency spectrum (Pierson & Moskowitz, 1964; Davidan et al., 1969, 1978; K. Hasselmann et al., 1973) and energy angular distribution (Longuet-Higgins et al., 1963; Krylov et al., 1976; Davidan et al., 1978; Yefimov & Solovyev, 1979; D. Hasselmann et al., 1980).

From the mid-1970s, intense development of the modern spectral theory of wind waves began. The theory of wind wave energy input was analysed in detail and partially supplemented in numerous publications (Snyder & Cox, 1966; Gent & Taylor, 1976; Zaslavskii & Krasitskii, 1976; Yefimov, 1981; Makin & Chalikov, 1986). In recent years some publications were devoted to the investigation of the interaction process between air flow and wave in order to determine the energy flux from wind to waves. The most important results were obtained by elaborating the model of the boundary layer above finite-amplitude waves (Chalikov, 1986; Makin 1987, 1989).

Studies of non-linear wave interaction in a spectrum were made in the following directions: analytical estimation and parameterisation of the collision integral; elaboration of numerical algorithms and calculation programs; and solution of the non-linear evolution of the spectrum. It was interesting from the theoretical point of view to find the stationary forms of the spectra that made the collision integral equal to zero. K. Hasselmann (1965) showed that the Rayleigh–Jeans distribution was isotropic, transforming to zero not only the collision integral, but also the integrand. Other stationary distributions were obtained (Zakharov & Smilga, 1981; Zakharov & Zaslavskii 1982, 1983a,b). Attempts to obtain non-stationary analytical solutions were also undertaken (Zaslavskii, 1989a,b; 1997, 2000).

In view of the importance of the non-linear energy redistribution mechanism in the wave spectrum and the need to take it into account in wind wave models for operational use, attempts were undertaken to approximate the collision integral by simpler analytical ratios than the accurate integral expression. For example, first attempts at approximating the collision integral can be found in the models of Barnett (1968); Ewing (1971); Theoretical Bases and Methods for Wind Sea Calculation (1988); S. Hasselmann and K. Hasselmann (1981). Among the most accurate approximations the results of Hasselmann et al. (1985); Polnikov (1988); and Zakharov & Pushkarev (1999), should be mentioned. They give a description of the non-linear energy exchange between wind waves and swell.

Among various results in determining the wave spectrum transformation in a non-uniform medium, representing a horizontal-non-uniform current and an uneven seabed, the best were obtained in the framework of the geometrical optics approximation. Wave element evolution, reflection and blocking in a current were investigated (Basovich & Talanov, 1977; Basovich et al., 1982; Hurghes & Grant, 1978; Theoretical Bases and Methods for Wind Sea Calculation, 1988; Phillips, 1980; Pokazeyev & Rosenberg, 1983). A review of these studies was made by D. Peregrine (1976), G. Kantargi, N. Tsivtsivadze and Kh. Akmuratov (1984). Further research in this field was continued by Dreyzis et al. (1986); Lavrenov (1984, 1985b, 1986, 1988a,b, 1989a,b, 1991a,b); Lavrenov & Ryvkin (1986, 1990); and Lavrenov et al. (1992), where the frequency-angular spectrum evolution at horizontally non-uniform non-stationary current at a non-uniform depths was described in the geometrical optics approximation.

Wind wave and ice interaction is one of the problems discussed in this monograph. It is rather complicated to describe theoretically wind waves in water partially covered by ice, as the dynamic processes in the "water–air–ice" three-phase medium in the case of the presence of ice should be taken into account. Both random wave movements and the probabilistic distribution of the characteristics of the ice cover make the solution rather complicated. The theory of this problem remains less researched, although nowadays some intensive studies have been done in this field (Bukatov & Bukatova, 1993; Lavrenov & Novakov, 2000; Liu et al., 1991; Marchenko, 1988; Masson &

LeBlond, 1989; Meylan & Squire, 1994; Shuchman & Rufenach, 1994; Wadhams et al., 1986). Analysing the modern state of the theory, three principal research trends can be pointed out: research into water surface gravity waves in the presence of ice cake (Bukatov & Bukatova, 1993; Lavrenov 1998a, Kheysin, 1967; Zubakin, 1976; Smirnov, 1996); investigation of surface gravity and internal wave propagation under an elastic layer simulating an ice sheet (Bukatov & Cherkesov, 1971; Marchenko, 1988); and study of wave diffraction caused by an ice floe (Kheysin, 1967; Mylan & Squire, 1994). In most cases these problems are solved with the help of the deterministic approximation. An attempt to solve the problem of wave development caused by wind action, dissipation due to wave breaking, non-linear energy transfer in the wave spectrum and wave scattering by ice floes without taking into consideration their elastic properties was undertaken by Masson & LeBlond (1989). The approach, elaborated in this monograph, allows the description of wind wave spectrum evolution in a spatial non-uniform medium using the geometrical optics approximation theory. It may also be applied to wave transformation in water with spatially non-uniform ice fields.

In the past few decades, many mathematical wind wave models have been developed. According to data from the World Meteorological Organisation (WMO) there were more than 20 models in 1985 without taking into account the models in the process of development (Marine Meteorology and Related Oceanographic Activities, 1986). The Meteorological Services in the USA, Great Britain, Norway, Germany, Japan and other countries used 11 models in their forecasting practice. In 1988 the National Services already used 16 models in their operational practice, and another 14 models were reported to be used in practical activities (Guide to Wave Analysis and Forecasting, 1988). In 1991, according to WMO information, the National Meteorological Services in 17 countries, including France, Germany, Greece, Hong Kong, India, Ireland, Japan, Malaysia, Holland, New Zealand, Norway, Saudi Arabia, Sweden, Great Britain, the USA and Russia used 22 models for official wind wave prediction. The total number of models used for operational purposes in different countries was more than 40 (WMO Wave Programme, 1991).

In Russia, the use of numerical modelling of wind waves began after some delay in comparison with Western countries. This was connected with a lagging behind in computer manufacturing. However, at present there are more than 20 wind wave models (Abuzyarov, 1981; Theoretical Bases and Methods for Wind Sea Calculation, 1988; Zaslavskii, 1989b; Catalogue of Applied Computer Programs for Shore Protection, 1989; Matushevskiy & Kabatchenko, 1989, 1991; Yefimov & Polnikov, 1991). These models are significantly different in their objectives, types, characteristics, verification using natural observations, etc.

In the early 1990s the variety of models increased. There were models describing wind waves not only on a regional, but also on a global scale (The WAM Model – A Third Generation Ocean Wave Prediction Model, 1988; Lavrenov & Pasechnik, 1989; Lavrenov & Ryvkin, 1989; Tolman, 1991;

Komen et al., 1994). There were more models taking into account the effect of an uneven seabed (Hasselmann & Collins, 1968; Collins, 1972; SWIM Group, 1985; Gutshabash & Lavrenov, 1987, 1988; Khandekar, 1989; Tolman, 1991) and spatially non-uniform currents (Hurghes & Grant, 1978; Kato & Sato, 1978; Sakai & Iwagaki, 1983; Booij et al., 1984; Gutshabash & Lavrenov, 1986; Holthuijsen et al., 1989; Khandekar, 1989). However, it should be noted that there were considerable discrepancies even between the results of the most advanced models.

This variety of mathematical models indicates that wind waves remain a very complicated object for investigation, and its theory is far from being completed. It is difficult to create one universal model optimal for all cases. That is why different approaches and methods have a right to exist. On the other hand, a large variety of models help us to better understand wind waves and develop model implementations using modern computers.

In 1985, due to a large number of different models, the SWAMP International Working Group (Sea Wave Modelling Project) proposed its own classification having made a comparison between the earlier developed models (Ocean Wave Modelling, SWAMP group, 1985). This classification was based on differentiation into types of source function presentation in the spectral energy density equation and, mainly, on the accuracy of the non-linear energy transfer estimation in the wind wave spectrum.

The participants of this group decided to initiate the development of a more advanced wind wave model. Thus, a new International Working Group WAMDI (Wave Modelling Group) was created with G. Komen, L. Cavaleri, M. Donelan, K. Hasselmann, S. Hasselmann, P. Janssen, etc. The first paper was published in 1988 (The WAM Model – A Third Generation Ocean Wave Prediction Model, 1988). The decision to combine the efforts of scientists from different countries for developing a wind wave model revealed, on the one hand, the importance of the problem and, on the other hand, its complexity.

It was decided to develop a new model based on the following principles. Firstly, it was necessary to use the most accurate approximation of the integral of non-linear energy transfer preserving the same cubic operator structure as the original expression (S. Hasselmann & K. Hasselmann, 1981). The proper formula was named a "discrete interaction approximation" (Komen et al., 1994). Secondly, the source function was to be supplemented with a more precise dissipation mechanism. The dissipation function (Komen et al., 1984) was obtained from a series of numerical computations of the energy balance equation with accurate collision integral estimation. The wind wave energy input mechanism was taken according to experimental data (Snyder et al., 1981). The mathematical problem was solved in spherical variables, allowing the use of this model as a global one. In addition, there was an attempt to generalise the model for the shallow water case. The model took into account refraction and bottom friction. The non-linear energy transfer function took into account the corresponding correction describing non-linear wave interaction in a finite depth basin. It was classified as a third-generation wind wave model.

This description was similar to the SWAMP group terminology. It meant that the first-generation models did not take into account non-linear wave interaction; the second-generation models included simple approximation of non-linear energy transfer. It should be noted that non-linear transfer was approximated in the WAM model, thus helping select the model into the more accurate next class. In the fourth-generation models, an attempt was made to estimate the self-consistent motion of the atmospheric boundary layer and rough sea surface (Komen et al., 1994). At present, the WAM model is being improved, tested and widely used both for global scale and local water areas. The operational WAM model variant assimilates satellite information for updating the wave forecast.

The publication of the monograph *Dynamics and Modelling of Ocean Waves* (Komen et al., 1994) by the WAMDI international working group was an important event. The monograph generalised theoretical and experimental wind wave investigations carried out in Western countries. Practically at the same time (at the beginning of 1995) the monograph of Russian researchers *Problems of Research and Mathematical Modelling of Wind Waves* (1995) was published. It presented some new results of wind wave research obtained in Russia and formulated unsolved problems. Thus, theoretical and experimental research of wind waves was summarised.

When the WAMDI group's work was over, a new international project WISE (Waves In Shallow Environments) appeared on the initiative of L. Holthuijsen, L. Cavaleri, et al. Its aim was to continue researching and developing a more advanced wind wave model for shallow sea areas. This model was created and named SWAN (Simulating Waves Nearshore) published by Ris, (1997), Booij et al., (1999). This was a third-generation model. In addition to parameterisation of the physical mechanisms forming a wave spectrum in deep water, it also included the effects of refraction, three-wave interactions and wave energy dissipation connected with waves breaking in shallow water.

It should be noted that the theory and methods of numerical modelling are being constantly improved: there are new papers, results (6th International Workshop on Wave Hindcasting and Forecasting, 2000) and models (WAVEWATCH, Tolman, 1991; PHIDIAS, Van Vledder et. al., 1994; TOMAWAC, Benoit et al., 1996). It should be noted that a new, so-called Narrow Angle Approximation Model (NAAM), has been developed recently in Russia (Zaslavskii, 2000), based on a precise estimation of non-linear energy transfer in the wind wave spectrum with narrow angular distribution (Zaslavskii, 1989b).

There is a new monograph by S. Massel (1996), in which an attempt is made to generalise the results of wind wave studies not only in Western countries, but also in Russia. It is a pity that only a restricted review of Russian publications without recent results was given in it.

This monograph is a logical continuation of the above-mentioned papers. An attempt to answer a number of questions, concerning a complex approach towards wind wave numerical simulation in the world's oceans under the condition of its spatial non-uniformities is undertaken. It should be pointed out that the non-uniformities are understood as the effect of currents, non-uniform ocean depths, ice, the influence of the Earth's sphericity, etc. on wind waves. This monograph considers wind waves in the framework of one general problem formulation as a probabilistic hydrodynamic process. Its spatial variability ranges between global scales of the oceans comparable to the Earth's radius, to regional sea and local coastal water area scales, where oceanic waves terminate their existence having travelled thousands of miles.

1 General Problem Formulation of Wind Wave Modelling in a Non-Uniform Ocean

1.1 Hydrodynamic Problem of Surface Wave Generation by Air Flow

The evolution of wind waves should be considered as the self-consistent motion of a two-layer water–air system under the corresponding dynamic and kinematic conditions in the two-media interface. The media motion is assumed to be determined by the laws of mass and momentum conservation. The former (i.e., the law of mass conservation) is written as:

$$\frac{d\rho_i}{dt} + \rho_i \operatorname{div}(\boldsymbol{U}_i) = 0 , \tag{1.1}$$

where ρ_i is the air ($i = 1$) or water ($i = 2$) density, respectively; $\boldsymbol{U}_i$ is the velocity of the medium motion.

If the fluid density remains constant, (1.1) is simplified to:

$$\operatorname{div}(\boldsymbol{U}_i) = 0 . \tag{1.2}$$

The equation of momentum conservation, referred to axes immovably connected with the rotating Earth, has the form:

$$\rho_i \frac{d\boldsymbol{U}_i}{dt} + \rho_i[\boldsymbol{\Omega U}_i] + \operatorname{grad}(P_i) - \rho_i \boldsymbol{g} = \boldsymbol{F}_i . \tag{1.3}$$

The first term is the inertia force of the mass acceleration; the second one, containing the rotation vector $\boldsymbol{\Omega}$ or the double angular velocity of the Earth's rotation, is the Coriolis force.

The absolute value of this vector is $|\boldsymbol{\Omega}| = 2\pi/12h = 1.46 \times 10^{-4}\,\text{s}^{-1}$, where h is the time in hours. In the term describing gravity force effects, the vector $\boldsymbol{g} = \{0, 0, -g\}$ characterises the acceleration due to gravity as $g = 9.81\,\text{m s}^{-2}$. The direction of the vector $\boldsymbol{g}$ determines a local vertical line.

The term $\boldsymbol{F}_i$ in the right hand-side of (1.3) is the resultant of all forces acting on a unit fluid volume. In almost all cases when viscosity effects are significant, water may be considered as an isotropic incompressible fluid and the stress tensor may be written in the form:

$$P_{ij} = -p\delta_{ij} + 2\mu e_{ij} , \tag{1.4}$$

where δ_{ij} is a single tensor ($\delta_{ij} = 1$ for $i = j$, otherwise $\delta_{ij} = 0$); μ is the viscous fluid coefficient, and

$$e_{ij} = \frac{1}{2}\left(\frac{\partial U_i}{\partial x_j} + \frac{\partial U_j}{\partial x_i}\right), \qquad (1.5)$$

where e_{ij} is the deformation velocity tensor. Consequently, if the incompressibility condition (1.2) is satisfied, the friction force per unit volume is equal to:

$$F_{ij} = 2\mu \frac{\partial e_{ij}}{\partial x_{ij}} = \mu \frac{\partial U_{ij}}{\partial x_{ij}}. \qquad (1.6)$$

Now the two-layer model with discontinuities in the density ρ and the kinematic coefficient of viscosity μ in the mobile interface surface $\eta(r,t)$ is considered as:

$$\rho = \begin{cases} \rho_a = 1.2 \times 10^{-3}\,\text{g s m}^{-3} \\ \rho_w = 1.0\,\text{g s m}^{-3} \end{cases} \quad \mu = \begin{cases} \mu_a = 1.5 \times 10^{-1}\,\text{s m}^2\text{s}^{-1} \text{ with } z > \eta \\ \mu_w = 1.0 \times 10^{-2}\,\text{s m}^2\text{s}^{-1} \text{ with } z < \eta, \end{cases} \qquad (1.7)$$

where z is the height. The lower fluid is assumed to be immovable at the initial time moment:

$$U(r,z,t=0) = 0, \quad \eta(r,t=0) = 0. \qquad (1.8)$$

The Cartesian coordinate system $\{r,t\}$ is chosen in such a way that the axis $z = x_3$ is directed vertically upwards. The plane $z = 0$ coincides with the undisturbed interface surface ($r = \{x,y\}$).

Due to significant differences in the values ρ_a, μ_a and ρ_w, μ_w, natural simplifications in the general equations (1.1)–(1.3) at $z > \eta$ and $z < \eta$ are different. When $\mu_w \rho_w / \mu_a \rho_a > 100$, it is assumed that the air flow at the initial stage of its development is the same as the usual turbulent boundary layer above a rigid wall. Further, the usual assumptions for the theory of a logarithmic boundary layer on a wall should be considered to be fulfilled, so that the friction velocity U_* can be assigned at a distance from the mobile underlying surface.

The situation with the lower fluid (water) is quite different. Due to the significant difference in the dynamic viscosity coefficients of water and air, momentum transfer by viscous stresses through the interface η are inefficient.

The velocity field should be written in the form $U = \text{grad}(\phi) + V$, where ϕ is the velocity potential and $V = \text{rot}(A)$ is its solenoidal (vorticity) component $[\text{rot}(U) = -\Delta(A)]$.

Then $\text{div}(U) = \Delta(\phi) = 0$ and $\Delta(U) = \Delta(V)$, i.e. a viscous force is determined only by the vorticity component. It is typically important only within thin boundary layers above the water surface and near the bottom. It can be taken into account only with some small corrections to the potential

1.1 Hydrodynamic Problem of Surface Wave Generation by Air Flow

approximation $U = \mathrm{grad}(\phi)$. In this approximation the water motion can be considered as potential and the dynamic equations for $z < \eta$ are as follows:

$$\frac{\partial \phi}{\partial t} + \frac{P}{\rho} + gz + \frac{1}{2}\left[(\nabla\phi)^2 + \left(\frac{\partial\phi}{\partial z}\right)^2\right] = 0 \ ; \qquad (1.9)$$

$$\left(\Delta + \frac{\partial^2}{\partial z^2}\right)\phi = 0 \ , \qquad (1.10)$$

where ∇ and Δ are the horizontal differential operators.

The velocity potential ϕ in (1.10) is determined by the solution of the Laplace equation (1.10) with boundary conditions at the free surface $z = \eta(x,y,t)$:

$$\frac{\partial \phi}{\partial n} = \frac{\partial \eta}{\partial t}[1+(\nabla\eta)^2]^{-1/2} \qquad (1.11)$$

and at the bottom at $z = H(x,y)$

$$\frac{\partial \phi}{\partial n} = 0 \ , \qquad (1.12)$$

where $\partial\phi/\partial n$ is the normal derivative to the surface η or to the bottom H, respectively.

However, it should be noted that the potential approximation is relatively rough to describe wind wave evolution. Unlike the water motion description, the viscous terms and flow vorticity are essential in the equations of motion of the atmospheric boundary layer. In this case, the initial equation (1.3) is solved, neglecting the Coriolis force for the boundary layer problem. The air flow velocity U is represented in the form of three items:

$$U = U_1 + U_2 + U_3 \ ,$$

where U_1 is the averaged flow velocity; U_2 is the deviation from U_1, created by waves at the water surface; U_3 is the random turbulent velocity fluctuation determined by the closing equations (Phillips, 1980).

The problem of self-consistent motion of the water–air, (i.e. the two-layer medium) is solved using a kinematic marginal condition and the condition of normal stress continuity at the interface $z = \eta$

$$U_\mathrm{a} = U_\mathrm{w} = \frac{\partial \eta}{\partial t}[1+(\nabla\eta)^2]^{-1/2} \ , \qquad (1.13)$$

$$P_\mathrm{a} = P_\mathrm{w} = -\gamma\rho\{\nabla\eta[1+(\nabla\eta)^2]^{-1/2}\} \ , \qquad (1.14)$$

where $\gamma \sim 10\,\mathrm{cm}^3\mathrm{s}^{-2}$ is the coefficient of surface tension at the water–air interface normalised by ρ. The value P_a (at $z = \eta$) should be determined in (1.14) using the solution of equations for random hydrodynamic fields U_a

and P_a of the boundary layer in the atmosphere. But the pressure P_w (at $z = \eta$) can be directly expressed by the potential velocity derivatives (1.9).

The complete system of equations (1.3), (1.9)–(1.14) for determining the surface evolution η with the initial conditions (1.8) presents considerable difficulty for analysis. Unlike in the usual classic theory of potential waves with the given pressure distribution P_a at the surface η, either the surface itself or the pressure are not determined in wind wave theory. These two unknown functions are not independent, so a joint solution of both equations (1.9)–(1.12) for wave disturbances at $z < \eta$ and complicated equations of vortex current above the wave boundary at $z > \eta$ are required.

1.2 Geometric Optics Approximation

The problem of mathematical wind wave modelling is aggravated by the fact that there are different horizontal and vertical non-uniformities significantly influencing the propagation and generation of surface gravity waves in the ocean under natural conditions. The most typical non-uniformities include spatial and temporal variability of current, turbulent motion and uneven bottom relief in oceanic areas. That is why it is interesting to consider the effect of non-uniformity on wave propagation and generation.

This is quite a complicated problem to be solved in the general formulation. First of all, it seems reasonable to consider wind wave propagation, with their length and period of the wave much less than the typical spatial and temporal medium variation scales, i.e. 1–100 km and 1–10 h. If the latter are assumed to be typical for a wide class of oceanic motions, it is possible to consider this problem using the method of geometric optics.

The geometric optics method is based on the assumption of the existence of plane waves. Plane waves are assumed to be described by the same propagation direction, length and amplitude everywhere. Natural waves obviously do not have such properties. But they can often be considered as plane in each local space area. In this case it is necessary for the wave amplitude a, the wave vector $\boldsymbol{k}$ and the frequency ω to remain almost unchanged at a distance of about the wavelength and over the time of about the wave period. Changes of these parameters are connected with variations of the wave propagation media, requiring small changes of these parameters in the media variability scale. Large-scale currents and the uneven bottom are considered to be non-uniform media. Thus, if the typical horizontal scale of the bottom relief variation is M_1, the spatial current scale is M_2 and T is its temporal scale variation, the necessary condition for the applicability of the geometric optics method: conditions:

$$(M_1 k)^{-1} \ll 1, \quad (M_2 k)^{-1} \ll 1, \quad (T\omega)^{-1} \ll 1. \tag{1.15}$$

If these conditions are satisfied, the so-called wave surfaces may be introduced in which the wave phase is the same at every point at a given moment.

The direction of wave propagation can be considered to be normal to the wave surface in each small area.

The notion of rays (i.e. lines with tangents to them coinciding at every point with the direction of wave propagation) is now introduced.[1] In geometric optics a ray is considered as wave propagation, digressing from their wave nature. The geometric optics approximation corresponds to a limited case of the small parameter ε (here $\varepsilon = \max\{(M_1 k)^{-1}, (M_2 k)^{-1}, (T\omega)^{-1}\}$).

The principal equations of geometric optics, describing the ray propagation, are now presented. Let the value $\eta(\boldsymbol{r}, t)$ be the free surface deviation from equilibrium. Thus, the value η for a plane monochromatic wave can be written as:

$$\eta = a \, \mathrm{e}^{\mathrm{i}(\boldsymbol{k}\boldsymbol{r}-\omega t)} = a \, \mathrm{e}^{\mathrm{i}\psi} \,. \tag{1.16}$$

In the case of a wave different from the plane one, but with geometric optics being applicable, the amplitude a is a function of the coordinates and time $a = a(\boldsymbol{r}, t)$. The phase is written in a more complicated form than by (1.16). However, it is essential for the phase to be of sufficiently large value $\psi \gg 1$ due to its change of 2π within the wavelength.

The expression (1.16) describes local sinusoidal waves. In small space areas and a short time interval the phase ψ can be expanded in the series:

$$\psi = \psi_0 + \boldsymbol{r} \frac{\partial \psi}{\partial \boldsymbol{r}} + t \frac{\partial \psi}{\partial t} + \dots \,. \tag{1.17}$$

Thus, the phase ψ is connected with the local wave vector $\boldsymbol{k}$ and the local frequency ω:

$$\boldsymbol{k} = \frac{\partial \psi}{\partial \boldsymbol{r}} = \mathrm{grad}(\psi) \,; \tag{1.18}$$

$$\omega = -\frac{\partial \psi}{\partial t} \,. \tag{1.19}$$

It follows from the ratio (1.18) that:

$$\mathrm{rot}(\boldsymbol{k}) = 0 \,, \tag{1.20}$$

i.e. the field of local wave vectors is non-vortical. This can be obtained from (1.19):

$$\frac{\partial \boldsymbol{k}}{\partial t} + \mathrm{grad}(\omega) = 0 \,, \tag{1.21}$$

which is a kinematic equation of wave density conservation (Phillips, 1980).

---

[1] This determination corresponds to the case of wave propagation in isotropic media (Kravtsov & Orlov, 1980). The surface gravity waves in non-uniform currents are attributed to dispersion waves in non-isotropic media. In this case the ray will be more precisely defined below.

Free waves can exist in the wave medium, not with the arbitrary frequency ω and wave vector $\boldsymbol{k}$, but those meeting some specific conditions. In this case the frequency is a function of the wave vector $\omega = F(\boldsymbol{k})$. The function type is dependent on the wave motion type under consideration and the corresponding force balance. However, in the case of a non-uniform and non-stationary medium the frequency ω depends not only on the wave vector $\boldsymbol{k}$, but also on the coordinate $\boldsymbol{r}$ and time t. In the case of slow medium parameter variation the dispersion ratio is local and can be written in the form (Kravtsov & Orlov, 1980):

$$\omega = F(\boldsymbol{k}, \boldsymbol{r}, t) , \quad \boldsymbol{k} = \boldsymbol{k}(\boldsymbol{r}, t) . \tag{1.22}$$

Using the equations (1.18) and (1.19), the local dispersion ratio can be rewritten in the form:

$$\frac{\partial \psi}{\partial t} + F\left(\frac{\partial \psi}{\partial \boldsymbol{r}}, \boldsymbol{r}, t\right) = 0 . \tag{1.23}$$

However, the phase equation (1.23) is significantly different from the dispersion ratio (1.22) in its content. It presents, not a simple algebraic correlation between the frequency and wave vector ratio, but a differential equation in partial derivatives relative to the unknown function ψ.

The equation (1.23) results in a remarkable analogy between geometric optics and the mechanics of a material particle. The phase equation (1.23) has the form of the Hamilton–Jacobi equation (Landau & Lifshits, 1973). It is solved in mechanics relative to the particle action D, connected with the momentum of the particle $\boldsymbol{P}$ and the Hamiltonian function $\mathcal{H}$:

$$\boldsymbol{P} = \mathrm{grad}(D) , \quad \mathcal{H} = -\frac{\partial D}{\partial t} .$$

By comparing these formulas with the ratios (1.18) and (1.19), it can be seen that the action of the material particle D in mechanics plays the role of the phase ψ in geometric optics. The particle momentum $\boldsymbol{P}$ is analogous to the wave vector $\boldsymbol{k}$, and the Hamiltonian function $\mathcal{H}$ plays the role of the frequency ω (Landau & Lifshits, 1973). Thus, there is an analogy between the behaviour of the material particle and the wave packet (i.e. the wave representing a superposition of monochromatic waves with frequencies within some small range and occupying a finite space area). The particle momentum corresponds to the wave vector, and the particle energy is similar to the wave packet frequency.

The characteristics of the (1.23) are determined by the system of ordinary differential equations:

$$\frac{\mathrm{d}\boldsymbol{r}}{\mathrm{d}t} = \frac{\partial F}{\partial \boldsymbol{k}} ; \quad \frac{\mathrm{d}\boldsymbol{k}}{\mathrm{d}t} = -\frac{\partial F}{\partial \boldsymbol{r}} ; \quad \frac{\mathrm{d}\omega}{\mathrm{d}t} = \frac{\partial F}{\partial t} . \tag{1.24}$$

The equations (1.24) are a system of Hamiltonian equations. The solution $\{r(t), t\}$ of (1.24) determines the spatial–temporal rays in three-dimensional space $\{x, y, t\}$. The spatial rays $r = r(t)$ are projections of spatial–temporal rays onto the coordinate space $r = \{x, y\}$.

It follows from (1.24) that a wave packet moves at the group velocity:

$$\frac{\partial F}{\mathrm{d}k} = C_\mathrm{g} \, . \tag{1.25}$$

The second equation (1.24) characterizes the change of the wave vector along the ray. The third equation (1.24) describes the frequency variation. The frequency remains constant along the ray: $\omega =$ constant in a stationary medium, with a dispersion ratio (1.22) without being explicitly dependent on time.

An expression for the wave phase ψ along the characteristics can be obtained similarly. Using the determination of the wave action D as an integral by the Lagrangian function L, it can be written as follows:

$$D = D_0 + \int_{t_0}^{t} L \, \mathrm{d}t = D_0 + \int_{t_0}^{t} \left(P \frac{\partial \mathcal{H}}{\partial P} - \mathcal{H} \right) \mathrm{d}t \, . \tag{1.26}$$

Thus, the following wave phase expression is obtained:

$$\psi = \psi_0 + \int_{t_0}^{t} (k C_\mathrm{g} - \omega) \, \mathrm{d}t \, , \tag{1.27}$$

where ψ_0 is the initial phase value.

The second term in the expression (1.27) is reduced to zero in a medium without dispersion. In this case the group velocity C_g coincides with the phase velocity $C = k\omega/k^2$. Thus, the phase is the preserved value $\psi = \psi_0$ in spatial–temporal rays. In the dispersion medium the so-called speed group delay, determined by the second item in the expression (1.27), arises (Kravtsov & Orlov, 1980). The group speed delay means a displacement of the wave packet propagation velocity in comparison with the phase velocity.

All these facts remain in force in the medium moving with speed V, as long as it changes slowly enough. The velocity V can be pointed out in the equations as follows. Let the value r be a spatial vector in the immovable coordinate system, and the value r_1 be a proper local vector in the coordinate system, moving alongside the medium. Then $r_1 = r - Vt$.

As a result of the transition to the new variable r_1, the Hamilton–Jacobi equation determining the phase (1.23) is written in the form:

$$\frac{\partial \psi}{\partial t} + F_1 \left(\frac{\partial \psi}{\partial r_1}, r_1, t \right) = 0 \, ,$$

where the Hamiltonian function F_1 is connected with the function F (1.22) by the ratio:

$$F_1 = F - V \frac{\partial \psi}{\partial r} .$$

The group velocity c_g in the moving coordinate system is expressed by the stationary coordinate system with ratio $c_g = C_g - V$.

Thus, it is sufficient to use the aforesaid formulae, to pass from the moving coordinate system to the stationary one and vice versa.

1.3 Wave Action Conservation Law

The kinematic equations (derived in the previous section on the basis of the geometric optics method) determine a non-vortical field of wave vector k in space and time. In order to obtain the distribution of the dynamic wave characteristics, such as the energy density, it is necessary to get information about the wave dynamics and wave interactions with the medium. If the wavelength and period are assumed to be small in comparison with the medium parameter variations, the amplitude evolution of gravity wave propagation over the oceanic surface in the presence of spatial-non-uniform currents and an uneven bottom can be considered using the geometric optics approximation.

The ocean is assumed to be an incompressible heavy homogeneous fluid. Its hydrodynamic equations are written in the form (1.1)–(1.3), whereas the Coriolis force for wind waves is not essential. The velocity vector U is assumed to consist of the horizontal V and the vertical W components.

The boundary conditions at the free surface at $z = \eta(r, t)$ can be written in the form:

$$P = P_a = 0 ; \quad W = \frac{\partial \eta}{\partial t} + (V\nabla)\eta , \qquad (1.28)$$

where P_a is the atmospheric pressure.

At $z = H(r, t)$ the bottom condition is written as:

$$W + (V\nabla)H = 0 . \qquad (1.29)$$

The small parameter ε introduced above is assumed to characterize the slowness of the main motion change along the horizontal coordinates and in time. No such slowness of variation along the vertical coordinate is assumed. All hydrodynamic fields are presented in the equations in the following form:

$$\tilde{\Phi}(r, z, t) = \Phi_0(r_e, z, t_e) + a\Phi(r, z, t) , \qquad (1.30)$$

where $\tilde{\Phi}$ is any kind of hydrodynamic function; Φ_0 is an averaged "background" field; Φ is a perturbation propagating in the background; $r_e = \varepsilon r$

and $t_e = \varepsilon t$ are the slowly changing horizontal coordinates and time; and a is a small amplitude parameter. As soon as $\boldsymbol{V}_0 = \boldsymbol{V}_0(\boldsymbol{r}_e, z, t_e)$ it follows from the continuity equation (1.2) that $W_0 \sim \varepsilon |\boldsymbol{V}_0|$. The slowness of the bottom surface variation $H = H(\boldsymbol{r}_e)$ is also assumed.

By substituting the expression (1.30) into (1.1)–(1.3) the main equations of the "background" motion can be derived:

$$\frac{\partial \boldsymbol{V}_0}{\partial t_e} + (\boldsymbol{V}_0 \nabla) \boldsymbol{V}_0 = -\frac{1}{\rho} \nabla_r P_0 ; \tag{1.31}$$

$$\nabla \boldsymbol{V}_0 = 0 ; \tag{1.32}$$

$$g\rho = -\frac{\partial P_0}{\partial z} . \tag{1.33}$$

The boundary conditions for the system (1.31)–(1.33) coincide with the expressions (1.28), (1.29), assigning a "0" index to all values. The solution of the equation is derived by the WKB expansion for the disturbance Φ:

$$\Phi = \{\Phi_1(\boldsymbol{r}_e, z, t_e) + \varepsilon \Phi_2(\boldsymbol{r}_e, z, t_e) + \ldots\} \exp[(i/\varepsilon)\psi(\boldsymbol{r}_e, t_e)] . \tag{1.34}$$

Substituting the expansion (1.34) into the equations for disturbances and equaling values of the order a in the expansion (1.30), the equation and boundary conditions for the vertical velocity of the first-order disturbance W_1 can be obtained (the index (1) is omitted below):

$$W'' + \left(-\frac{\sigma''}{\sigma} - k^2\right) W = 0 . \tag{1.35}$$

$$\left(\frac{W}{\sigma}\right)' = g\frac{k^2}{\sigma^3} \quad \text{with} \quad z = \eta_0, \ W = 0 \quad \text{with} \quad z = -H(\boldsymbol{r}_e) , \tag{1.36}$$

where $\sigma = \omega - (\boldsymbol{k}, \boldsymbol{V})$ is the Doppler frequency depending on the vertical coordinate z. A prime means a derivative with respect to z.

The boundary problem produces a set of dispersion ratios for different modes:

$$\omega = F(\boldsymbol{k}, \boldsymbol{r}_e, t) \tag{1.37}$$

and the eigenfunctions $W = W(\boldsymbol{r}_e, z, t_e)$, dependent on $\boldsymbol{r}_e$ and t_e, are given with the help of the marginal problem (1.35), (1.36).

The other wave values are expressed using W by the formulas:

$$\begin{aligned} \boldsymbol{V} &= \frac{i\boldsymbol{k}}{k^2}\sigma\left(\frac{W}{\sigma}\right)' - \left(\frac{iW}{\sigma}\right)\frac{\partial \boldsymbol{V}_0}{\partial z} ; \\ P &= \frac{i\sigma}{k^2}\left(\frac{W}{\sigma}\right)' , \quad \eta = i\frac{W}{\sigma} . \end{aligned} \tag{1.38}$$

By further separating the terms of order $a\varepsilon$ in the main equations and boundary conditions and taking into account the expressions (1.30), (1.34), the equation and boundary conditions for the value W_2 can be obtained:

$$[W_2]'' - \left(\frac{\sigma''}{\sigma} + k^2\right) W_2 = Q \; ;$$

$$\sigma \frac{\partial W_2}{\partial z} \left(\frac{gk^2}{\sigma} + \sigma'\right) W_2 = Q_1 \quad \text{with} \quad z = \eta_0 \; ; \tag{1.39}$$

$$W_2 = -(\boldsymbol{V}\nabla)H = Q_2 \quad \text{with} \quad z = -H \; ,$$

where Q, Q_1 and Q_2 are the functions expressed by Φ_0 and Φ_1 (the explicit form of these functions is given by Voronovich (1976)). In order to solve the non-uniform marginal problem (1.39), it is necessary for the functions Q, Q_1, Q_2 to be orthogonal eigenfunctions of the corresponding homogeneous marginal problem (the solvability condition). This results in:

$$\int_{-H}^{\eta_0} Q \frac{iW}{k^2} \, dz - \frac{iW}{k^2\sigma} Q_1 \bigg|_{z=\eta_0} - \frac{iW'}{k^2} Q_2 \bigg|_{z=-H} . \tag{1.40}$$

After bulky transformations (1.40) can be reduced to the form of the adiabatic invariant conservation law:

$$\frac{\partial A}{\partial t_e} + \nabla_{r_e}(\boldsymbol{C}_{\mathrm{g}} A) = 0 \; , \tag{1.41}$$

where:

$$A = -\int_{-H}^{\eta_0} \frac{\sigma''}{2\sigma^2 k^2} W^2 \, dz + \left(\frac{g}{\sigma^3} + \frac{\sigma'}{2\sigma^2 k^2}\right) W^2 \bigg|_{z=\eta_0} ; \tag{1.42}$$

$$\boldsymbol{C}_{\mathrm{g}} A = \int_{-H}^{\eta_0} \left[-\boldsymbol{V}_0 \frac{\sigma''}{2\sigma^2 k^2} + \frac{1}{2\sigma k^2} \frac{\partial^2 \boldsymbol{V}_0}{\partial z^2} - \frac{\boldsymbol{k}}{k^2}\right] W^2 \, dz$$

$$+ \left[\boldsymbol{V}_0 \left(\frac{g}{\sigma^3} + \frac{\sigma''}{2\sigma^2 k^2}\right) - \frac{1}{2\sigma^2 k^2} \frac{\partial \boldsymbol{V}_0}{\partial z} + \frac{g\boldsymbol{k}}{\sigma^2 k^2}\right] W^2 \bigg|_{z=\eta_0} . \tag{1.43}$$

It follows from the marginal problem properties (1.35) that the a ratio of the expressions (1.42) and (1.43) can be shown to be the actual group velocity $\boldsymbol{C}_{\mathrm{g}} = \partial F/\partial \boldsymbol{k}$.

It should be noted that the adiabatic invariant conservation law (1.41) is valid not only for arbitrary velocity fields, but also for those described by the equations of hydrodynamics (1.1)–(1.3).

The case of the mean current velocity, independent of the vertical coordinate z, should be considered. Using the ratios (1.35), (1.36) the dispersion relation $\sigma^2 = gk\,\text{th}(kH)$ can be easily obtained. The adiabatic invariant transfer rate C_g is equal to:

$$C_g = V_0 + \frac{\partial \sigma}{\partial k} = V_0 + \frac{1}{2}\frac{k}{k}\left(\frac{g\,\text{th}(kH)}{k}\right)^{1/2}\left(1 + \frac{2kH}{\text{sh}(2kH)}\right). \quad (1.44)$$

Using the expressions (1.41)–(1.44) it follows that:

$$A = \frac{E}{\sigma}, \quad (1.45)$$

where E is the wave energy density.

The expression (1.45) is widely known in science the wave action density. The law of wave action density preservation (1.41) with (1.44) is the simplest and the most common expression in wave dynamics. It was established on the basis of the variation principle by B. Whitham (1965) and developed by F. Bretherton and C. Garret (1968) and F. Bretherton (1971), and A. Voronovich and V. Goncharov (1982). It should be noted that the equation of adiabatic invariant preservation law (1.41)–(1.43) represents a more general law as compared to the principle of wave action conservation, because it takes into account the vertical non-uniformity of the mean current velocity.

The equation (1.41) indicates that the local velocity of wave action change is balanced by the action flow divergence transferred with the group velocity C_g relative to moving medium. If the mean velocity V does not remain constant, then the wave vector k and the frequency σ can be varied in space and time according to (1.24). In this case the wave energy density is not conserved with preservation of the wave action A. There is some energy exchange between the waves and the current.

An important consequence of solving the problem is that the characteristics of (1.41) coincide with (1.24), being in turn characteristics of the phase equation (1.23).

Now the problem with initial conditions should be considered. It is necessary to determine the initial surface $\tilde{S}$, where the starting values are prescribed. The equations of the surface $\tilde{S}$ are written in the parametric form $\boldsymbol{r} = \boldsymbol{r}_0(\xi, \zeta)$, where ξ and ζ are curvilinear coordinates in the surface $\tilde{S}$. Let the wave field $\eta^0(\xi, \zeta)$ be determined by the initial value of the wave phase $\psi|_{\tilde{S}} = \psi^0(\xi, \zeta)$ and let the amplitude $a|_{\tilde{S}} = a^0(\xi, \zeta)$ be prescribed in the surface $\tilde{S}$ at $\tau = 0$ (τ is a parameter varying along a ray, for example, time, i.e. $\tau = t$). If a wave propagates along a ray, then the existing ray point $\boldsymbol{r}(\tau_0) = \boldsymbol{r}_0(\xi, \zeta)$ on the surface $\tilde{S}$ is a natural initial condition for the ray trajectory $\boldsymbol{r} = \boldsymbol{r}(\tau)$. The solutions of differential equations for rays (1.24) corresponding to the initial conditions can be given in the form of $\boldsymbol{r} = \boldsymbol{r}(\xi, \zeta, \tau)$, $\boldsymbol{k} = \boldsymbol{k}(\xi, \zeta, \tau)$. In this case the parameters ξ and ζ indicate rays leaving the surface $\tilde{S}$, with the parameter τ pointing at a fixed ray position. The value combination ξ, ζ, τ is called the ray coordinates. In the general case these coordinates are not orthogonal.

The equation $r = r(\xi, \zeta, \tau)$ determines a set of rays generated by the prescribed field distribution in the initial surface $r(\tau_0) = r_0(\xi, \zeta)$. The equation for a set of rays describes a relation between the ray coordinates and the Cartesian coordinates. If the Jacobian $J_1 = \partial(x, y, z)/\partial(\xi, \zeta, \tau)$ is not equal to zero in the considered area, then the equation $r = r(\xi, \zeta, \tau)$ can be solved unambiguously relative to the ray coordinates corresponding to the point of observation $\xi = \xi(r)$, $\zeta = \zeta(r)$, $\tau = \tau(r)$.

The results obtained in this chapter give the solution of wave propagation in a water surface with a horizontal non-uniform current and uneven bottom as follows:

$$\eta(r, t) = a_0 J_1^{-1/2} J_2^{-1/2} e^{i\Psi}, \tag{1.46}$$

where the wave phase ψ is determined by the initial conditions according to (1.26): $\psi(r, t) = \psi(r_0) + \int_0^t (k\,C_g - \omega)\,dt$.

Unlike the classical case (Kravtsov & Orlov, 1980) an additional multiplier $J_2 = \sigma_0/\sigma$ appears in the expression (1.46) due to the influence of the spatial non-uniform currents, so long as the equation of wave action density conservation (1.41) is solved instead of that for the energy.

An important result of the solution (1.46) is that the following equality holds along the characteristics (Kravtsov & Orlov, 1980):

$$|C_g A|\,dl = \text{const}, \tag{1.47}$$

where dl is the distance between two infinitely close projections of the characteristics towards the coordinate space $\{x, y\}$. It follows from (1.24) that the ratio (1.47) establishes the preservation law of wave action flow along a ray tube. Another simple consequence of (1.24) and (1.47) should be noted. If the medium properties are not dependent on time t, then the frequency ω is preserved. Additionally the value of the wave vector component k_x is also preserved along the characteristics in the spatial-cylindrical case (i.e. the wave medium properties depend only on one coordinate y). These characteristics are parallel lines (see Fig. 1.1). The ratio (1.47) can be written in the simple form: $C_{gy} A = \text{const}$. These kinds of ratios are used for solving a wide range of problems. For example, they describe wave propagation in shallow water with a depth varing only along one direction, i.e. at parallel isobaths or in the presence of horizontal-shear currents. A number of such problems will be considered below.

The results presented in this chapter allow us to consider in general wave propagation with slow temporal and weak horizontal non-uniformity in the average medium state of the ocean.

The sphere of application of this theory should be made clear. The aforementioned methods describing waves in water are based on the assumption of waves being locally plane. However, this assumption is not always fulfilled. Sometimes there are situations in which changes, small in comparison with the wavelength, are accumulated. This may lead to the phenomenon where

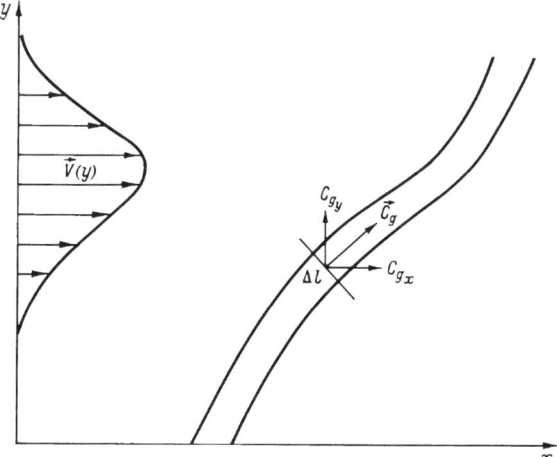

Fig. 1.1. Wave rays in the background of a non-uniform current

the wave field can be significantly different from a local plane one. Thus, if the Jacobian becomes zero, $J_1 = 0$, in the solution (1.46), then a singularity (caustic) occurs due to the ray tube width being decreased to zero. An infinite narrowing of the ray tube width occurs due to ray intersection in the ratio (1.47). In this case the estimate of the wave height becomes unrealistically large in the vicinity of the caustic. These are considered below when the use of diffraction approaches is necessary.

An alternative solution method can be developed using the spectral approach developed in this monograph. The advantage of this method is that the ray set $r = r(\xi, \zeta, \tau)$ in physical space can have a sufficiently complicated form. This results in complexity of constructing the smooth solution for the entire space. However, there can be only one phase trajectory, crossing a given point in the phase space $\{k, r\}$, i.e. the phase trajectories do not intersect each other in the spectral solution. In fact, this quality is a consequence of the uniqueness of solutions for ordinary differential equation systems with given initial conditions.

1.4 Statistical Wind Wave Description

An obvious feature of wind waves is their random character. As wind waves present a non-stationary probabilistic hydrodynamic process, the ideas and methods of random process theory are widely used for their theoretical and experimental investigation. The displacement of the water–air interface $\eta(r, t)$ is the main experimentally observed characteristic of wind waves. That is why it is necessary to consider $\eta(r, t)$ as a random moving surface in the probabilistic description of wind waves. Probabilistic distributions of the values η in finite space and time sets $\{r_n, t_n\}$ ($n = 1, 2$) become the objects of study. As the measurement data indicate, the distribution of probabilities for a value η at a fixed point is approximated by the Gaussian distribution.

A theoretical description of wind waves in the terms of finite-dimensional densities is associated with significant difficulties. It allows investigating only the simplest statistical characteristics of the value η, with the second moment or the correlation function being the most important:

$$K(\Delta r, \Delta t) = \langle \eta(r, t) \eta(r + \Delta r, t + \Delta t) \rangle , \quad (1.48)$$

where angle brackets denote averaging by a statistical ensemble.

The spatial–temporal correlation function $K(\Delta r, \Delta t)$ is connected with the spectrum $S(k, \omega)$ of a random process by the Fourier transformation:

$$S(k, \omega) = \frac{1}{(2\pi)^3} \int K(\Delta r, \Delta t) \exp[-i(k\Delta r - \omega \Delta t)] \, d\Delta r \, d\Delta t . \quad (1.49)$$

The dispersion of surface elevation $\langle \eta^2 \rangle$ is calculated by integrating the spectrum $S(k, \omega)$ over the two-dimensional wave vector k and the frequency ω.

The two-dimensional spatial wave spectrum $S(k)$ is determined by (1.49) according to the formula:

$$S(k) = \int S(k, \omega) \, d\omega = \frac{1}{(2\pi)^2} \int K(\Delta r, 0) \exp(-ik\Delta r) \, d\Delta r \quad (1.50)$$

and the frequency spectrum $S(\omega)$ is determined as:

$$S(\omega) = \int S(k, \omega) \, dk = \frac{1}{2\pi} \int K(0, \Delta t) \, e^{i\omega \Delta t} \, d\Delta t . \quad (1.51)$$

The second moments or their corresponding spectra are known to give complete statistical information about a random field, in the case of a Gaussian process (Davidan et al., 1978). This determines the importance of the spectral wave characteristic information, as soon as the experimental data of the distribution function allow us to consider a random field of the level disturbances η as being approximately Gaussian. For the given spectrum the Gaussian surface model can be a basis for obtaining statistical information about the geometric characteristics of a moving random surface: the mean number of stationary points (maxima, minima, hyperbolic points, etc.) per unit surface area, statistical distributions of maximum and minimum heights, etc. Numerous results have been obtained by W. Pierson, Yu. Krylov, and by others. The methods of statistical geometry for random surfaces were most consistently developed by M. Longuet-Higgins in the 1960s and later by V. Rozhkov and Yu. Trapeznikov (1990).

The first empirical estimates for wind waves were based on the relation between their simplest characteristics and wind speed. The main purpose of wind wave theory is to determine this relation with a help of the dynamic equations describing the "water–air" system. Sea waves and the wind speed field are of a random character. That is why the main problem of wind wave theory (on the level of second moments) can be formulated specifically as the problem of determining the surface wave spectrum by the statistical characteristics of a random wind speed field of the atmospheric turbulent boundary layer.

1.5 Kinetic Equation of the Evolution of the Wind Wave Spectrum

The description of the formulation of the hydrodynamic problem of wind waves is given in Sect. 1.1. Apart from the complexity of solving this problem, there is an additional difficulty in wind wave modelling, connected with the random character of wind waves. That is why any attempt to solve the problem of the numerical simulation of wind waves on real ocean scales with the help of the deterministic hydrodynamic formulation is impractical. The number of degrees of freedom of the system is practically unlimited.

Significant achievements in the numerical simulation of wind waves are connected with the kinetic equation describing the evolution of the wave spectrum under the external field action with the wind field being an external field action. The phenomenological description of the equation is as follows. The statistically space-homogeneous, stationary, random water–air interface $\eta(r,t)$ is considered in the previous Sect. 1.4. For describing the random field evolution "slow" coordinates and time are introduced. Field variability scales essentially exceed the typical lengths and periods of waves.

The corresponding generalization of the statistically homogeneous and stationary field can be achieved by considering the local spectra, depending on the slow coordinates r_e and the time t_e (the index "e" is omitted below):

$$S = S(\boldsymbol{k}, \omega, \boldsymbol{r}, t) \,. \tag{1.52}$$

Similarly, the wave action spectrum can be written as:

$$N = N(\boldsymbol{k}, \omega, \boldsymbol{r}, t) = S/\sigma \,. \tag{1.53}$$

Now the general phenomenological equation for the spectral density evolution of the wave action can be formally written as the transfer equation:

$$\frac{\partial N}{\partial t} + \frac{\partial N}{\partial \boldsymbol{r}}(N\dot{\boldsymbol{r}}) + \frac{\partial}{\partial \boldsymbol{k}}(N\dot{\boldsymbol{k}}) + \frac{\partial}{\partial \omega}(N\dot{\omega}) = G \,. \tag{1.54}$$

If the derivatives $\dot{\boldsymbol{r}}, \dot{\boldsymbol{k}}, \dot{\omega}$ are written in the form of Hamiltonian equations:

$$\frac{d\boldsymbol{r}}{dt} = \frac{\partial \mathcal{H}}{\partial \boldsymbol{k}} \,; \quad \frac{d\boldsymbol{k}}{dt} = -\frac{\partial \mathcal{H}}{\partial \boldsymbol{r}} \,; \quad \frac{d\omega}{dt} = \frac{\partial \mathcal{H}}{\partial t} \,, \tag{1.55}$$

then equation (1.54) can be rewritten in the form of the full time derivative:

$$\frac{dN}{dt} = \frac{\partial N}{\partial t} + \frac{\partial N}{\partial \boldsymbol{r}}\frac{d\boldsymbol{r}}{dt} + \frac{\partial N}{\partial \boldsymbol{k}}\frac{d\boldsymbol{k}}{dt} + \frac{\partial N}{\partial \omega}\frac{d\omega}{dt} = G. \quad (1.56)$$

The equations (1.54) and (1.56) are called kinetic equations. They are well-known in theoretical physics. It is a generalization of Liouville's theorem of the conservation of the distribution function of the gas as a particle system moving in phase space (Landau & Lifshits, 1973). The value in the right-hand side of (1.56), is called the collision integral. The integral-differential equation (1.56) with the collision integral describing a molecule collision in phase space is called the Boltzmann equation, which was proposed in 1872.

As noted in Sect. 1.2, the ratios (1.55) present equations of motion for wave packets in terms of the variables $\boldsymbol{r}$ and $\boldsymbol{k}$ (it follows that $F = \mathcal{H}$, Sect. 1.3). These equations coincide in form with the Hamiltonian equations, being the central point in classic mechanics (Lantsosh, 1965; Landau & Lifshits, 1973). They are solved in terms of the particle momentum p and its coordinate q. The canonical Hamiltonian equations represent a system of $2s$ (in our case $s = 3$) first-order differential equations for $2s$ unknown functions $p(t)$ and $q(t)$. They substitute s of the second-order equations for the Lagrange motion simulation method.

The total time derivative of the Hamilton function $\mathcal{H}$ is written as:

$$\frac{d\mathcal{H}}{dt} = \frac{\partial \mathcal{H}}{\partial t} + \sum_i \frac{\partial \mathcal{H}}{\partial q_i}\dot{q}_i + \sum_i \frac{\partial \mathcal{H}}{\partial p_i}\dot{p}_i. \quad (1.57)$$

Substituting $\dot{q}_i$ and $\dot{p}_i$ into (1.57) from (1.55), the last two terms are cancelled reciprocally:

$$\frac{d\mathcal{H}}{dt} = \frac{\partial \mathcal{H}}{\partial t}. \quad (1.58)$$

In particular, if the Hamiltonian function is not dependent on time explicitly, then $d\mathcal{H}/dt = 0$, i.e. $\mathcal{H}$ is conserved.

If the Hamiltonian function is not dependent on the coordinates, the corresponding component of the generalized momentum is preserved while the system is moving. This can be written as:

$$\dot{p}_i = -\frac{\partial \mathcal{H}}{\partial q} = 0. \quad (1.59)$$

Such a coordinate is called cyclic.

Let f be some function of the coordinates q, momentum p and time t. Its full time derivative is as follows:

$$\frac{df}{dt} = \frac{\partial f}{\partial t} + \sum_j \frac{\partial f}{\partial q_j}\dot{q}_j + \sum_j \frac{\partial f}{\partial p_j}\dot{p}_j. \quad (1.60)$$

1.5 Kinetic Equation of the Evolution of the Wind Wave Spectrum

Substituting the values $\dot{q}_j$ and $\dot{p}_j$ from the Hamiltonian equation (1.55) gives

$$\frac{df}{dt} = \frac{\partial f}{\partial t} + \{\mathcal{H}\,f\}\,, \tag{1.61}$$

where the designation:

$$\{\mathcal{H}\,f\} = \sum_j \left(\frac{\partial \mathcal{H}}{\partial p_j} \frac{\partial f}{\partial q_j} - \frac{\partial \mathcal{H}}{\partial q_j} \frac{\partial f}{\partial p_j} \right) \tag{1.62}$$

is introduced.

The expression (1.62) is called the Poisson brackets for the values $\mathcal{H}$ and f. Thus, the kinetic equation (1.54) can also be considered as a sum of a non-stationary term $\partial N/\partial t$ with the corresponding Poisson brackets for N and ω.

The functions of dynamic variables that remain constant in the motion of the system are usually called motion integrals. It can be seen from (1.61) that the condition for f to be a motion integral ($df/dt = 0$) can be written in the form:

$$\frac{\partial f}{\partial t} + \{\mathcal{H}\,f\} = 0\,. \tag{1.63}$$

If the motion integral is not dependent explicitly on time, then $\{\mathcal{H}\,f\} = 0$, i.e. the Poisson brackets with the Hamiltonian function must be equal to zero. An important feature of the Poisson brackets is that if f and g are two motion integrals, then the brackets composed of them are also the motion integrals $\{fg\}$ (Poisson's theorem).

For the geometric interpretation of the dynamical system the notion of phase space is often used, as the space of $2s$ dimensions where coordinate axes are the values of the s coordinates and s momentum of the system. Point motion in the phase space depicts a corresponding line, called the phase trajectory. The entire area is moved, assuming that every point of the area of the given phase space propagates with time according to the equations of motion of the considered dynamical system. Its volume can be proved to be invariant $\int d\Gamma = $ const. (Lantsosh, 1965). This statement (the Liouville theorem) follows directly from the phase volume invariants under canonical transformations and from the fact that variations under motion themselves can be considered as canonical transformations.

In the mathematical modelling of wind waves the traditional transfer from the hydrodynamic equations (1.5)–(1.13) to the kinetic equation (1.54) is as follows (Hasselmann, 1979; Yefimov & Polnikov, 1991). The hydrodynamic fields are assumed to be random functions, written in the form of the Fourier (or Fourier–Stieltjes) integral. The equations of motion are presented for spectral components of water surface elevation according to the hydrodynamic equations of the homogeneous field approximation. Using the statistical formulas for closing the higher moments of the Fourier components the solution leads to the evolution equation of the wind wave field spectrum S. It should be noted that the initial formulation of the hydrodynamic problem does not allow one to obtain correctly a different comprehensive type of physical mechanism for forming wind wave spectra. This is referred to dissipation connected with wave breaking.

It should be pointed out that the interpretation of the kinetic equation as the interaction of random wave fields became wellknown after K. Hasselmann's publications (1960, 1966, 1979). Using the method of Feymann diagrams the description of the non-linear interaction is generalised (taking into account the non-conservative interaction of wave fields and external fields) by means of the methods of theoretical physics for the wind wave case. The function in the right side of (1.54) describes the different physical mechanisms forming the wind wave spectrum. At present the right side of (1.54) is called "a source function". It is written as the sum of different approximations to the physical mechanism:

$$G = \sum_i G_i \,. \tag{1.64}$$

According to wind wave theory, the source function should be assumed to include, at least, the following components (Davidan et al., 1985): G_1 is the mechanism describing the wind-to-wave energy flux due to the influence of the turbulent pressure variation field; G_2, G_3, G_4 are the energy flux to waves due to wave interaction (G_2 – linear, G_3 – non-linear) with averaged air flow and atmospheric turbulence (G_4); G_5 is the energy exchange due to the wave interaction with water turbulence; G_6 is the energy dissipation due to bottom friction; G_7 is the energy dissipation due to wave breaking; G_8 is the non-linear energy transfer in the wind wave spectrum. The list of possible mechanisms forming the wind wave spectrum, including, for example, the interaction between waves and the ice cover G_9 could be extended. These are the main components of the source function; they are still insufficiently studied.

1.6 General Problem Formulation for Determining the Wave Action Spectral Density in the Ocean

The kinetic equation can be used to describe wind wave field evolution in the ocean. It can be written in the most general form similarly to (1.56) in the spherical coordinates $\{\varphi, \vartheta, R\}$ relative to some value $\tilde{N}$, whose connection to waves is be specified below:

$$\begin{aligned}
\frac{\partial \tilde{N}}{\partial t} &= \frac{\partial}{\partial \varphi}(\tilde{N}\,\dot{\varphi}) + \frac{\partial}{\partial \vartheta}(\tilde{N}\,\dot{\vartheta}) + \frac{\partial}{\partial R}(\tilde{N}\,\dot{R}) \\
&\quad + \frac{\partial}{\partial k_\varphi}(\tilde{N}\,\dot{k}_\varphi) + \frac{\partial}{\partial k_\vartheta}(\tilde{N}\,\dot{k}_\vartheta) + \frac{\partial}{\partial k_R}(\tilde{N}\,\dot{k}_R) + \frac{\partial}{\partial \omega}(\tilde{N}\,\dot{\omega}) \\
&= G \,,
\end{aligned} \tag{1.65}$$

1.6 General Problem Formulation

where $\tilde{N}$ is a function depending on time t, latitude φ, longitude ϑ, radius R, corresponding values of the generalized momentum $\{k_\varphi, k_\vartheta, k_R\}$ and frequency ω.

The Hamiltonian function $\mathcal{H}$ is assumed to allow the writing of the equations of motion in the spherical coordinate system $\{\varphi, \vartheta, R\}$ in the following form:

$$\frac{dR}{dt} = \frac{\partial \mathcal{H}}{\partial k_R} \; ; \quad \frac{d\varphi}{dt} = \frac{\partial \mathcal{H}}{\partial k_\varphi} \; ; \quad \frac{d\vartheta}{dt} = \frac{\partial \mathcal{H}}{\partial k_\vartheta} \; ; \tag{1.66}$$

$$\frac{dk_R}{dt} = -\frac{\partial \mathcal{H}}{\partial R} \; ; \quad \frac{dk_\varphi}{dt} = -\frac{\partial \mathcal{H}}{\partial \varphi} \; ; \quad \frac{dk_\vartheta}{dt} = -\frac{\partial \mathcal{H}}{\partial \vartheta} \; ; \tag{1.67}$$

$$\frac{d\mathcal{H}}{dt} = \frac{\partial \mathcal{H}}{\partial t} \; . \tag{1.68}$$

Assuming that the motion occurs on a spherical surface, then $dR/dt = dk_R/dt = 0$.

Substituting the ratios (1.66)–(1.68) into (1.65):

$$\frac{\partial \tilde{N}}{\partial t} + \frac{\partial \tilde{N}}{\partial \varphi}\dot{\varphi} + \frac{\partial \tilde{N}}{\partial \vartheta}\dot{\vartheta} + \frac{\partial \tilde{N}}{\partial k_\varphi}\dot{k}_\varphi + \frac{\partial \tilde{N}}{\partial k_\vartheta}\dot{k}_\vartheta + \frac{\partial \tilde{N}}{\partial \omega}\dot{\omega} = G \; . \tag{1.69}$$

Now the relation between (1.65) or (1.69) and the problem of numerical simulation of the wave in the ocean will be determined. The equations of motion for the wave packet on the ocean surface are obtained assuming that the ocean depth H and current velocity $\boldsymbol{V}$ depend on the latitude φ and the longitude ϑ, i.e. $H = H(\varphi, \vartheta)$, $\boldsymbol{V} = \boldsymbol{V}(\varphi, \vartheta, t)$. Based on the geometric optics approximation (see Sects. 1.2 and 1.3), the Hamiltonian function of the wave packet motion is written in the form:

$$\mathcal{H} = \sqrt{gk\,\text{th}(kH)} + \boldsymbol{V}\boldsymbol{k} \; . \tag{1.70}$$

An additional factor of the Hamiltonion function influencing wave propagation is an effect connected with the Earth's rotation. However, as shown by Backus (1962), its effect is so small that it can be completely neglected for wind waves.

The wave equations of motion on the spherical surface are written in the form:

$$\frac{d\varphi}{dt} = c_g \frac{\partial k}{\partial k_\varphi} + \frac{V_\varphi}{R} \; ; \tag{1.71}$$

$$\frac{d\vartheta}{dt} = c_g \frac{\partial k}{\partial k_\vartheta} + \frac{V_\vartheta}{R\cos(\varphi)} \; ; \tag{1.72}$$

$$\frac{dk_\varphi}{dt} = -\left\{ c_g \frac{\partial k}{\partial \varphi} + f\frac{\partial H}{\partial \varphi} + \frac{\partial V_\varphi}{\partial \varphi}\frac{k_\varphi}{R} + \frac{\partial V_\vartheta}{\partial \varphi}\frac{k_\vartheta}{R\cos(\varphi)} + \frac{V_\vartheta}{R}\frac{k_\vartheta \sin(\varphi)}{\cos^2(\varphi)} \right\} \; ; \tag{1.73}$$

$$\frac{dk_\vartheta}{dt} = -\left\{ f \frac{\partial H}{\partial \vartheta} + \frac{\partial V_\varphi}{\partial \vartheta} \frac{k_\varphi}{R} + \frac{\partial V_\vartheta}{\partial \vartheta} \frac{k_\vartheta}{R\cos(\varphi)} \right\} ; \qquad (1.74)$$

$$\frac{d\mathcal{H}}{dt} = \frac{d\omega}{dt} = \frac{k_\varphi}{R} \frac{\partial V_\varphi}{\partial t} + \frac{k_\vartheta}{R\cos(\varphi)} \frac{\partial V_\varphi}{\partial t} , \qquad (1.75)$$

where

$$k = \sqrt{\frac{k_\varphi^2}{R^2} + \frac{k_\vartheta^2}{R^2 \cos^2(\varphi)}} , \qquad (1.76)$$

and

$$\frac{\partial k}{\partial k_\varphi} = \frac{k_\varphi}{kR^2} , \quad \frac{\partial k}{\partial k_\vartheta} = \frac{k_\vartheta}{kR^2 \cos^2(\varphi)} ; \qquad (1.77)$$

$$\frac{\partial k}{\partial \varphi} = \frac{k_\varphi}{kR^2} \frac{\tan(\varphi)}{\cos^2(\varphi)} ; \quad f = \sqrt{\frac{gk}{\text{th}(kH)} \frac{k}{\text{ch}^2(kH)}} ; \qquad (1.78)$$

$$c_g = \frac{1}{2} \sqrt{\frac{g \, \text{th}(kH)}{k}} \left(1 + \frac{2kH}{\text{sh}(2kH)}\right) ,$$

where V_ϑ, V_φ are the zonal and meridional components of the current velocity. Equations (1.71)–(1.75) describe the motion of the wave packet in a spherical surface under the influence of the non-uniform current velocity $\boldsymbol{V}(\varphi,\vartheta,t)$ and depth $H(\varphi,\vartheta)$.

The wave number $k = |\boldsymbol{k}|$ (or the frequency ω) and the angle β between the wave vector direction and the x axis of local rectangular coordinates are usually used in numerical simulations of wind waves. The wave number k is connected with the former variables k_φ and k_ϑ by the ratio (1.76), and the angle β can be determined as:

$$\tan \beta = \frac{k_\varphi \cos(\varphi)}{k_\vartheta} . \qquad (1.79)$$

Using the ratios (1.73), (1.74), it can be shown that variations of the new variables k and β in time are connected with the previous variables by:

$$\dot{k} = \frac{1}{kR^2} \left[\dot{k}_\varphi k_\varphi + \frac{\dot{k}_\vartheta k_\vartheta}{\cos^2 \varphi} \right] ; \qquad (1.80)$$

$$\dot{\beta} = \frac{\cos \varphi \cos^2 \beta}{k_\vartheta^2} [k_\vartheta \dot{k}_\varphi - \dot{k}_\vartheta k_\varphi] . \qquad (1.81)$$

The relation between the value $\tilde{N}$, introduced by the aforementioned method, and the spectral density of the wave action $N(\boldsymbol{k})$, usually used in local rectangular plane coordinates $\{x,y\}$, will now be determined. It should be noted that the typically used spectral density $N(\boldsymbol{k})$ is determined as

1.6 General Problem Formulation

a wave action referred to the phase volume element $dk_x\, dk_y\, dx\, dy$, whereas the $\tilde{N}$ value in the initial kinetic equation (1.65) is referred to the phase volume element $dk_\varphi\, dk_\vartheta\, d\varphi\, d\vartheta$. In order to define the spectral density of the wave action N, the corresponding values can be equated taking into account their phase volumes. As a result, the following ratio is derived:

$$\tilde{N}(k_\varphi, k_\vartheta, \varphi, \vartheta) = J\, k N(k, \beta, x, y)\,, \qquad (1.82)$$

where J is the Jacobian transfer function from $\tilde{N}$ to N:

$$J = \frac{\partial(k, \beta, x, y)}{\partial(k_\varphi, k_\vartheta, \varphi, \vartheta)}\,. \qquad (1.83)$$

In order to define the Jacobian J it is necessary to calculate a fourth-order determinant. Using the relation $dx\, dy = R^2 \cos\varphi\, d\varphi\, d\vartheta$ (1.76) and omitting the intermediate calculations, the Jacobian function can be shown to be equal to $J = 1/k$. This could have been expected directly using the Liouville theorem (Landau & Lifshits, 1973).

Thus, (1.69) describes the usual spectral density evolution of the wave action $N(k, \beta, \varphi, \vartheta)$. Transferring to new variables and using the ratios (1.76)–(1.82), this equation can be written as:

$$\frac{\partial N}{\partial t} + \frac{\partial N}{\partial \varphi}\dot\varphi + \frac{\partial N}{\partial \vartheta}\dot\vartheta + \frac{\partial N}{\partial k}\dot k + \frac{\partial N}{\partial \beta}\dot\beta + \frac{\partial N}{\partial \omega}\dot\omega = G\,, \qquad (1.84)$$

where N is a function of the latitude φ, longitude ϑ, wave number k, angle β, frequency ω and time t.

In the case of a traditional spectral density of the wave energy $S = S(\sigma, \beta)$, depending on the eigenfrequency σ (the intrinsic frequency measured in the coordinate system related to the current) and the angle β, its relation to the wave action density $N(k, \beta)$ is determined as (Lavrenov, 1986):

$$S(\sigma, \beta) = N(k, \beta)\, k\sigma\, \frac{\partial k}{\partial \sigma}\,. \qquad (1.85)$$

Thus, if the solution of (1.84) is found, the ratio (1.85) allows the determination of the spectral energy density. The most important feature of (1.84) is that its left-hand side can be expressed in the form of a full time derivative. It should be noted that this fact was not noticed by the authors of the WAM model (Komen et al., 1994). It follows that the wave action density is preserved along the characteristics in the case of the source function being equal to zero $G = 0$.

The equations of motion (1.71)–(1.75) using new variables can be written in the following form:

$$\frac{d\varphi}{dt} = c_g \frac{\sin(\beta)}{R} + \frac{V \sin(\gamma)}{R}\,; \qquad (1.86)$$

$$\frac{d\vartheta}{dt} = c_g \frac{\cos(\beta)}{R\cos(\varphi)} + \frac{V \cos(\gamma)}{R\cos(\varphi)}\,; \qquad (1.87)$$

$$\frac{dk}{dt} = -k\frac{\tan(\varphi)\cos(\beta)}{R}V\sin(\beta-\gamma) - k\frac{1}{R}\left\{\sin(\beta)\left[\frac{\partial V}{\partial \varphi}\cos(\gamma-\beta)\right.\right.$$

$$\left.+V\sin(\beta-\gamma)\frac{\partial \gamma}{\partial \varphi}\right] + \frac{\cos(\beta)}{\cos(\varphi)}\left[\frac{\partial V}{\partial \vartheta}\cos(\gamma-\beta) + V\sin(\beta\gamma)\frac{\partial \gamma}{\partial \vartheta}\right]\Bigg\}$$

$$-\frac{1}{R}f\left[\sin(\beta)\frac{\partial H}{\partial \varphi} + \frac{\cos(\beta)}{\cos(\varphi)}\frac{\partial H}{\partial \vartheta}\right] ; \qquad (1.88)$$

$$\frac{d\beta}{dt} = -\frac{\tan(\varphi)\cos(\beta)}{R}[c_g + kV\cos(\gamma-\beta)] - \frac{k\cos(\beta)}{R}\left[\frac{\partial V}{\partial \varphi}\cos(\gamma-\beta)\right.$$

$$\left.+V\sin(\beta-\gamma)\frac{\partial \gamma}{\partial \varphi}\right] + \frac{k\sin(\beta)}{R\cos(\varphi)}\left[\frac{\partial V}{\partial \vartheta}\cos(\gamma-\beta) + V\sin(\beta-\gamma)\frac{\partial \gamma}{\partial \vartheta}\right]$$

$$+\frac{1}{R}f\left[-\cos(\beta)\frac{\partial H}{\partial \varphi} + \frac{\sin(\beta)}{\cos(\vartheta)}\frac{\partial H}{\partial \vartheta}\right] ; \qquad (1.89)$$

$$\frac{d\omega}{dt} = k\cos(\beta-\gamma)\frac{\partial V}{\partial t} + kV\sin(\beta-\gamma)\frac{\partial \gamma}{\partial t} , \qquad (1.90)$$

where $V_\vartheta = V\cos(\gamma)$; $V_\varphi = V\sin(\gamma)$.

Thus, the problem of determining the spectral density of the wind wave action is reduced to solving the equation system (1.84), (1.86)–(1.90) under the given initial (or boundary) conditions. It should be noted that the equation system contains functions depending on the variable parameters such as the depth field $H(\varphi,\vartheta)$, the current velocity field $\boldsymbol{V} = \{V_\varphi(\varphi,\vartheta,t), V_\vartheta(\varphi,\vartheta,t)\}$, and also the wind speed field $\boldsymbol{U} = \{U_\varphi(\varphi,\vartheta,t), U_\vartheta(\varphi,\vartheta,t)\}$. The latter is included in the source function G and determines the wind wave input.

In the general case the process of obtaining a numerical solution of the problem (1.84)–(1.90) is a very complicated task demanding significant computer resources. A great variety of different physical factors, and their different spatial and temporal scales make the numerical implementation of the problem very difficult.

1.7 Spatial–Temporal Scale Considerations in the Analysis of Problem Solution

Estimations of different components in the right-hand side of the equation system (1.84), (1.86)–(1.90) show different spatial–temporal scales of the physical mechanisms forming a wind wave field in the ocean. It should be noted that the wave number and the change in the angle β is due to the current spatial and temporal variations and the non-uniformity of the depth in shallow water. At the same time the angle β is changed without any current or depth influence due to the ocean surface sphericity effect.

1.7 Spatial–Temporal Scale Considerations

In order to obtain quantitative estimates of various factors, the functions in the right-hand side of (1.86)–(1.90) are transformed into their non-dimensional form:

$$\dot{\tilde{\varphi}} = \frac{\dot{\varphi}R}{\langle|c_g|\rangle} \ ; \quad \dot{\tilde{\vartheta}} = \frac{\dot{\vartheta}R}{\langle|c_g|\rangle} \ ; \quad \dot{\tilde{k}} = \frac{\dot{k}R}{\langle|c_g|\rangle} \ ; \quad \dot{\tilde{\beta}} = \frac{\dot{\beta}R}{\langle|c_g|\rangle} \ ,$$

where $\langle|c_g|\rangle$ is a mean group velocity estimate. In this case the right-hand sides of the equations also become non-dimensional with the new parameters: $\alpha = \langle|V|\rangle/\langle|c_g|\rangle$ is the ratio of the mean current velocity to the wave propagation speed; $\chi = \langle|\Delta V|\rangle/\langle|kV|\rangle$ is the ratio defining the effectiveness of non-uniform current on the wave kinematics, where $\langle|k|\rangle$ is the mean estimate of the wave number; and $\delta = \langle|k|\rangle\langle\Delta H\rangle\,\mathrm{e}^{-2\langle|k|\rangle\langle H\rangle}$ is a parameter characterizing the refraction effect in shallow water. It is clear that a comparison of the non-dimensional parameters α, χ, δ determines the quantitative significance of the different mechanisms. Thus, for waves with a period $\tau = 6$ s in shallow water at $kH = 1$ and depth gradient $\Delta H/\Delta L \approx 10^{-3}$ the parameter δ is of order 10^3–10^4. The parameter β is of a smaller value. For $\Delta V/\Delta L \sim 10^{-4}\,\mathrm{s}^{-1}$ the value β is estimated as 10^2. This is significantly greater than the effects due to sphericity of the ocean surface (in this case they are about 1).

Thus, even the roughest estimates indicate that the shallow-water and the large-gradient current effects are of the greatest influence on the change of the wave elements at comparatively small distances. The sphericity is practically of no importance for describing such effects. It is manifested at global distances. In this case the small-gradient currents with global scales (Kenyon, 1981) can play their role.

It is reasonable to carry out a study of the influences of different effects on the solution of the problem, isolating the spatial–temporal scales. This allows the simplification of the analysis of the problem. It reveals the most effective mechanisms forming the wind wave spectrum, taking into account the spatial–temporal scale of wave development in a specific geographical area. It should be noted that the geometric aspect of the problem is analysed by a description of wave packet propagation in phase space. This is described not only by the right-hand side, but also by the left-hand side of the kinetic equation.

Thus, it is possible to investigate the problem at the following spatial–temporal scales:

1. The global scale (for the spatial scale $L_1 \sim 10^6$–10^7 m and the temporal scale $T_1 \sim 10^6$ s) taking into account the Earth's surface curvature and global currents for modelling wind waves. In this case the typical spectrum-forming mechanisms are effective under deep water conditions (G_2, G_5, G_8, etc.). The spatial scale of wave field non-uniformity is determined by a typical scale of atmospheric perturbations (cyclones). An example of such a model is given in Chap. 2.

2. The regional scale I ($L_2 \sim 10^5$–10^6 m, $T_2 \sim 10^5$ s) with the wind waves modelled under deep water conditions in seas, large lakes, basins, etc; the

effective mechanisms are the same (G_2, G_5, G_8, etc.) and the sphericity is of no importance (see Chap. 4).

3. The regional scale II ($L_3 \sim 10^3$–10^5 m, $T_3 \sim 10^3$–10^5 s) is a typical one with medium spatial non-uniformity being taken into account: currents (Chap. 5) and bottom relief (Chap. 6). It is the most complicated case for both mechanisms, typical for deep water conditions and mechanisms connected with wave transformation under the influence of non-uniform currents as well as in shallow water with bottom dissipation (G_6). Such models describe the wave evolution in shallow and tidal seas, extensive continental shelf areas and open ocean areas with strong currents.

4. The local scale. This is a spatial–temporal scale of wave transformation in non-uniform currents with large velocity gradients, as well as in sea coastal shallow water areas (with typical local scales of the coastal areas being $L_4 \sim 10^2$–10^4 m and typical temporal wave evolution scale $T_4 \sim 10^4$ s). The mechanisms of refraction, transformation and bottom friction can be significantly larger in these cases than a four resonance non-linear wave interaction (see Sect. 4.1) in the spectrum (G_8) and wind wave energy input (G_2). These examples are given in Chaps. 5 and 6. The models can also be assumed to include wave and ice floe interaction mechanisms (G_9) in the Arctic seas (see Chap. 7).

5. The small scale. This is the scale of wave transformation in the surf and marginal ice zone ($L_5 \sim 10$–10^2 m, $T_5 \sim 10$–10^2 s) with intensive wave dissipation due to waves breaking in shallow water or in the marginal zone (see Chap. 7).

The above differentiation of the effects by scales does not exclude solving the problem as a whole, i.e., creating general model complexes, whose successive implementation (with a smaller-scale model using the results of a larger-scale calculation as the boundary or initial data) allows numerical simulation of wind wave evolution in the general case, including all physical and geometrical factors.

2 Mathematical Simulation of Wave Propagation at Global Distances

2.1 Problem Formulation of Ocean Wind Wave Modelling Using Spherical Variables

Chapter 1 presents a general problem formulation of the evolution of the wind wave spectrum considering various determining factors. Due to the complexity of formulating the general problem, sphericity solution effects are considered in this chapter. The investigation is limited to the deep-water case, taking into consideration that the average ocean depth essentially exceeds the characteristic lengths of wind waves and swell.

It should be noted that traditionally most wind wave models are based on the wave energy balance equation, written in the usual rectangular coordinates $r = \{x, y\}$ (it is not used in the WAM model (Komen et. al., 1994) and some of its modifications (Tolman, 1992)). The problems are solved in the horizontal plane. This is justified by considering relatively limited water areas, for example, lakes. However, it is doubtful that we can use this approach for waves in oceans with their sizes comparable to the Earth's radius. A specific correction for wave estimation can be introduced taking into account the Earth's curvature, first of all describing swell propagation over large distances that can exceed $90°$ along a great circle arc (Snodgrass et al., 1966).

The spherical shape of the oceanic surface results in another geometrical divergence of wave energy compared to that on a plane surface. This leads to a smaller (or greater) wave height decrease with waves propagating from the generation area. The Earth's rotation may have some influence on the waves resulting in deviation of their propagation trajectory from a geodesic line. However, these effects are so insignificant that they can be completely neglected (Backus, 1962).

Generally speaking, waves travelling directly from the source and waves having curled around the Earth and returning back from the other side can be observed due to the Earth's round shape. This situation is observed, for example, in seismology in the case of Rayleigh wave propagation. It is excluded for hydrodynamic waves in the ocean, as the water surface on the Earth is actually completely overlapped by continents. Therefore, only waves moving directly from the source are considered in this problem (see Fig. 2.1).

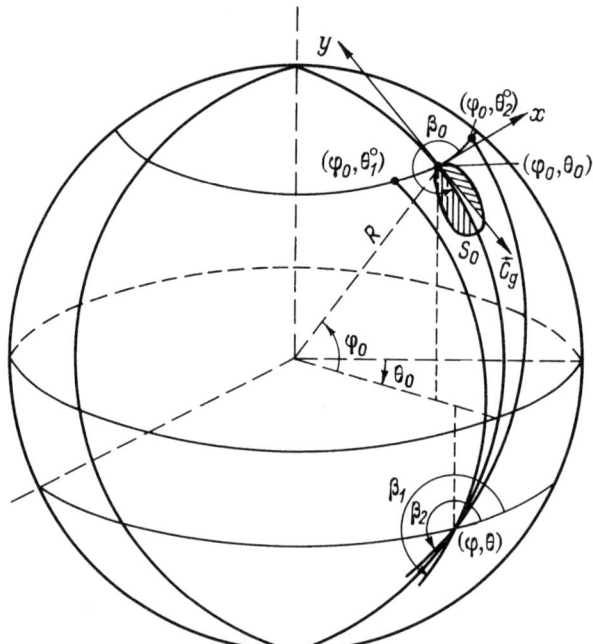

Fig. 2.1. Geometry of the formulation of the problem: $\{x, y\}$ are local rectangular coordinates; $\{\varphi, \theta\}$ are spherical coordinates; c_g is the group velocity vector; β is the angle between the wave vector direction and the x axis; $\theta_1^\circ, \theta_2^\circ$ are wave generation area margins at $\varphi = \varphi_0$; β_1, β_2 are limiting angle values at which the source can be "seen" at the observation point. The "petal" denotes the angular distribution of the initial spectrum

It follows from the general formulation (1.86)–(1.90) of the problem in the case of deep water and the absence of a current, that the wave number $|\boldsymbol{k}|$ and the frequency σ remain constant along the wave propagation trajectory. The equation of the conservation of the wave action density (1.84) can be expressed in terms of the frequency–angle energy spectrum $S(\sigma, \beta)$. Using the relation between the wave action density and the energy spectrum: $N(k, \beta) = S(\sigma, \beta) \cdot (\partial\sigma/\partial k)/k\sigma$, the wave energy balance equation is:

$$\frac{\partial S}{\partial t} + \frac{\partial S}{\partial \varphi}\dot{\varphi} + \frac{\partial S}{\partial \vartheta}\dot{\vartheta} + \frac{\partial S}{\partial \beta}\dot{\beta} = G \ . \tag{2.1}$$

It should be noted that the source function $G(\sigma, \beta, \varphi, \vartheta, t)$ is of a local character and it can be approximated by ratios used in the usual plane coordinate system, as mentioned in Chap. 1. Taking into account wave energy advection, the left part of (2.1) is determined by the surface shape in which the waves propagate.

In the case under consideration, the wave packets move along a geodesic line, which is the shortest distance between the two nearest points in the

2.1 Wind Wave Modelling Using Spherical Variables

surface. The equations of wave packet propagation (1.86)–(1.88) can be expressed as the geodesic line equation (Dubrovin et al., 1986):

$$\ddot{q}^m + \Gamma^m_{ij} \dot{q}_i \dot{q}^j = 0 ,\qquad(2.2)$$

where $\Gamma^m_{ij} = \frac{1}{2} g^{km} \left(\frac{\partial g_{jk}}{\partial z^i} + \frac{\partial g_{ik}}{\partial z^j} + \frac{\partial g_{ij}}{\partial z^k} \right)$ are the Christoffel symbols; q is the generalized coordinate: $q^1 = \varphi$, $q^2 = \vartheta$; and g_{ik} is the metric tensor of the Riemannian surface, prescribed on the spherical surface.

Based on (2.2), the wave packet propagation equation can be written as:

$$\ddot{\varphi} + \frac{1}{2} \dot{\vartheta}^2 \sin(2\varphi) = 0 ;\qquad(2.3)$$

$$\ddot{\vartheta} \cos^2(\varphi) - \dot{\varphi}\dot{\vartheta} \sin(2\varphi) = 0 .\qquad(2.4)$$

The angle β can be determined by the vector projection of the wave group velocity $\mathbf{c}_g$ onto the crossed parallel:

$$\cos(\beta) = \frac{\cos(\varphi)\dot{\vartheta}}{\sqrt{\dot{\varphi}^2 + \cos^2(\varphi)\dot{\vartheta}^2}} = \frac{2\sigma R}{g} \cos(\varphi)\,\dot{\vartheta} .\qquad(2.5)$$

By separating the variables, (2.3)–(2.4) can be easily integrated to give

$$\varphi = \arcsin\left\{ \sqrt{1-\alpha^2} \sin\left[\frac{g}{2R\sigma}(t-t_0) + \arcsin\left(\frac{\sin(\varphi_0)}{\sqrt{1-\alpha^2}} \right) \right] \right\} ;\qquad(2.6)$$

$$\vartheta = \vartheta_0 + \arctan\left(\frac{\sin(\varphi)}{\sqrt{\cos^2(\varphi)/\alpha^2 - 1}} \right) - \arctan\left(\frac{\sin(\varphi_0)}{\sqrt{\cos^2(\varphi_0)/\alpha^2 - 1}} \right) ,\qquad(2.7)$$

where $\alpha = \cos(\varphi_0)\cos(\beta_0)$. The integration constants in (2.6) and (2.7) are obtained using the initial conditions: $\varphi = \varphi_0$, $\vartheta = \vartheta_0$ at $t = t_0$.

Based on the ratio (2.5), it can be shown that:

$$\cos(\varphi)\cos(\beta) = \cos(\varphi_0)\cos(\beta_0) .\qquad(2.8)$$

This means that the product of $\cos(\varphi)$ and $\cos(\beta)$ remains constant along the trajectory for the wave packet propagating on a sphere.

Thus, if the wave packet propagates to the northeast starting from the latitude φ_0, crossing this parallel at angle β_0 ($\beta_0 \leq \pi/2$), the angle β is decreased, attaining zero value, $\beta_0 = 0$, at latitude $\varphi = \arccos[\cos(\varphi_0)\cos(\beta_0)]$. Then the wave packet continues propagating in the reverse direction and returns to the initial latitude φ_0 at an angle $\beta = -\beta_0$, moving to the southern hemisphere. The parallel, in which the ray "turn" occurs, seems to be a peculiar caustic line, relative to which the direction of the wave packet motion is changed. Actually, there is no turn at all. The effect is created by the geometry of the spherical surface, on which the wave packet propagates along the arc of the Earth's great circle.

Thus, the problem of wind wave energy evolution under oceanic surface sphericity results in integrating the spectral equation (2.1) with the characteristics (2.6)–(2.8) for the prescribed boundary or initial conditions.

2.2 Correspondence of Local and Global Coordinate Systems

The transformation of the spectral energy balance equation (2.1), written with the use of the spherical variables φ, ϑ, β, to a local plane rectangular coordinate system $\{x, y\}$ is now considered. The local coordinates $\{x, y\}$ around the point $\{\varphi_0, \vartheta_0\}$ (Fig. 2.1) are introduced as follows (Kamenkovich & Monin, 1978):

$$x = R\cos(\varphi_0)(\vartheta - \vartheta_0) \ ; \quad y = R(\varphi - \varphi_0) \ . \tag{2.9}$$

Then

$$\frac{dx}{dt} = R\cos(\varphi)\frac{d\vartheta}{dt} \ , \quad \frac{dy}{dt} = R\frac{d\varphi}{dt} \ .$$

Substituting the variables (2.9) into (2.1) gives

$$\frac{\partial S}{\partial t} + \frac{\partial S}{\partial x}\frac{dx}{dt} + \frac{\partial S}{\partial y}\frac{dy}{dt} + \frac{\partial S}{\partial \beta}\frac{d\beta}{dt} = G \ . \tag{2.10}$$

The equation (2.10) coincides with the spectral density equation, written in the usual problem formulation in the plane rectangular coordinate system $\{x, y\}$, with the exception of the fourth term in its left part. Using the expression (2.8), this can be written as:

$$\frac{\partial S}{\partial \beta}\frac{d\beta}{dt} = -\frac{g}{2\sigma}\tan\left(\frac{y}{R} + \varphi_0\right)\cos(\beta)\frac{\partial S}{\partial \beta} \ . \tag{2.11}$$

The expression (2.11) can be interpreted as a local error arising in the energy spectral density estimation using the rectangular coordinate system $\{x, y\}$ due to the fact that the Earth's surface sphericity is not taken into account. The value of the expression (2.11) is considerably less than the other items in the left side of (2.10) for regional spatial–temporal scales. However, in the case of $|y/R + \varphi_0| \to \pi/2$, the expression (2.11) becomes sufficiently large, and it cannot be neglected. Since the local coordinate system (2.9) is introduced for a small area around the point $\{\varphi_0, \vartheta_0\}$, (i.e. at $y/R \ll 1$), the term (2.11) is sharply increased close to the poles ($\varphi_0 = \pm\pi/2$). In this case the largest errors occur in (2.10), due to neglecting the expression (2.11). The question arises: how to make the optimal choice of the sphere projection onto the plane under the conditions of the problem.

The characteristics of the spectral energy balance equation are typically written in the rectangular coordinate system $\{x, y\}$ as:

$$\frac{dx}{dt} = \frac{g}{2\sigma}\cos(\beta) \; ; \quad \frac{dy}{dt} = \frac{g}{2\sigma}\sin(\beta) \; . \tag{2.12}$$

The angle β is assumed to be constant along the wave packet propagation trajectory, i.e. $\beta = \beta_0$. However, the angle β varies, depending on the latitude φ, according to (2.8) with wave propagation on a sphere. An additional error is introduced in omitting this fact in the definition of the spectral energy in the local coordinate system (2.11). An estimate of the change in the angle β in the first approximation can be presented as:

$$\Delta\beta = c\tan(\beta_0)\tan(\varphi_0)\Delta\varphi \; .$$

The variation of the angle β is essential even for middle latitudes and relatively limited water areas (for example, for the Black Sea, $\varphi \approx 45°$, $\Delta\varphi \approx 5°$ and $\Delta\beta \approx 5°$). The variation of the angle β is increased as $\beta_0 \to 0, \pi$, and a higher-order approximation should be used for this estimation. This effect is intensified with enlargement of the water area.

Thus, the errors appear due to the following two reasons in the transformation to the local plane coordinate system. The first error is due to using a spherical projection onto the plane (Vakhrameyeva et al., 1986). For example, it is done in the NEDWAM model (Burgers, 1990). The second reason is determined by the fact that the angle β is changed with the wave packet propagation in a spherical surface. But this change is not taken into account in local problem formulation.

2.3 Numerical Simulation of Swell Propagation Using the Method of Characteristics

An elementary solution of the energy balance equation (2.1) on a spherical surface can be obtained by neglecting the source function ($G = 0$). This can be justified in the case of swell propagation from wave generation. The following problem with the boundary conditions is considered as:

$$\frac{dS}{dt} = 0 \; ; \quad S(\omega, \beta, \varphi, \vartheta, t)\Big|_{\varphi=\varphi_0} = S_0(\omega, \beta, \varphi, \vartheta, t) \; . \tag{2.13}$$

It follows from (2.1) that the spectral density S remains constant along the trajectory of wave packet propagation. Using (2.6) and (2.7), the solution of the problem (2.13) can be written as:

$$S(\omega,\beta,\varphi,\vartheta,t) = S_0\left\{\omega,\beta_0,\vartheta - \arctan\left(\frac{\sin(\varphi)}{\sqrt{\cos^2(\varphi)/\alpha^2 - 1}}\right)\right. \tag{2.14}$$

$$+ \arctan\left(\frac{\sin(\varphi_0)}{\sqrt{\cos^2(\varphi_0)/\alpha^2 - 1}}\right),$$

$$\left. t - \frac{2R\omega}{g}\left[\arcsin\left(\frac{\sin(\varphi)}{\sqrt{1-\alpha^2}}\right) - \arcsin\left(\frac{\sin(\varphi_0)}{\sqrt{1-\alpha^2}}\right)\right]\right\}.$$

The solution (2.14) for the stationary case $\partial S/\partial t = 0$ will be found. The initial wave energy spectral density is assumed to be independent of time t and is given as:

$$S_0(\omega,\beta_0,\vartheta,t) \tag{2.15}$$

$$= \begin{cases} S_0(\omega)\frac{8}{3\pi}\sin^4(\beta_0[H(\vartheta - \vartheta_1^\circ) - H(\vartheta - \vartheta_2^\circ)]) & \text{with } \pi < \beta_0 < 2\pi, \\ 0 & \text{with } \pi < \beta_0 < \pi, \end{cases}$$

where $H(\vartheta)$ is the Heaviside function.

This problem formulation corresponding to the spectral density source of wave energy is given at the parallel $\varphi = \varphi_0$ between two meridians $\vartheta_1^\circ < \vartheta < \vartheta_2^\circ$ (see Fig. 2.1). The maximum of the angular energy distribution is emitted along a meridian to the south. It is possible to assume that this problem formulation describes the swell evolution propagating in the Pacific Ocean. This could be a result of persistent north winds blowing in the Bering Sea (see Fig. 2.1).

Substituting (2.15) into (2.14), the angle β can be limited by the condition:

$$\sin^2\beta_0 = 1 - \left(\frac{\cos(\varphi)}{\cos(\varphi_0)}\cos(\beta)\right)^2 \geq 0. \tag{2.16}$$

In fact, this means the change in the angular distribution width of the spectral energy density due to the surface sphericity in which the waves propagate. If, at some surface point, the initial wave energy is distributed in the interval of the angles $\beta_1 \leq \beta \leq \beta_2$, corresponding to two arcs of the great circle O_1 and O_2, then the energy would not scatter over the entire spherical surface, but remain between the arcs O_1 and O_2, beyond which "a shadow zone" is observed.

Besides the condition (2.16), the energy range of the change of the angle β at the calculated point is limited by the finite linear width of the initial spectral energy density source (2.15), i.e. $\beta_2 \leq \beta \leq \beta_1$. The values β_1 and β_2 at a given point $\{\varphi,\vartheta\}$ can be found by solving the transcendental equation (2.7). The spectral zero moment is presented as:

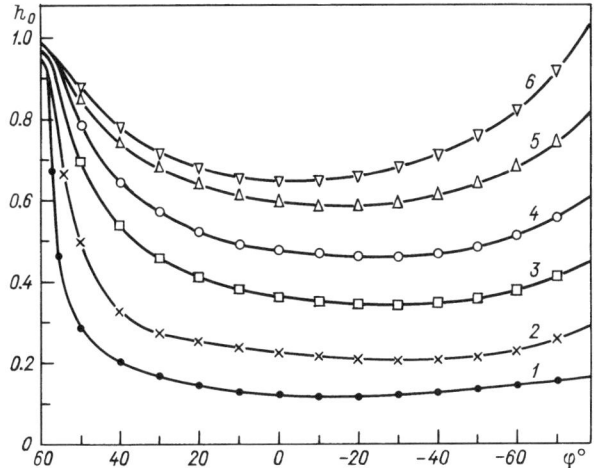

Fig. 2.2. Change of relative average wave heights along a meridian for different generation area $\Delta\vartheta$ at latitude $\varphi_0 = 60°$: $1 - 1°$; $2 - 5°$; $3 - 15°$; $4 - 30°$; $5 - 60°$; $6 - 80°$

$$m_0 = \int_0^\infty \int_{\beta_2}^{\beta_1} S \, d\omega \, d\beta = \int_0^\infty S_0(\omega) \, d\omega \int_{\beta_2}^{\beta_1} \frac{8}{3\pi} \left(1 - \frac{\cos^2(\varphi)}{\cos^2(\varphi_0)} \cos^2(\beta)\right)^2 d\beta$$

$$= \tilde{m}_0 \frac{8}{3\pi} \left\{ \beta - 2\frac{\cos^2(\varphi)}{\cos^2(\varphi_0)} \left(\frac{\beta}{2} + \frac{\sin(\beta)}{4}\right) \right.$$

$$\left. + \frac{\cos^4(\varphi)}{\cos^4(\varphi_0)} \left(\frac{3\beta}{8} + \frac{\sin(2\beta)}{4} + \frac{\sin(4\beta)}{32}\right) \right\} \Bigg|_{\beta_2}^{\beta_1}, \quad (2.17)$$

where $\tilde{m}_0$ is the zero moment for the boundary value of the spectrum (2.15).

It should be noted that the average wave period calculated with the help of the second-order moment ($\tau = 2\pi\sqrt{m_2/m_0}$) coincides with its value at the boundary. The function value in the brackets of (2.17) is changed with increasing distance between the observation point and the boundary. It includes equally both the second-order moment and the zero-order moment. Thus, the average wavelength, calculated using appropriate moments, can be different.

The angle β_1 is determined numerically using the ratio (2.7) for the case $\varphi = 60°$ and $\vartheta = (\vartheta_1° + \vartheta_2°)$ at $\beta_2 = 3\pi - \beta_1$. Then the numerical values of the expression (2.14) are determined. The calculated results for different angles $\varphi_0 - \varphi$ (or distances from the source) and different source widths $\Delta\vartheta$ are shown in Fig. 2.2. As it is seen, the height is decreased with the distance from the source and wave approaching the Equator. Furthermore, the heights of waves propagating from small non-extended sources, are decreased

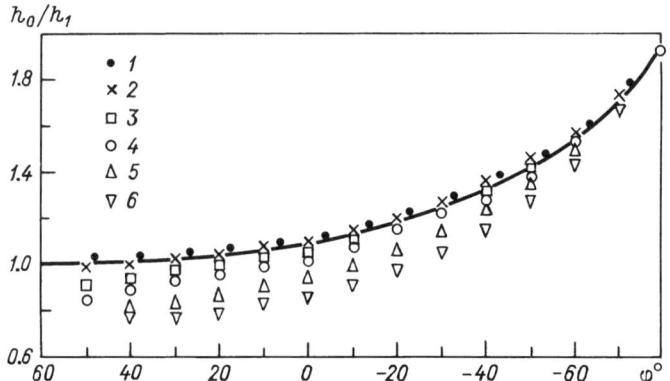

Fig. 2.3. Ratio of average wave heights h_0/h_1 for different coordinates φ. Designations 1–6 are the same as in Fig. 2.2. The *dotted line* denotes approximation (2.16)

to a greater extent. After passing the Equator the wave height is gradually increased with distance. This effect becomes prominent for extended sources ($\Delta\vartheta = 60$–$90°$) and is a direct manifestation of the surface sphericity effect on wave propagation.

The solution should be compared with the similar one obtained for the plane case without taking into account surface sphericity. In order to do that, the triangle with tops $(\varphi_0, \vartheta_1°)$, $(\varphi_0, \vartheta_2°)$, (φ, ϑ) (where $\vartheta = (\vartheta_1°+\vartheta_2°)/2$) should be "stretched" in a plane. The linear width of the source and the distance l from the source to the given point (φ, ϑ) coincide with those considered above.

In this case the final expression for the relative wave height h_1 can be presented in the analytical form:

$$h_1^2 = \frac{m_{01}}{m_0} = \frac{16}{3\pi} \frac{\cos(\varphi_0)}{\cos(\varphi)} \left[\frac{3}{8} \left(\frac{\pi}{2} - \beta_1 \right) + \frac{\sin(2\beta_1)}{4} - \frac{\sin(4\beta_1)}{32} \right], \quad (2.18)$$

where

$$\beta_1 = \arctan\left[\frac{2(\varphi_0 - \varphi)}{\Delta\vartheta \cos(\varphi)} \right].$$

The solution (2.18) is asymptotically described as $h_1 \sim 1/\sqrt{l}$ for large distance l from the source. This means that the wave height is decreased from the source at a large distance in the plane case according to the cylindrical law.

The results presented as the ratio of wave height on a spherical surface to the plane surface h_0/h_1 are given in Fig. 2.3. As can be seen, the ratio h_0/h_1 is always more than that obtained for small sources ($\Delta\vartheta \sim 1°$). As for extended sources ($\Delta\vartheta = 15$–$90°$), the opposite situation can also be observed. The wave height on the sphere is decreased sufficiently quickly for distances

2.3 Swell Propagation Simulation by Characteristics Method

comparable to the source size. This can be less than that in the plane problem. This effect is increased for a larger source size comparable with the Earth's radius. Thus, for small distances from the source, the wave energy is spread to a greater degree in various directions on the sphere, than in the plane. With increasing distance, a point appears on the sphere with the wave height being the same as in the plane, whereas the greater the size is, the further this point is from the source. The wave heights on a spherical surface become larger than on a plane surface with an increase in distance. The increase in wave height is explained by the fact that the spectral components leaving the source and propagating along the great circle arcs gather at a diametrically opposite point, where their trajectories intersect.

The ratio h_0/h_1 for small values $\Delta\vartheta = 1-5°$ can be approximated by the dependence:

$$\frac{h_0}{h_1} \approx \left(\frac{\varphi_0 + 90°}{\varphi + 90°}\right)^{0.246} \quad (-90° < \varphi < 90°), \qquad (2.19)$$

indicating the difference between the calculated wave heights for the sphere and the plane.

A comparison of the wave heights calculated with and without sphericity yields quantitative estimates showing how the result can be changed in this or that case. In particular, the estimated wave height taking into account the sphericity can be twice as large as the wave height calculated using a plane model for wave propagation from north to south in the Pacific Ocean. It should be noted that a comparison of two estimation results is performed for the most favourable case of wave calculation in the plane. Thus, an estimate of the "error from below" is obtained, as soon as the general direction of wave propagation coincides with the meridian, taken as the vertical coordinate axis in the plane model calculation. The distortion of the projection of the cylindrical sphere onto a plane along the meridian is absent, and the angular energy distribution is assumed to be sufficiently narrow. But in the more general case there would always be a problem of choosing an optimal spherical projection onto a plane, with the wave propagation rays not being direct lines.

The aforementioned theoretical estimates can be confirmed by the following facts. First, it should be noted that a stationary problem is considered. The wave generation source is taken in a simplified form without wave dissipation. The observation data testify to a sufficiently weak dissipation of low-frequency swell in the ocean (Davidan et al., 1985). It is noted (Hasselmann et al., 1973) that swell waves can propagate without any significant damping through contrary wind areas. But formulation of the stationary problem with the generation source taken in a simplified form might be rather a rough consideration. This reveals the spherical influence on the wave height distribution in the ocean surface in its simplest form. The picture observed in natural conditions can be distorted without taking this factor into account. As a rule, storm areas move, being changed in time and space. This, as well as wave dispersion, i.e. the difference in propagation velocity of various spectral components, should result in "blurring" of some surface sphericity effects.

One of the first basic papers (Snodgrass et al., 1966) investigating swell propagation over global distances should be mentioned. Six wave stations located along the great circle arc between New Zealand and Alaska had been making swell observations in the Pacific Ocean for two and a half months. The observation results had large scatter, although the average values of the wave height revealed a noticeable decrease of swell travelling over relatively small distances compared to the area of the storm. The wave height decrease coefficient δ (according to the formula $h_0 \sim e^{-\delta l}$, where l is the travelled distance) is estimated as $\delta \sim 2.1 \times 10^{-7}$ m^{-1} for the frequency 0.07 Hz. The height decrease actually ceases with further swell propagation, and the average decrease coefficient makes up no more than 0.2×10^{-7} m^{-1}.

These data can be compared with our calculation. The generation area size is assumed to be 2200 km. This corresponds approximately to the storm horizontal scale estimated by Snodgrass et al. (1966). According to the data, the wave height makes up 0.79 of its initial value at a distance of 1100 km from the storm. The wave height is 0.63 of the initial value at a distance of 12,000 km, being approximately equal to the distance between Alaska and New Zealand. According to our wave calculations for a spherical surface, the wave heights are 0.82 and 0.65 of the initial value, while the wave calculations for the plane surface are 0.95 and 0.38, respectively. As can be seen from the comparison of the aforementioned estimates, the wave heights on a sphere are generally in satisfactory agreement with full-scale observations. The present study points out the importance of taking into account the Earth's spherical surface in the numerical simulation of swell propagation in the ocean.

2.4 The Influence of Current on the Evolution of Waves at the Global Scale

The influence of the Earth's spherical surface on wave propagation in the ocean are considered in this chapter. The influence of current and shallow water was not taken into account previously. This is quite understandable as in most cases the influence of these factors on waves is of a more local character. These problems are considered in more detail in the following chapters. However, the influence of global-scale currents on waves probably with no large gradients, but with some contribution comparable to spherical effects, is considered in this section. For example, the circumpolar current (or the so-called west wind current) is a typical current influencing wave propagation in the Southern ocean, located approximately at latitude 40–60° S. Its current distribution is zonally directed westward.

It should be noted that it was already mentioned in some papers (Snodgrass et al., 1966) that waves due to Pacific Ocean storms near Antarctic shores changed their direction from that calculated with the help of weather

2.4 The Influence of Current on the Evolution of Waves at the Global Scale

maps. It was also pointed out that there were waves observed at places completely in the shadow zone (Munk et al., 1963). An interesting explanation of the problem concerning wave refraction in a current was given by K. Kenyon (1981). He calculated analytically wave propagation rays with current velocity component $V_y(x)$, linearly dependent on the coordinate x. Kenyon described these phenomena of wave refraction in a current and offered explanation of the cause of such anomalies by the Antarctic subpolar current. However, he solved the problem in a plane coordinate system, without taking into account global scales of the current.

It is quite clear that in order to define the wave action spectral density in the most general case, it is necessary to solve (1.84) together with the ray equations (1.86)–(1.90) written with the help of spherical variables. The equations can be solved numerically. However, the influence of the current global scale on waves can be estimated analytically in elementary situations as well.

Thus, the problem should be solved in the deep-water case assuming that the current velocity V is stationary. This varies depending only on the latitude φ:

$$\boldsymbol{V} = \{0, V_\vartheta(\varphi)\} . \tag{2.20}$$

The equations of the characteristics (1.86)–(1.90) with the stationary current (2.20) are as follows:

$$\frac{d\varphi}{dt} = c_g \frac{\sin(\beta)}{R} ; \tag{2.21}$$

$$\frac{d\vartheta}{dt} = c_g \frac{\cos(\beta)}{R\cos(\varphi)} + \frac{V_\vartheta}{R\cos(\varphi)} ; \tag{2.22}$$

$$\frac{dk}{dt} = -k \frac{\tan(\varphi)\cos(\beta)}{R} V_\vartheta \sin(\beta) - k\frac{1}{R}\sin(\beta)\cos(\beta)\frac{\partial V_\vartheta}{\partial \varphi} ; \tag{2.23}$$

$$\frac{d\beta}{dt} = -\frac{\tan(\varphi)\cos(\beta)}{R}[c_g + kV_\vartheta\cos(\beta)] - \frac{k\cos(\beta)}{R}\cos(\beta)\frac{\partial V_\vartheta}{\partial \varphi} . \tag{2.24}$$

However, instead of solving (2.21)–(2.24) directly, an attempt at finding the motion integrals of the system (2.21)–(2.24) is undertaken. It should be noted that the coordinate ϑ is cyclic in this problem formulation. That is why the generalized momentum component k_ϑ (as follows from (1.59) and (1.70)) remains constant along the trajectory of wave packet propagation, i.e. $k_\vartheta = $ const. Using (1.76) and (1.78), the motion integral (i.e. the constant value along the trajectory of wave packet propagation) can be written as:

$$k\cos(\beta)\cos(\varphi) = k_\vartheta/R = \text{const} . \tag{2.25}$$

It can be seen that this motion integral is a generalization of the relation (2.8) obtained for the case without any current.

The second motion integral is a constant of the frequency ω (1.90). It can be written for the prescribed form of current change as:

$$\sqrt{gk} + kV_\vartheta(\varphi)\cos(\beta) = \omega \ . \qquad (2.26)$$

The ratios (2.25) and (2.26) are sufficient to determine the wave number k and the angle β along the trajectory dependent on changing the current velocity $V_\vartheta(\varphi)$ and the latitude φ.

Thus, it is sufficient to solve (2.21) and (2.22) including the dependencies (2.25) and (2.26) instead of the set of equations (2.21)–(2.24). However, even this simplified equation system cannot always be solved analytically. Nevertheless, the ratios (2.25) and (2.26) reveal a number of factors of the wave element transformation in a current. Thus, if a wave packet propagates along the trajectory from the point $\{\varphi_0, \vartheta_0\}$, where the wave number is k_0, and the wave vector direction is β_0, then the wave number k and the angle β can be easily calculated with the help of the ratios (2.25) and (2.26) at the trajectory point $\{\varphi, \vartheta\}$:

$$\frac{gk}{\omega^2} = \left(1 - \frac{k_\vartheta V_\vartheta(\varphi)}{R\omega\cos(\varphi)}\right)^2 \ ; \qquad (2.27)$$

$$\cos(\beta) = \frac{k_\vartheta g}{R\omega^2\cos(\varphi)}\left(1 - \frac{k_\vartheta V_\vartheta(\varphi)}{R\omega\cos(\varphi)}\right)^{-2} , \qquad (2.28)$$

where $k_\vartheta = Rk_0\cos(\varphi_0)\cos(\beta_0)$.

The changes of the wave number k and the angle β, depending on the latitude φ and the current velocity V_ϑ, are described by the ratios (2.27) and (2.28). A fair current reduces the wave number (i.e. increases the wavelength) and the angle β. The current does not influence waves in the specific case when the current velocity changes as $V_\vartheta \sim \cos(\varphi)$.

In some values of the parameters the absolute value of the right-hand side (2.28) can become more than 1. For such values of the latitude φ and the current velocity $V(\varphi)$ there exists no solution for a wave with parameters ω and k. The critical velocity limiting the area where waves exist is determined by the ratio:

$$V_\vartheta = \frac{\cos(\varphi)R\omega}{k_\vartheta}\left(1 \pm \sqrt{\frac{|k_\vartheta|g}{R\omega^2\cos(\varphi)}}\right) \ . \qquad (2.29)$$

When the current velocity is equal to this value, the angle β is equal to zero or π. This corresponds to wave propagation parallell to the current velocity. The plus sign in (2.29) with $\beta > 0$ refers to waves propagating from sub-Antarctic areas towards the Equator, and the minus sign to the opposite direction. As seen from (2.21)–(2.24), the wave packet trajectory makes a turn when the wave packet reaches a point where the ratio (2.29) is satisfied. The latitude of the point can be determined with the help of (2.27) and (2.28):

2.4 The Influence of Current on the Evolution of Waves at the Global Scale

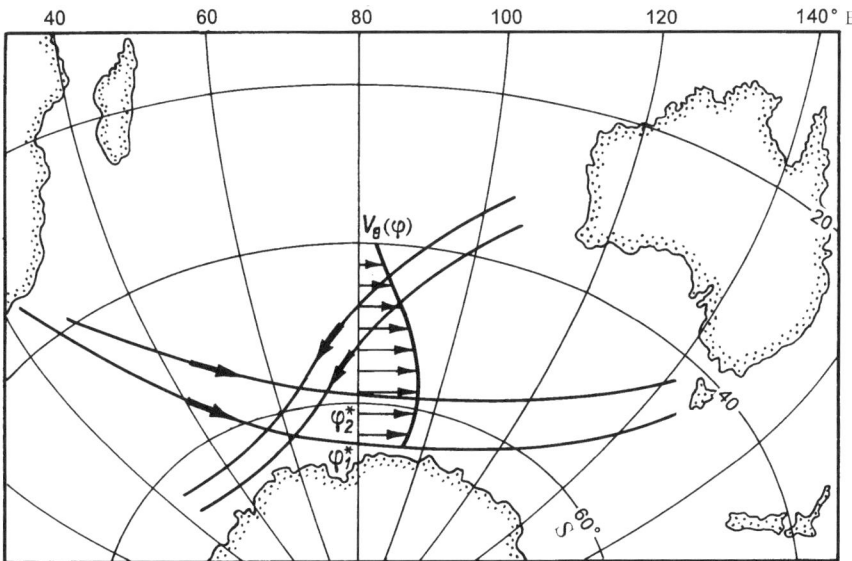

Fig. 2.4. Wave propagation rays in the Southern Ocean under circumpolar current conditions: φ_1^* – ray "turning" points without any current; φ_2^* – ray "turning" points with current

$$\cos(\varphi^*) = \frac{2k_\vartheta V_\vartheta^2}{Rg\left(1 + \frac{2\omega}{g}V_\vartheta \pm \sqrt{1 + \frac{4\omega}{g}V_\vartheta}\right)} \ . \qquad (2.30)$$

Unlike wave propagation without currents, considered in the previous section, in this case the turning point depends not only on the latitude φ and the initial propagation angle β_0 (see (2.8)), but also on the current velocity.

For waves originally propagating in the southern direction, and small values of the ratio $\omega V/g \ll 1$, the expression (2.30) can be written as:

$$\cos(\varphi^*) \approx \frac{k_\vartheta g \cos(\varphi_0)\cos(\beta_0)}{\omega^2\left(1 - \frac{2\omega}{g}V_\vartheta\right)} \ . \qquad (2.31)$$

According to (2.31) for $V_\vartheta = 1\,\mathrm{m\,s^{-1}}$ and the wave period $\tau = 11\,\mathrm{s}$ (with $\omega \sim 0.5\,\mathrm{rad\,s^{-1}}$), the angle φ^* is changed additionally due to current, by approximately 10°, i.e. the turning point is shifted along a latitude over 1000 km south or northward depending on the current velocity direction (see Fig. 2.4).

Estimates of the influence of the current on waves obtained in this section reveal that the current is reflected and the waves cannot reach the Antarctic shores (see Fig. 2.4). This occurs in the case of propagation of northwest waves typically generated by westerly and northwesterly winds. Thus, superposition of incident and reflected waves should be observed to the north of this zone, resulting in increased regional wave density. At the same time, northeast waves deviate from the great circle arc and moved more intensively towards Antarctica due to the currents. The problem of wave spectrum transformation in non-uniform currents is considered in more detailed in the following chapters. Only the kinematic evaluation revealing the influence of small-gradient non-uniform currents on waves propagating over global distances has been presented in this section. The influence of currents appears to be comparable with the effect of the Earth's surface sphericity.

3 Numerical Implementation of the Wave Energy Balance Equation

3.1 Statement of the Problem

Successful solution of the problem of hindcasting and forecasting a sea wind wave depends on the quality of the physical model, the numerical implementation of the wave energy balance equation and the accuracy of the wind field data. It should be noted that in general most modern wind wave models are not characterized by their high-level numerical implementation. The numerical imeplementation of the wave energy balance equation is principally important for accuracy and model efficiency (Lavrenov & Ryvkin, 1989; Ryvkin, 1990; Ryabinin, 1991a,b; Tolman, 1992). Numerical implementation errors are comparable with those obtained due to inaccuracy of wind data and imperfect parameterization of the mechanisms of the physical model.

The method of characteristics used for solving (2.1) allows one to obtain a solution accurately reproducing wave propagation. In order to define the frequency-angular spectrum at one point, all spectral components arriving from the entire water basin should be collected.

However, it is unreasonable to use the method of characteristics in the previous form to estimate wind waves over vast oceanic areas. Firstly, it does not allow the calculation of the non-linear wave interaction, since it is necessary to obtain information about all spectral harmonics at every grid point, whereas they are gathered only at one point. Secondly, it is important to have complete information of the integral wave parameters and the spectrum at all grid points for the prediction of ocean operational conditions. The application of the method of characteristics for large areas is not justified, since all rays must be collected from the entire area at each calculating point at every time step (there are about 400 points in the North Atlantic for a numerical grid with $2.5° \times 2.5°$).

For numerically solving (2.1), the widely-used finite-difference method can be employed. A review of these methods for the plane case is given by Rogers et al. (1999). However, numerical implementation of (2.1) on a sphere is rather difficult. Unlike the solution of the similar problem in a plane, it is necessary to approximate the additional term $\partial S / \partial \beta \cdot \mathrm{d}\beta / \mathrm{d}t$, increasing the dimension of the equation. The problem becomes more complicated when it is implemented in full form, taking into account a spatially non-uniform current and an uneven bottom.

50   3 Numerical Implementation of the Wave Energy Balance Equation

Among the different numerical methods used in wind wave models, the numerical implementation of the WAM model of the wave energy balance equation should be pointed out (Komen et. al., 1994). An attempt was made to solve this equation for a spherical surface, taking into account wave refraction in shallow water and currents. The WAM model is probably the first model to implement the equation most completely. A scheme of directed first- and second-order differences is used in the WAM model to approximate the advective terms in (2.1). The central-difference second-order scheme is applied for approximating the term characterizing the spectral density variation as the direction function β.

However, the numerical scheme chosen in the WAM model for solving (2.1) does not seem to be sufficiently optimal. The problem is that due to large numerical diffusion the error of estimating the term $\partial S/\partial \beta \cdot \mathrm{d}\beta/\mathrm{d}t$ becomes significant when calculating the wave propagation of a narrow-directed spectrum. Thus, an additional source of errors appears as compared with the balance equation for the plane. In order to minimize them it is necessary to increase considerably the number of calculations.

It is interesting to mention the numerical implementation of the wave energy balance equation for a spherical surface proposed by V. Ryvkin (1990). He suggested that it is possible to choose specially discrete directions at each latitude to achieve the maximum approximation in order to get a precise solution. This can be done if a special set of characteristics for the entire grid area is prescribed, so that these characteristics pass through the latitude at different grid points. The sets of directions at different latitudes are not the same. The solutions are determined in terms of the characteristics. That is why there is no need for interpolation in order to calculate the spectral density S depending on the angle β. It should be pointed out that in spite of some advantages of this method, it is rather difficult to generalize the proposed choice of the special direction grid for non-uniform current and uneven bottom.

Another most typical cause of errors in wave energy propagation should be noted. In numerical calculations, the continuous frequency-angular wave spectrum is prescribed as a specific number of spectral components. The finite width of the spectral frequency and angular bands introduces some complications for the numerical simulation of wave propagation. Ideally, the initial wave energy contained originally within some area should propagate quite smoothly over the ocean surface in time. However, in most wind wave models the spectral resolution is so rough that it induces the so-called "garden-sprinkler effect" (Booij & Holtuijsen, 1987). This results in the energy spreading from the source along the directions prescribed in advance by the discrete spectrum representation in the initial area. At some distances from it, an anomalous increased concentration of wave energy is manifested in these directions, but is explicitly insufficient in the other directions. Thus, a limited angular resolution of the model introduces artificial anisotropy in the spatial wave energy distribution. As a consequence of this phenomenon, the model prediction of swell propagation from a distant storm is unsatisfactory.

The existence of this problem was mentioned by many scientists (Ocean Wave Modeling (SWAMP group), 1985; Booij & Holtuijsen, 1987; The WAM model, 1988; Ryabinin, 1991a,b; Tolman, 1992), but no sufficiently simple solution was found. It appears that the most natural way to solve this problem is to increase the spectral resolution. According to the estimations of Booij & Holtuijsen (1987), a typical width of the frequency ($\Delta\omega$) and the angular ($\Delta\beta$) bands for wave calculation in the North Atlantic area should be $\Delta\omega \equiv 0.03\omega$ and $\Delta\beta \equiv 1.5°$. However, the use of such a fine resolution is unlikely to be advisable in practical applications.

The second solution of the "garden-sprinkler" problem was proposed by Booij & Holtuijsen (1987) for the case of wave propagation in a plane surface. They suggest adding two terms to the wave energy balance equation, to correct the effects connected with the finite width of the frequency and angular bands. The inclusion of the correction terms into the numerical wave propagation schemes requires the solution of a more complex equation than the traditional wave energy balance equation. This includes additional terms with second-order partial derivatives and the solution of an additional equation for estimating the wave age, which is not determined locally. The solution of this problem requires additional computational time.

That is why it is necessary to develop an alternative method, which would be no worse than the finite difference method in accuracy and requires less computational time. This approach can be elaborated by combining the method of characteristics and a simple polynomial interpolation allowing us to solve (1.84) at the grid points on a sphere.

Thus, an attempt is undertaken to solve the well-known "philosophical problem" of the wave description – whether waves are considered to be a field description or particle propagation. The method of characteristics considers the propagation of wave packets (particles), and at the same time the finite-difference and interpolation methods appear to describe the field. In our opinion, the proposed numerical method is different from ordinary finite-difference methods because it describes more precisely the correspondence with wave propagation physics.

It should be noted that the alternative approach proposed below removes the "garden-sprinkler" effect without a noticeable increase in computation time. Moreover, application of this approach in the semi-Lagrangian numerical method (hereinafter referred to as the interpolation-ray or the INTERPOL method) allows the use of much larger integration time steps in comparison with the Courant condition (CLF) without losing numerical accuracy.

3.2 Formulation of the Wave Propagation Problem

Main equation. The evolution of a two-dimensional sea wave spectrum $S(\omega, \beta, \varphi, \vartheta, t)$, being a function of the frequency ω, direction β (measured counterclockwise from the parallel), latitude φ, longitude ϑ and time t is described by the equation used in the WAM model (The WAM model, 1988; Komen et al., 1994) in the form:

$$B(S) = \frac{\partial S}{\partial t} + \frac{1}{\cos\varphi}\frac{\partial(\dot\varphi\cos\phi S)}{\partial\varphi} + \frac{\partial(\dot\vartheta S)}{\partial\vartheta} + \frac{\partial(\dot\beta S)}{\partial\beta} = G, \quad (3.1)$$

where $B(S)$ is the differential operator and G is the source function. Based on the set (1.86)–(1.90), the equations of motion for a wave packet along the arc of a great circle can be written as follows:

$$\dot\varphi = C_g \frac{\sin\beta}{R}; \quad (3.2)$$

$$\dot\vartheta = C_g \frac{\cos\beta}{R\cos\varphi}; \quad (3.3)$$

$$\dot\beta = -C_g \tan\varphi \frac{\cos\beta}{R}, \quad (3.4)$$

where C_g is the group velocity and R is the Earth's radius.

Further, a solution of (3.1)–(3.4) for swell wave propagation is investigated, assuming that the source function G is equal to zero.

Initial conditions. In order to formulate more or less real initial conditions, the problem of satellite data assimilation might be taken into consideration. The data are used to improve the results of calculating and predicting wind waves (Bauer et al., 1992; Burgers et al., 1992; Lionello et al., 1992).

It is assumed that wind waves of large height are revealed with the help of satellite altimeter observations in the northern area of the North Sea. Initially, the perturbation centre is located at the point $\vartheta_0 = 0°$, $\varphi_0 = 72°$. The initial perturbation is distributed over space in accordance with the approximation $\exp(-\delta r/L_{\max})$, where δr is the distance between the centre of the initial perturbation and the considered point, and $L_{\max}$ is the correlation radius of the initial perturbation. This can be taken as $L_{\max} = 150$ km for the North Sea scale, according to Burgers et al. (1992). It is assumed that the waves propagate southward at the initial moment of time. This means that the general direction of wave propagation makes up the angle $\beta_0 = -90°$. A significant wave height at the centre would be 10 m and a mean period 15 s.

The spectrum of the initial perturbation is approximated by the formula:

$$S_0(\omega,\beta,\varphi,\vartheta,t=0) = S_0(\omega,\beta)F(\varphi,\vartheta) = S_0(\omega)Q_0(\beta)F(\varphi,\vartheta), \quad (3.5)$$

where $F(\varphi,\vartheta)$ is a spatial distribution function; $Q_0(\beta)$ is a function of the angular energy distribution; and $S_0(\omega,\beta)$ is the initial wave frequency-angular spectrum. The spatial distribution function is assumed to be:

$$F(\varphi,\vartheta) = \exp(-\alpha\sqrt{[(\vartheta-\vartheta_0)^2\cos^2\varphi_0 + (\varphi-\varphi_0)^2]}), \quad (3.6)$$

where $\alpha = 2(R/L)$ and $L = 150$ km, being the constants determining the extent of the decrease of the initial perturbation with distance.

The angular energy distribution function is assumed to be equal to:

$$Q_0(\beta) = \begin{cases} \frac{8}{3\pi}\cos^4(\beta - \beta_0) & \text{with } |\beta - \beta_0| \leq \frac{\pi}{2}, \\ 0 & \text{with } |\beta - \beta_0| > \frac{\pi}{2}. \end{cases} \quad (3.7)$$

The frequency spectrum is described by the dependence:

$$S_0(\omega) = (n+1)\, m_0 \frac{\omega_{max}^n}{\omega^{n+1}} \exp\left[-\frac{n+1}{n}\left(\frac{\omega_{max}}{\omega}\right)^n\right], \quad (3.8)$$

where m_0 is the zero spectrum moment, ω_{max} is the frequency of the spectral maximum and n is a parameter characterizing the frequency bandwidth. For swell, it can be assumed that $n = 5$ (Davidan et al., 1985).

The propagation of the initial perturbation over the water area during 48 hours is considered. The problem is solved analytically and then numerically using two different methods. As the first method, a numerical scheme implemented in the WAM model is used. The interpolation-ray method is used as the second one. The numerical calculation errors are estimated by comparison with the analytical solution.

For obtaining an analytical solution of the problem the ratios (3.2)–(3.4) are substituted in (3.1), written in the advective form (2.1), i.e. in the form of a complete time derivative. As noted above, the spectral energy density is preserved along the characteristics in the case of the source function equal to zero, $G = 0$. This analytical solution (2.14) of (2.1) was obtained in the previous section.

3.3 Standard Numerical Propagation Schemes

Two numerical propagation schemes should be mentioned among the usually used ones. They are implemented into the WAM model (1988). The first-order upwind scheme, applied for (3.1), can be written as:

$$S_j^{n+1} = S_j^n - \sum_k \frac{\Delta t}{\Delta x_k \cos\varphi_j}\left[(u\cos\varphi S^n)_{k_+} - (u\cos\varphi S^n)_{k_-}\right]. \quad (3.9)$$

The second-order leapfrog scheme is written in the following form:

$$S_j^{n+1} = S_j^{n-1} - \sum_k \frac{\Delta t}{2\Delta x_k \cos\varphi_j}\left[(u\cos\varphi S^n)_{k_+} - (u\cos\varphi S^n)_{k_-}\right]$$

$$+ \text{diffusion}, \quad (3.10)$$

where the index n is a time step number, and indices k_- and k_+ are referred to the neighbouring grid points in the upstream and downstream propagation directions, respectively, relative to the reference grid point j. The index k runs over the three propagation directions β, φ, θ and the values u_k, Δx_k denote the velocity components β, φ, θ, (3.2)–(3.4) and the grid spacing in the relevant directions, respectively.

The first-order scheme is characterized by a higher numerical dispersion, with the effective diffusion coefficient $D \sim \Delta x^2/\Delta t$ (The WAM model, 1988). For numerical stability the time step should satisfy the CLF condition $\Delta t < \Delta x/u$, so $D > \Delta x u$. There is a smaller inherent numerical dispersion in the advection term of the second-order scheme. Its disadvantage is an unphysical negative energy generation in sharp gradient regions. This drawback can be eliminated by including explicit diffusion terms, as indicated in (3.10). The explicit diffusion necessary for removal of the negative side lobes in (3.10) should be of the same order as the implicit numerical diffusion in (3.9) in order to make the effective dispersion comparable for both schemes.

Further, the numerical results of the above scheme are compared with the analytical and interpolation-ray method solutions of the problem.

3.4 Elimination of the "Garden Sprinkler Effect"

Kinetic equation corrections determined due to frequency-angular discretization. In order to obtain a numerical solution of the wave energy balance equation (3.1), a representation of the continuous spectrum in discrete frequency-angular form is required. Let the value $S^n(\omega_k, \beta_l)$ be a spectral component at time step n corresponding to frequency ω_k and angle β_l.

The averaged energy $\bar{S}(\omega_k, \beta_l)$ in the range

$$\{(\omega_k - 0.5\,\Delta\omega \leq \omega \leq \omega_k + 0.5\,\Delta\omega), (\beta_l - 0.5\,\Delta\beta \leq \beta \leq \beta_l + 0.5\,\Delta\beta)\}$$

can be determined by integrating the continuous spectrum within this range:

$$\bar{S}(\omega_k, \beta_l) = \frac{1}{\Delta\beta} \frac{1}{\Delta\omega} \int_{\beta_l - \frac{\Delta\beta}{2}}^{\beta_l + \frac{\Delta\beta}{2}} \int_{\omega_k - \frac{\Delta\omega}{2}}^{\omega_k + \frac{\Delta\omega}{2}} S(\omega, \beta)\,d\beta\,d\omega \,. \tag{3.11a}$$

The usual way to estimate the integral (3.11a) is based on the rectangle or trapezium method (Ryvkin, 1990; Ocean Wave Modeling, 1985). However, a more accurate algorithm can be applied, assuming that the value S is a continuous and twice differentiable function. In order to estimate the integral (3.11a), bi-quadratic interpolation including the function values at the grid points $\omega_{k-1}, \omega_k, \omega_{k+1}$ and $\beta_{l-1}, \beta_l, \beta_{l+1}$ can be used. The cubature formula for energy assessment is obtained by "multiplying" the quadrature formulas of each variable (Krylov & Shulgina, 1966):

3.4 Elimination of the "Garden Sprinkler Effect"

$$\bar{S}(\omega_k, \beta_l) = \sum_{i,j=-1}^{1} a_j b_i S(\omega_{k+i}, \beta_{l+j}), \qquad (3.11b)$$

where a_j, b_i are the interpolation coefficients: $a_{-1} = a_1 = b_{-1} = b_1 = 1/24$; $a_0 = b_0 = 11/12$. The averaged energy within this interval contains not only the component $S(\omega_k, \beta_l)$ but the spectra values of the neighbouring components as well.

In a similar way the wave energy balance equation (3.1) can be transformed to describe the mean energy evolution taking into account the frequency $\Delta\omega$ and the angular resolution $\Delta\beta$. This is achieved using an integral operator of the type (3.11a) in (3.1).

The traditional expansion of trigonometric functions valid for small values $|\Delta\beta| < 1$ can be estimated as:

$$\cos\beta_{l\pm1} = \cos(\beta_l \pm \Delta\beta) = \cos\beta_l(1 - (\Delta\beta)^2/2) \\ \mp \sin\beta_l \Delta\beta(1 - (\Delta\beta)^2/6) + O((\Delta\beta)^4); \qquad (3.12a)$$

$$\sin\beta_{l\pm1} = \sin(\beta_l \pm \Delta\beta) = \sin\beta_l(1 - (\Delta\beta)^2/2) \\ \pm \cos\beta_l \Delta\beta(1 - (\Delta\beta)^2/6) + O((\Delta\beta)^4). \qquad (3.12b)$$

Taking into account that the group velocity in (3.1)–(3.4) is inversely proportional to the frequency $C_g \sim 1/\omega$, the following expansion can be applied:

$$\omega_{i\pm1}^{-1} = \omega_i^{-1}\left[1 \mp \frac{\Delta\omega}{\omega} + \left(\frac{\Delta\omega}{\omega}\right)^2 \mp \left(\frac{\Delta\omega}{\omega}\right)^3 + O\left(\frac{\Delta\omega}{\omega}\right)^4\right]. \qquad (3.12c)$$

Thus, (3.1) is written as:

$$\frac{\partial \bar{S}}{\partial t} + \left[\frac{1}{\cos\varphi}\frac{\partial(\dot\varphi(1+\varepsilon/2+\delta)\cos\varphi\bar{S})}{\partial\varphi} + \frac{\partial(\dot\vartheta(1+\varepsilon/2+\delta)\bar{S})}{\partial\vartheta}\right. \\ \left. + \frac{\partial(\dot\beta(1-\varepsilon/2+\delta)\bar{S})}{\partial\beta}\right] + \varepsilon\left[\frac{C_g}{R}\frac{\partial}{\partial\beta}\left(\cos\beta\frac{\partial\bar{S}}{\partial\varphi} - \frac{\sin\beta}{\cos\varphi}\frac{\partial\bar{S}}{\partial\vartheta}\right) + \frac{\partial\dot\beta}{\partial\beta}\frac{\partial^2\bar{S}}{\partial\beta^2}\right] \\ - \delta\frac{C_g\omega}{R}\frac{\partial}{\partial\omega}\left[\left(\sin\beta\frac{\partial\bar{S}}{\partial\varphi} + \frac{\cos\beta}{\cos\varphi}\frac{\partial\bar{S}}{\partial\vartheta} - \tan\varphi\frac{\partial}{\partial\beta}(\cos\beta\bar{S})\right)\right] \\ + O(\varepsilon^2) + O(\delta^2) = 0, \qquad (3.13)$$

where $\varepsilon = (\Delta\beta)^2/12 \ll 1$, $\delta = (\Delta\omega/\omega)^2/12 \ll 1$.

Deriving (3.13), it is assumed that $\varepsilon \sim \delta$. The first two terms (the second is in the first square brackets) in (3.13) are similar to the appropriate terms of the main wave energy balance equation (3.1). In addition they contain the correction terms of order δ and ε, describing changes in the propagation velocity of spectral components connected with the frequency and angular discretizations. It should be noted that the third and fourth terms of (3.13), included in square brackets, are of most interest. They make up the correction of the equation due to the finiteness of the angular and frequency spectral resolution. These terms contain the derivatives on the spatial, angular and frequency variables. They are proportional to the values ε and δ and the group velocity C_g. The last terms in (3.13) are higher-order corrections $O(\varepsilon^2)$ and $O(\delta^2)$.

The correction terms in (3.13) are comparable to a similar expression derived for the case of wave propagation over a plane surface (Booij & Holtuijsen, 1987). This depends on the "wave age", which is not defined locally. In order to determine it, an additional equation must be solved. The main advantage of the correction term (3.13), derived in the present study, consists in determining it locally for wave propagation over a spherical surface. For a plane surface, the kinetic equation can be written in a similar way:

$$\frac{\partial \bar{S}}{\partial t} + C_{gx}\frac{\partial \bar{S}}{\partial x} + C_{gy}\frac{\partial \bar{S}}{\partial y} + \varepsilon C_g \frac{\partial}{\partial \beta}\left\{\frac{\partial(\cos\beta \bar{S})}{\partial y} - \frac{\partial(\sin\beta \bar{S})}{\partial x}\right\}$$

$$+ C_g \omega \frac{\partial}{\partial \omega}\left\{\frac{\partial(\cos\beta \bar{S})}{\partial x} - \frac{\partial(\sin\beta \bar{S})}{\partial y}\right\} + O(\varepsilon^2) + O(\delta^2) = 0, \quad (3.14)$$

where $C_{gx} = (1 + \varepsilon/2 + \delta)\,C_g\cos\beta$ and $C_{gy} = (1 + \varepsilon/2 + \delta)\,C_g\sin\beta$ are components of the group velocity. Thus, additional terms in (3.13) and (3.14) are dependent on the frequency-angular resolution, group velocity and spatial and angular non-uniformity of the wave field.

Investigation of a special case. The additional terms in the left-hand side of the wave energy balance equation (3.13) significantly increase the computation time in wind wave models. However, this problem can be considerably simplified in some special cases.

It should be noted that the strong influence of the "garden-sprinkler effect" is caused by a coarse angular resolution. There are 12 directions usually used in most models (Ryvkin, 1990; Ryabinin, 1991a,b; Booij and Holtuijsen, 1987; Ocean Wave Modeling, 1985), but, actually, this number is not sufficient. At the same time frequency spectral descritization plays a less important role.[1] Our consideration is limited only by the angular descritization effect.

The value $\mathcal{L}$ can be assumed as a typical spatial scale of wave propagation in some basin. For the ocean, the value $\mathcal{L}$ makes up the order of several thousand kilometres while for shelf seas (such as the North Sea), it is the

---

[1] This fact can be quantitatively confirmed by actual assessment of the parameters ε and δ. For example, for the WAM model (1988) $\Delta\beta = \pi/6 \sim 0.5$ and $\Delta\omega/\omega \cong 0.1$, so $\varepsilon \gg \delta$.

3.4 Elimination of the "Garden Sprinkler Effect"

order of several hundred kilometres. Using the results of Booij and Holtuijsen (1987), it can be shown that:

$$\frac{\partial(\cos\beta\bar{S})}{\partial\varphi} - \frac{1}{\cos\varphi}\frac{\partial(\sin\beta\bar{S})}{\partial\vartheta} \cong \frac{R}{\mathcal{L}}\frac{\partial\bar{S}}{\partial\beta}. \qquad (3.15)$$

The factor $(R/\mathcal{L})$ is the order of 10 for a typical spatial sea scale, and it is about 1 for an ocean.

With an exception of the ocean subpolar areas, it can be assumed that the correcting term in the second square brackets of (3.13) is an order larger than the other ones. Keeping the main correction in (3.13), this is written as follows:

$$\frac{\partial\bar{S}}{\partial t} + \frac{1}{\cos\varphi}\frac{\partial(\dot{\tilde{\varphi}}\cos\varphi\bar{S})}{\partial\varphi} + \frac{\partial(\dot{\tilde{\vartheta}}\bar{S})}{\partial\vartheta} + \frac{\partial(\dot{\tilde{\beta}}\bar{S})}{\partial\beta} \cong \mathcal{A}\frac{\varepsilon C_\mathrm{g}}{R}\frac{\partial^2\bar{S}}{\partial\beta^2}, \qquad (3.16)$$

where

$$\mathcal{A} = R/\mathcal{L} - \tan(\varphi)\sin(\beta), \quad \dot{\tilde{\varphi}} = \dot{\varphi}(1+\varepsilon/2),$$

$$\dot{\tilde{\vartheta}} = \dot{\vartheta}(1+\varepsilon/2), \quad \dot{\tilde{\beta}} = \dot{\beta}(1-\varepsilon/2).$$

As seen from (3.16), its left-hand side is an ordinary diffusive operator describing a weak "energy exchange" between the nearest angular components. The parameter $\mathcal{A}$ is dependent on the latitude φ and the angular direction of the wave propagation β. The correcting term value in (3.16) is determined by two factors. The first one is dependent on the angular spectral discretization, the second is determined by the effects of wave propagation on a sphere. The parameter $\mathcal{A}$ decreases with wave propagation northward in the northern hemisphere and increases with wave propagation in the opposite direction. The dependence becomes more significant for the case of wave propagation over global distances.

In the general case the problem of solving (3.16) is connected with the correct estimation of the parameter $\mathcal{A}$. A large value of this parameter causes considerable angle smoothing resulting in anomalous isotropy of the angular energy distribution. But in the case of too small a value of this parameter, the "garden sprinkler" effect cannot be eliminated.

In order to get a clear understanding of the solution of the equation, a common diffusion equation with a simplified right-hand side ($\mathcal{A} = R/\mathcal{L}$) should be considered:

$$\frac{\partial\bar{S}}{\partial\tau} = \delta\frac{\partial^2\bar{S}}{\partial\beta^2}, \qquad (3.17)$$

where

$$\delta = \mathcal{A}(\Delta\beta)^2 C_\mathrm{g}/12\,R. \qquad (3.18)$$

The elementary β-periodical solution of (3.17) can be written as:

$$S_m(\beta,\tau) = B(m)\exp(\pm im\beta - m^2\tau\delta),\qquad(3.19)$$

where $B(m)$ is a function of the parameter m determined by the initial conditions of the problem. In time duration the solution of (3.19) tends to the isotropic spectrum:

$$\lim_{\tau\to\infty} S(\beta,\tau) = B(m) = \frac{1}{2\pi}\int_0^{2\pi} S(\beta,0)\,\mathrm{d}\beta.\qquad(3.20)$$

The characteristic time scale of the process relaxation is estimated as:

$$\Delta\tau = 1/(\delta m^2) = 12\,R/(m^2\mathcal{A}(\Delta\beta)^2 C_\mathrm{g}) = 48\,\pi R/(m^2\mathcal{A}(\Delta\beta)^2 gT),\qquad(3.21)$$

where T is the average wave period. Since in reality the initial non-isotropic wave spectrum should not become quasi-isotropic, it can be assumed that the validity of the approximation (3.16) is limited by the condition:

$$\Delta\tau \geq t_\mathrm{max},\qquad(3.22)$$

where t_max is the maximum duration of wave propagation. This condition gives an upper limit to the value $\mathcal{A}$:

$$\mathcal{A} \leq 48\,\pi R/(m^2(\Delta\beta)^2 gT t_\mathrm{max}).\qquad(3.23)$$

Thus, for an average wave period of 10 s, an angular resolution $\Delta\beta = \pi/12$, $m = 5$ and a maximum duration of 36 hours, it is estimated that $\mathcal{A} \leq 10^2$. A more precise value of the parameter $\mathcal{A}$ can be obtained by a numerical simulation.

Numerical algorithm. The numerical implementation of the proposed correction of the finite angular resolution effects can be presented by a simple finite difference approximation:

$$S^{n+1}(\omega_k,\beta_l) = \nu S^n(\omega_k,\beta_{l-1}) + (1 - 2\nu)S^n(\omega_k,\beta_l)$$
$$+ \nu S^n(\omega_k,\beta_{l+1}),\qquad(3.24)$$

where n is an indicator of the time step number, $\nu = \mathcal{A}C_\mathrm{g}\Delta t/12\,R$ and $\mathcal{A} = R/\mathcal{L}$ are dimensionless parameters depending on typical wave scales. As shown in (3.21), the value ν is not explicitly dependent on the angular resolution $\Delta\beta$. It should be noted that (3.16) describes an angular smoothing of the wave spectrum or a weak "energy exchange" between the angular components, taking place at every time step of wave propagation.

One of the algorithm testing methods consists in controlling whether the total wave energy is conserved or not. In order to estimate the total energy change for all directions, the spectrum is integrated over all angles:

$$\frac{2\pi}{L}\sum_{l=1}^{L}S^{n+1}(\omega_k,\beta_l)\Delta\beta = \frac{2\pi}{L}\sum_{l=1}^{L}\{\nu S^n(\omega_k,\beta_{l-1}) + (1-2\nu)S^n(\omega_k,\beta_l)$$
$$+\nu S^n(\omega_k,\beta_{l+1})\}\,\Delta\beta\,. \quad (3.25)$$

Since the spectrum $S(\omega_k,\beta_l)$ is a periodic function of the directions β_l, it is possible to show that the operator (3.25) preserves the full energy, as:

$$\frac{2\pi}{L}\sum_{l=1}^{L}S^{n+1}(\omega_k,\beta_l)\Delta\beta = \frac{2\pi}{L}\sum_{l=1}^{L}S^n(\omega_k,\beta_l)\Delta\beta\,. \quad (3.26)$$

3.5 Interpolation-Ray (INTERPOL) Method

As an alternative to the numerical scheme used in the WAM model, a numerical scheme, based on a semi-Lagrangian method (Lavrenov, 1992; Ryvkin, 1990; Ryabinin, 1991a,b) is considered. This can be modified for calculating wave propagation on a spherical surface. Further this scheme, in combination with the angular smoothing technique, is called the interpolation-ray (INTERPOL) method.

The wave energy balance equation is solved in the advective form (2.1) for the spectral components propagating along its characteristics (3.2)–(3.4) with the help of the semi-Lagrangian numerical method. Thus, the energy is preserved along the characteristics in the simple swell propagation case (i.e. with zero source function $G = 0$). It is necessary to determine the spectrum value at the initial position of the characteristics.

Now, the grid point (ϑ_i,φ_j) is considered with an incoming wave packet of frequency ω_k and propagation direction β_l. The coordinates of the initial location $(\vartheta_i^0,\varphi_j^0)$, where the wave packet is situated at the previous time step, can be found using (3.2)–(3.4). This position does not coincide with a grid point. Polynomial interpolation should be applied to define the initial value of the spectrum S^0 at the point $(\vartheta_i^0,\varphi_j^0)$ as:

$$S_{ij}^0(\omega_k,\beta_l^0) = \sum_{p=1}^{M}\sum_{q=1}^{L} a_{pq} S_{i+f(p)j+f(q)}^{n-1}(\omega_k,\beta_l)\,, \quad (3.27)$$

where $a_{pq} = a_{pq}(\omega_k,\beta_l^0,\Delta\vartheta,\Delta\varphi,\Delta t,\vartheta,\varphi)$ are interpolation coefficients; $f(p)$ and $f(q)$ are integer functions; and $S_{pq}^{n-1}(\omega_k,\beta_l)$ is the spectral value in the grid point (ϑ_p,φ_q) at the preceding time step $t = t_{n-1}$. The values $M = L = 2$ are used in the bilinear interpolation case which is optimal for obtaining the solution to the problem in the polynomial interpolation class (Ryvkin, 1990).

Another problem of spectral interpolation is that the propagation direction β is not constant along the characteristics with waves propagating in a spherical surface. Using the relation (2.8), the variation of the angle β

along the characteristics can be found as:

$$\cos \beta_l \cos \varphi_i = \cos \beta_l^0 \cos \varphi_i^0 . \qquad (3.28)$$

And again, the initial angle β_l^0 does not coincide with the angles prescribed by spectral representation in the grid point. The spectral density corresponding to β_l^0 may be approximated by a polynomial. The initial propagation angle at the grid point (ϑ_i, φ_j) is assumed to be β_l. It follows from (2.8) that the angle β is dependent only on the latitude φ, but neither on the frequency ω, nor on the longitude ϑ. That is why for all grid points, located at latitude φ_j, the value of the propagation angle is known and equal to β_j. The appropriate spectrum value is also defined. For the grid points located at another latitude, say φ_{j-1}, an equivalent propagation angle can be derived using the relation (2.8).

In order to determine the spectrum value for β_l^0 the following interpolation is used:

$$S_{i-1,j-1}(\beta_l^0) = \sum_{m=-1}^{1} a_m S_{i-1,j-1}(\beta_{l+m}) , \qquad (3.29)$$

where a_m are the interpolation coefficients.

The wave spectrum component is assumed to be zero in the wave shadow zone, which can take place in a spherical surface (in our case formally $|\cos \beta_l^0(\varphi_{j-1})| > 1$).

The smoothing operator (3.25) is applied at the next step of the interpolation-ray method.

The boundary conditions are numerically implemented as follows. If a regular grid point corresponds to land, the appropriate spectral component is assumed to be equal to zero in (3.27).

3.6 Comparison of Results of Numerical Wave Propagation Schemes

Numerical grid and determination of main value results. For numerical simulations of swell propagation over a spherical surface a grid stretching in longitude from $-12°$ to $12°$ and in latitude from 51 to $75°$ is chosen. The grid includes 25×49 points, with a spacing of $0.5°$ in latitude and $1.0°$ in longitude, comprising approximately 55 km between neighbouring points at the grid area centre. This can be considered as a simplified area of the Norwegian and North Seas. A simplified form of the area allows us to obtain an accurate analytical solution of the problem.

The problem of wave propagation starting from the initial condition (3.5)–(3.8) is solved analytically, and after that using in turn the numerical upwind scheme (3.9) and the INTERPOL method. The numerical results of different methods is compared by their integral values enumerated below.

3.6 Comparison of Results of Numerical Wave Propagation Schemes

The mean wave height $h(\varphi, \vartheta, t)$ is determined as:

$$h^2(\varphi, \vartheta, t) = 2\pi \iint S(\omega, \beta, \varphi, \vartheta, t)\, d\omega\, d\beta. \tag{3.30}$$

The normalized total energy of the propagating wave field is as follows:

$$\sum(t) = E(t)/E(t=0), \tag{3.31}$$

where:

$$E(t) = \iiiint S(\omega, \beta, \varphi, \vartheta, t) R^2 \cos\varphi\, d\omega\, d\beta\, d\varphi\, d\vartheta. \tag{3.32}$$

The latitude coordinate value referred to the propagation of the wave area centre (the time shift of the centre of mass coordinate) is determined:

$$\langle \varphi(t) \rangle = \frac{1}{E(t)} \iiiint \varphi(t) S(\omega, \beta, \varphi, \vartheta, t)\, R^2 \cos\varphi\, d\omega\, d\beta\, d\varphi\, d\vartheta. \tag{3.33}$$

The wave energy diffusion is evaluated in space with time. The parameter $\Theta(t)$, characterizing this value, is determined as the square root of the surface area. It contains waves with a height greater than $1/3$ of their maximum value at the current moment t:

$$\Theta(t) = \sqrt{T(t)/T(0)}, \tag{3.34}$$

where

$$T(t) = \iint F(\varphi, \vartheta, t) R^2 \cos\varphi\, d\varphi\, d\vartheta \tag{3.35}$$

and $F(\varphi, \vartheta, t)$ is the Heaviside function:

$$F(\varphi, \vartheta) = F\left(h(\varphi, \vartheta, t) - \frac{1}{3} h_{\max}(t)\right). \tag{3.36}$$

The root mean square (RMS) error of the calculation of the wave height over the water area can be evaluated as:

$$\mathrm{RMS}(t) = \sqrt{\sum_{i,j} \mathrm{ERR}^2(\varphi_i, \vartheta_j, t)/N}, \tag{3.37}$$

where N is the total number of grid points and ERR is the normalized local wave height error:

$$\mathrm{ERR}(t) = \frac{h_{\mathrm{model}}(\varphi, \vartheta, t) - h_{\mathrm{anal}}(\varphi, \vartheta, t)}{h(t)}, \tag{3.38}$$

where $h_{\mathrm{anal}}^{\max}(t)$ is the maximum value of the wave height calculated analytically over the entire area at time t.

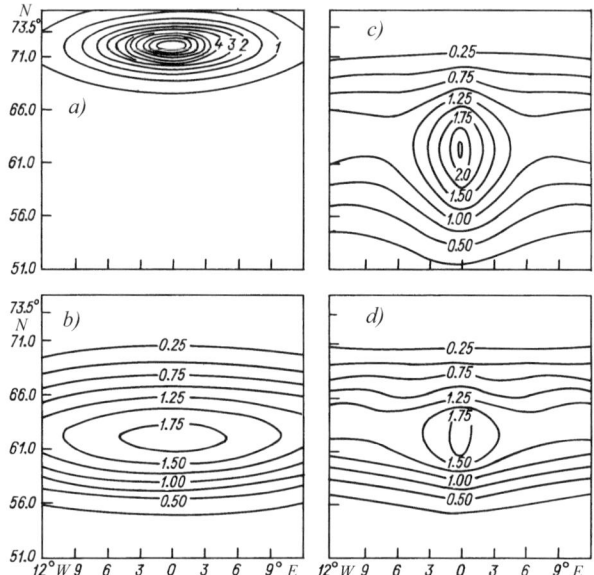

Fig. 3.1. Spatial wave height distribution at the initial time (**a**) and after 24 hours obtained analytically (**b**), by scheme (3.9) (**c**) and the INTERPOL method (**d**)

Numerical results. The initial spatial distribution of the wave height is shown in Fig. 3.1a. The results of wave height distributions after 24 hours of propagation obtained by the analytical solution, the upwind numerical scheme (3.9)[2] and the INTERPOL method are presented in Fig. 3.1b,c and d, respectively. There are 12 directions and a time step of 20 minutes was used in the numerical calculations. The characteristic shape of the numerical solution is presented in Fig. 3.1c,d. The spatial distribution of normalized wave height errors (map of errors) ERR (3.38) at $t = 24$ hours is shown in Fig. 3.2. The calculations using the first numerical scheme (3.9) and, to a lesser extent, the INTERPOL results show the tendency for the wave energy to concentrate along the basic directions prescribed initially by the spectral representation in the initial perturbation area. The wave heights at these directions are overestimated, while in the other directions, they are underestimated. This is a purely geometrical effect caused by the coarse angular resolution of the scheme, being a manifestation of the "garden-sprinkler effect".

The form of the spatial wave height distribution in a water area is determined by the spectral discrete presentation prescribed beforehand in the disturbance source. In Fig. 3.1, the general direction of wave propagation coincides exactly with the angular direction (base) prescribed by the discrete representation in the perturbation source. Several analogous numerical ex-

---

[2] The calculations using the WAM model were made by J. Onvlee (the Royal Netherlands Meteorological Institute, KNMI).

3.6 Comparison of Results of Numerical Wave Propagation Schemes   63

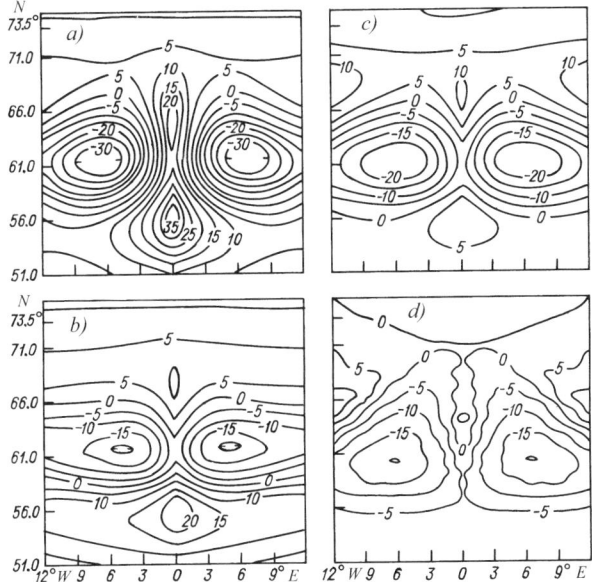

Fig. 3.2. ERR error spatial distribution (in per cent) at $t = 24$ hours: (**a**) by the scheme (3.9) with 12 directions, time step is 20 min; (**b**) by the scheme (3.9) with 24 directions, time step is 20 minutes; (**c**) by the INTERPOL method with 12 directions, time step is 20 minutes; (**d**) by the NTERPOL method with 12 directions, time step is 6 hours

periments have been carried out. Thus, the wave propagation direction is shifted by the angle $\Delta\beta/2$ with respect to the basic direction prescribed at the source. As shown in Fig. 3.1, most of the wave energy in the numerical solution is concentrated along one of these directions close to the general angle of wave propagation. The spatial wave height distribution is split into a double-peaked structure with the angle being precisely between two basic directions used in the numerical scheme. However, in this case the degree of anisotropy and the wave height error are not significantly different from the case in Fig. 3.1.

The difference between the numerical results of the upwind scheme (3.9) and the INTERPOL method and the analytical solution is shown quantitatively in Fig. 3.2. A high degree of anisotropy induced by the limited angular resolution (see Fig. 3.1b–d) is distinctly revealed in the distribution of the normalized wave height error. This is clearly seen for the results obtained by the numerical scheme (3.9) with 12 directions at time $t = 24$ hours. The local errors are 40 per cent overestimated in the general direction of wave propagation and 35 per cent underestimated in the other directions. These errors are increased with time, especially in the general direction of wave propagation. The numerical results are overestimated by 25 per cent at $t = 12$ hours, and in some points the error comprises 90 per cent at $t = 48$ hours. In the lateral directions, the underestimation of numerical values comprises 15–20 per cent at $t = 12$ hours to 40 per cent at $t = 48$ hours of wave propagation.

Increasing the number of directions to 24 reduces the error level approximately by a factor of two (see Fig. 3.2b) in the numerical scheme (3.9). The results of the INTERPOL method with 12 directions (see Fig. 3.2c) are in better agreement with the analytical solution compared to the results of the scheme (3.9). Its accuracy is comparable to that of the scheme (3.9) with 24 directions. The INTERPOL method with 24 directions further reduces the error level by a factor of two.

As mentioned above the time step is not limited by the CLF criterion in the INTERPOL method. Actually, its accuracy could be improved by applying larger time steps. This can be shown by comparing the error levels of the INTERPOL method for time steps of 20 minutes and 6 hours (see Fig. 3.2c,d).

The parameters of the integral solution (the temporal evolution of the total wave energy, position of the wave field centre, degree of energy diffusion and root mean square error (3.32)–(3.34), (3.37)) are presented in Fig. 3.3. As shown, both numerical schemes are able to reproduce the value of the solution of the first two mentioned parameters (see Fig. 3.3a,b). The results of the INTERPOL method show a slightly larger dispersion than the results of the scheme (3.9). The energy distribution obtained in the grid area by the INTERPOL method is closer to the exact solution, as can be seen from the results of the energy diffusion function $\Theta(t)$ (see Fig. 3.3c).

It is interesting to consider the root mean square (RMS) error (see Fig. 3.3d). There is a monotonic error increase for all numerical methods in the initial wave propagation stages (up to 24 hours). After that the error begins decreasing due to the wave area extending beyond the numerical grid. The RMS error (for the entire numerical area) obtained by the scheme (3.9) with 12 directions comprises 8 per cent in 12 hours and 20 per cent in 40 hours. The shift of the basic directions by $\Delta\beta/2$ leads to approximately the same error. A two-fold increase in the number of directions decreases the level of errors approximately by half for the middle stage of wave propagation. The INTERPOL method with 12 directions and the same 20 minute time step results in a 5 per cent error at $t = 12$ hours and 12.5 per cent at $t = 40$ hours. However, the increase of the time step to 3 hours gives a 3 per cent error at $t = 12$ hours and 11 per cent error at $t = 40$ hours. The 6-hour time step gives a 10 per cent error at $t = 40$ hours. Thus, the error level decreases with increasing time step.

In order to study the dependence of the model results on the spatial form of the initial perturbation, the calculations for a number of functions and the extent of their decrease from the initial perturbation centre are repeated. It turns out that in the prescribed numerical grid the form of the initial perturbation does not significantly influence the wave height distribution and evolution. Qualitatively, the characteristic features of the spatial wave height

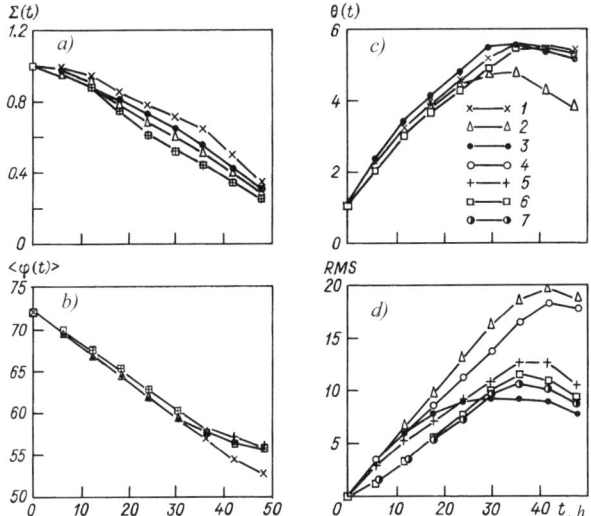

Fig. 3.3. Temporal variations of the integral parameters of the numerical solution: (**a**) normalized total energy $\Sigma(t)$; (**b**) latitude coordinate of wave centre area $\langle\varphi(t)\rangle$; (**c**) energy diffusion over space $\Theta(t)$; (**d**) root mean square RMS(t) error. 1 – analytical solution; 2 – scheme (3.9) with 12 directions, 20 minute time step; 3 – scheme (3.9) with 24 directions, 20 minute time step; 4 – scheme (3.9) with 12 directions, 20 minute time step and with the shift of basic directions at $\Delta\beta/2$; 5 – INTERPOL method with 12 directions, 20 minute time step; 6 – INTERPOL method with 12 directions, 3 hour time step; 7 – INTERPOL method with 12 directions, 6 hour time step

distribution and the evolution of the integral parameters are very similar for different forms of initial perturbations. However, quantitatively, the local error and the degree of the spatial wave height anisotropy are increased with decreasing spatial "spreading" of the initial wave perturbation.

3.7 Numerical Integration of the Source Function

Statement of the problem. The numerical solution of the wave energy balance equation when the source function is assumed to be zero was considered in the previous section. The INTERPOL method was shown to be effective. Now, it is interesting to consider the solution of the equation with a non-zero source function forming a wind wave spectrum under the influence of different physical mechanisms. The numerical solution of the wave energy balance equation can be divided into two stages: numerical implementation of the wave energy propagation and time integration of the source function. It is shown by Tolman (1992) that insufficient accuracy of the numerical solution can result in incorrect interpretation of the physical processes forming a wind wave spectrum.

To integrate numerically the right-hand side of the wave energy balance equation, a number of first and second order explicit and implicit numerical schemes (Ocean Wave Modeling, 1985; Ryvkin, 1990) are used, including different Runge–Kutta schemes. It should be noted that the use of high-order numerical schemes requires a lot of CPU time, so they are ineffective for wind wave operational models.

The problem is that different wind wave spectrum bands are developed with a different rate, depending on the wind speed and wave frequency. As a result, the processes of high-frequency spectrum band formation happen quite rapidly. Their numerical calculation requires a sufficiently small time step that leads to increasing the number of steps and computing time.

This problem is solved differently in numerical models. In most cases some definite limits are used to restrict the increase of the spectrum value in the high-frequency bands (Ocean Wave Modeling, 1985).

In the numerical implementation of the WAM model (The WAM model, 1988; Komen et al., 1994) the spectrum is divided into two ranges: a prognostic one, covering both the maximum area and the low-frequency band, and a diagnostic one, covering the spectral high-frequency tail. The two conditions are imposed on the spectral value in the second range. The first one consists in prescribing the dependence of the spectral energy density, such as $S(\omega) \sim \omega^{-5}$, starting with some definite frequency. In addition, the evolution rate of the wave energy spectral density (or the source function value) should not exceed a definite value. This is represented in the numerical implementation of the WAM model as follows:

$$G(S, \omega, \beta) = \text{sign}(G(S, \omega, \beta)) \min(|G|, G_{\max}) , \qquad (3.39)$$

where $G_{\max} = 0.62 \times 10^{-4} f^{-5}$, and f is the cyclic frequency $f = \omega/2\pi$.

As noted by Burgers (1990), the constraints for the spectral density value introduced in this manner indicate an unsatisfactory description of the wave energy dissipation and inefficient numerical integration algorithm of the wave energy balance equation. If the spectrum and source function constraints introduced in the numerical implementation are used, their influence on the solution of the spectrum is not quite clear. The question appears whether the wave energy preservation law is violated or not. Do the introduced constraints serve as additional energy sources?

The problem is that the wave energy dissipation function, being quasi-linear on the spectrum (4.30), is used in the WAM model. But this dissipation function does not balance the spectrum in its high-frequency band. That is why some additional physically unjustified constraints are introduced in the numerical implementation of the model at every time step.

Another important remark connected with practical wind wave calculations should be noted. The wind speeds are introduced into the model with some time step usually coinciding with the synoptic time interval. As a rule it comprises 12, 6 or 3 hours (in most favourable cases).

In this respect, the most optimal numerical scheme of the wave energy balance equation consists in obtaining the most accurate numerical solution in every synoptic term (i.e. at the time moment corresponding to the outcome of the wind wave result), using the least number of iterations. The "ideal" numerical scheme should produce the most accurate calculated value in every corresponding synoptical term in one numerical iteration. However, this is hardly possible due to numerical instability or insufficient accuracy of the calculations. Probably, it is necessary to find some compromise between the time step of the initial information input, and the number of numerical iterations and numerical errors.

Numerical schemes for integration of the source function of the wave energy balance equation. Different numerical schemes used for integrating the source function of the wave energy balance equation are considered, to investigate their accuracy and effectiveness. The following schemes can be identified as explicit, semi-implicit and implicit.

Explicit schemes. Two simple numerical schemes are considered as examples of explicit schemes. Euler's first-order explicit scheme is written in the form of:

$$S_{n+1} = S_n + \Delta t G(S_n, U_n) , \qquad (3.40)$$

where S_n, S_{n+1} are the energy spectral densities at time steps n and $n+1$, respectively; Δt is the time step; and G is the source function estimated using the spectral density S_n and the wind speed U_n.

The Adams explicit two-step scheme (or the predictor–corrector method) is executed in two consecutive steps:

$$S^{(1)}_{n+1} = S_n + \Delta t G(S_n, U_n) ; \qquad (3.41a)$$

$$S^{(2)}_{n+1} = S_n + \frac{1}{2}\Delta t \left(G(S_n, U_n) + G(S^{(1)}_{n+1}, U_{n+1}) \right) , \qquad (3.41b)$$

where $S^{(1)}_{n+1}$ is the preliminary spectral density value (predictor), and $S^{(2)}_{n+1}$ is its specified value (corrector) at the time step $n+1$. The source function G_n can be dependent on the wind speed U, having different values at time moments n and $n+1$.

Semi-implicit scheme of the WAM model. The numerical scheme used in the WAM model (Komen et al., 1994) is presented as an example of a semi-implicit scheme. This scheme is based on the trapezium implicit formula:

$$S_{n+1} = S_n + \frac{1}{2}(G(S_n, U_n) + G(S_{n+1}, U_{n+1}))\Delta t . \qquad (3.42)$$

To obtain the value S_{n+1} in explicit form, it is necessary to solve the equation:

$$S_{n+1} - \frac{1}{2}(G(S_{n+1}, U_{n+1}))\Delta t = S_{n+1} + \frac{1}{2}(G(S_n, U_n))\Delta t \, . \tag{3.43}$$

This can be done precisely only in some simple cases. But in general the source function $(G(S_{n+1}, U_{n+1}))$ is dependent on the spectral density S_{n+1} in a sufficiently complicated way.

To solve (3.43) relative to S_{n+1}, the source function G_{n+1} is expanded in the Taylor's series:

$$G_{n+1} = G_n + \frac{\partial G_n}{\partial S}\Delta S + \ldots \, . \tag{3.44}$$

The functional derivative in (3.44) is written in the form of diagonal Λ_n, and non-diagonal N_n matrices:

$$\frac{\partial G_n}{\partial S} = \Lambda_n + N_n \, . \tag{3.45}$$

Substituting (3.45) into (3.43), the following expression is obtained:

$$\left[1 - \frac{1}{2}(\Lambda_n(U_{n+1}) + N_n(U_{n+1}))\Delta t\right]\Delta S$$
$$= \frac{1}{2}(G(S_n, U_n) + G(S_n, U_{n+1}))\Delta t \, , \tag{3.46}$$

where $\Delta S = S_{n+1} - S_n$.

According to the papers (The WAM model, 1988; Komen et al., 1994), the contribution of non-diagonal terms to (3.46) is sufficiently small to be neglected. Thus, the change of spectral density at time step n is equal to:

$$\Delta S = \frac{1}{2}(G(S_n, U_n) + G(S_n, U_{n+1}))\Delta t \left[1 - \frac{1}{2}\Delta t \Lambda_n(U_{n+1})\right]^{-1} . \tag{3.47}$$

It is required to calculate not only the source function, but also its derivative (3.43) in order to utilize the semi-implicit scheme (3.47).

It should be noted that the explicit schemes (3.40)–(3.41) and the semi-implicit scheme (3.47) can be used for sufficiently general forms of the source function G.

Implicit schemes. The most effective method of numerical solution for the wave energy balance equation consists in implementation of implicit numerical schemes. However, there are no recommendations for their development in the general case. Every specific case requires additional investigation of the right-hand side function of the wave energy balance equation. The use of implicit schemes requires determining in advance the source function

3.7 Numerical Integration of the Source Function

in analytical form. An attempt can be undertaken to represent the source function in polynomial form or as a spectral density power expansion:

$$G(\omega, \beta, S, \boldsymbol{U}) = \sum_i A_i S^i(\omega, \beta, \boldsymbol{U}) \,, \qquad (3.48)$$

where A_i are the expansion coefficients, depending on the frequency ω, the direction of the spectral component propagation β and the wind speed $\boldsymbol{U}$.

In the simplest case, assuming that the main contribution to the source function is a linear term in the spectrum $G = A_1 S$ (where A_1 is a generalized coefficient), the trapezium method (3.42) results in the following formula:

$$S_{n+1} = S_n \frac{1 + \frac{1}{2}\Delta t A_n}{1 - \frac{1}{2}\Delta t A_{n+1}} \,. \qquad (3.49)$$

It should be noted that the numerical method based on using (3.49) is called the optimal order method (Arushanyan & Zaletkin, 1990).

The development of the optimal scheme seems to be based on the explicit form of the source function, taking into consideration its non-linear expansion terms. For example, it is effective to use the non-linear dissipation function of the spectrum. In cases of known explicit form of the source function there is no need to use its expansion in Taylor series, as in the semi-implicit scheme (3.47).

The source function is assumed to have the following form:

$$G(S) = BS(1 - cS^\alpha) \,, \qquad (3.50)$$

where B, c and α are generalized coefficients. The value B may be dependent on time t as well. For a wind wave, for example, it can be assumed that B is an increment of the wave energy increase due to the wind wave energy input (according to Miles' theory). The parameter c limits the energy increase due to wave dissipation by some definite value. If the equilibrium interval $S_\infty(\omega, \beta)$ is assumed to be a limited value, then $c = S_\infty^{-\alpha}$. This means that the dissipation function depends non-linearly on the spectrum. For example, Phillips (1985) suggested that the wave energy dissipation function is cubically dependent on the spectrum for the wave-breaking dissipation mechanism.

In this case the determination of the formula for the implicit numerical scheme results in solving the algebraic equation:

$$S_{n+1}\left[1 - \frac{1}{2}B_{n+1}(1 - cS_{n+1}^\alpha)\Delta t\right] = S_n + \frac{1}{2}(G(S_n, U_n))\Delta t \,. \qquad (3.51)$$

This can easily be solved, for example, with $\alpha = 1$ or $\alpha = 2$. The solution of (3.51) with $\alpha = 1$, which makes physical sense, can be presented in the convenient form for numerical calculations:

$$S_{n+1} = \frac{2A_2}{b + \sqrt{b^2 + 4a_2 A_2}} \,, \qquad (3.52)$$

where

$$b_2 = \frac{1}{2}B_{n+1}\Delta t - 1 \; ; \quad a_2 = \frac{1}{2}cB_{n+1}(\Delta t)^2 \; ; \quad A_2 = S_n + \frac{1}{2}(G(S_n, U_n))\Delta t \; .$$

It should be noted that a similar algorithm can be developed for the case of the non-linear interaction in the source function (see Sect. 4.1). However, in this case some additional difficulties can appear. Not only should the spectral density for the considered wave component $\{\omega_i, \beta_j\}$ be taken into account, but also the analogous values for other components $\{\omega_{i+k}, \beta_{j+n}\}$. That is why it is necessary to solve the corresponding non-linear set of algebraic equations rather than one equation (3.51).

Test results of different numerical schemes. Some of the above-mentioned numerical schemes have been tested and compared with the numerical and analytical solutions. These can be obtained in the constant wind case for the source function (3.50). The precise analytical solution is written as follows:

$$S(t) = S_0(cS_0^\alpha \pm |1 - cS_0^\alpha|\, e^{-\alpha Bt})^{-1/\alpha} \; , \tag{3.53}$$

where the $(+)$ sign is used for the case where the value $1 - cS^\alpha$ is greater than zero, and the $(-)$ sign is for the opposite case.

The energy spectral density values calculated with the help of different numerical algorithms and the analytical solution are presented in Fig. 3.4, 18 minutes after starting the evolution. The $15\,\mathrm{m\,s^{-1}}$ wind speed value and the Miles wind input formula for the parameter B are used. In the numerical integration a 3-minute time step is implemented.

As seen in Fig. 3.4, there are essential differences between the results, in spite of a sufficiently small time step for numerical integration. A comparison of the numerical results with the analytical solution shows that the least accurate are the values calculated with the help of the Euler explicit method. It should be noted that although the results obtained according to the Adams explicit method are in satisfactory agreement with the analytical solution, there is an anomalous negative value at the frequency $4\,\mathrm{rad\,s^{-1}}$. The values calculated by the semi-implicit method (3.47) and the implicit numerical scheme (3.52) are mostly close to the analytical solution.

Further calculations were made with increased numerical integration steps. It should be noted that the results obtained with a help of the Euler and Adams explicit schemes are unstable, especially in the high-frequency band. That is why they are no longer used.

The results using an analytical and three numerical schemes with a 20-minute integration step at the time moment of 3 hours are presented in Fig. 3.5. The initial stage of the calculations is characterized by the overestimated spectral density value calculated using the implicit scheme (3.52). The underestimated values are calculated by the semi-implicit method (3.47) and

3.7 Numerical Integration of the Source Function   71

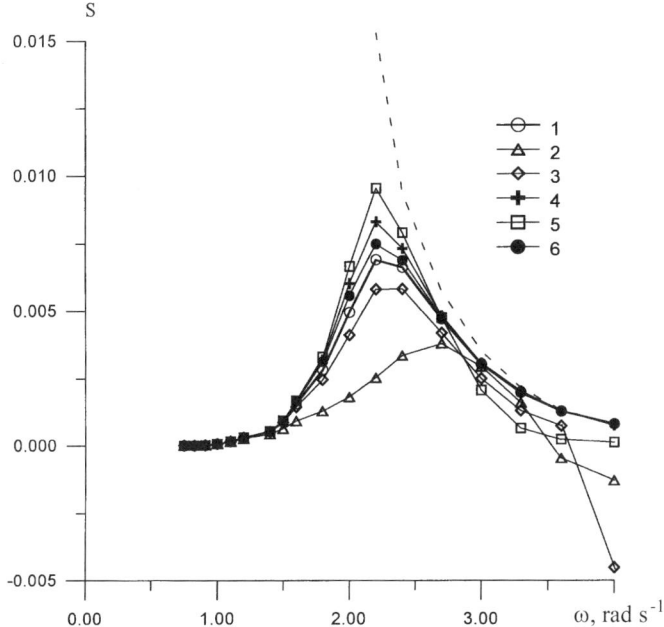

Fig. 3.4. Spectral energy density values calculated by different numerical algorithms at time moment $t = 18$ minutes with integration step $t = 3$ minutes: 1 – analytical solution; 2 – Euler's method (3.40); 3 – Adams' method (3.41); 4 – semi-implicit method (3.47); 5 – optimal order method (3.49); 6 – implicit method (3.52). The *dotted line* denotes the equilibrium interval values

the scheme (3.49), the latter even attaining negative values. To a different extent all three numerical solutions approach the analytical one during further calculations. Being the most accurate, the implicit method (3.52) is quite close to the analytical method for all frequency bands after 4–6 iterations.

Similar calculations are repeated with the numerical integration steps equal to 1, 3, 6 and even 12 hours. The results carried out with a help of all numerical schemes are stable. The results obtained by the implicit numerical scheme (3.52) are the most acceptable for large numerical integration steps.

The analytical solution (3.53) gives a quantitative estimate of the relative error of the numerical solutions. It can be determined at every time step as the total difference between the numerical and analytical solutions. They are normalized by the maximum value of the analytical solution (root mean square error RMSE):

$$\text{RMSE}_i(t) \qquad (3.54)$$
$$= \sum_{j=1, k=1}^{N,M} \sqrt{\left(\frac{S_i(\omega_j, \beta_k, t) - S_{\text{anal}}(\omega_j, \beta_k, t)}{S_{\text{anal}}(\omega_j, \beta_k, t)|_{\max}}\right)^2 \bigg/ (N-1)(M-1)},$$

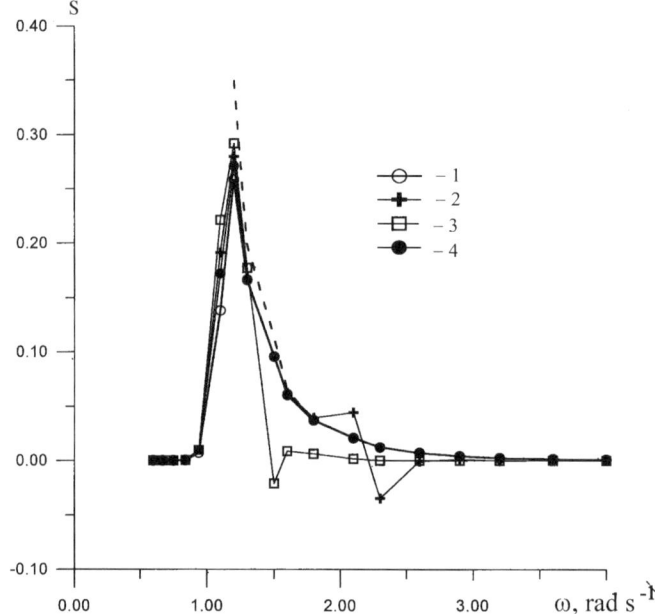

Fig. 3.5. Calculated spectral density values by different numerical algorithms with integration step $\Delta t = 20$ minutes at 3 hours: 1 – analytical solution; 2 – semi-implicit method (3.47); 3 – optimal order method (3.49); 4 – implicit scheme (3.52). *Dotted line* denotes the equilibrium interval values

where the sum is calculated for all spectral components: frequencies ($j = 1, \ldots, N$) and directions ($k = 1, \ldots, M$). The index i denotes the results for different numerical schemes.

It should be noted that normalization by the maximum spectral density of the analytical solution in (3.54) yields "an estimate from below". The calculation error would be much greater if it were normalized by the current spectral density value.

The changes of calculation errors in time for different methods and the integration time step of 1 hour are shown in Fig. 3.6.

The general tendency of decreasing calculation error is related with numerical values, approaching the analytical solution and increase of the maximum spectral density, used for normalizing the calculation error in (3.54).

The relative changes in the calculation error turned out to be the same in comparison with the previous calculations. The most accurate calculations came from the implicit numerical scheme (3.52). With the exception of the first integration steps, there are obvious advantages of the scheme (3.52) compared with the other ones within the entire calculation range.

The numerical experiments reveal that the Adams scheme (3.41) is the most accurate among the explicit schemes (3.40, 3.41). However, these

3.7 Numerical Integration of the Source Function 73

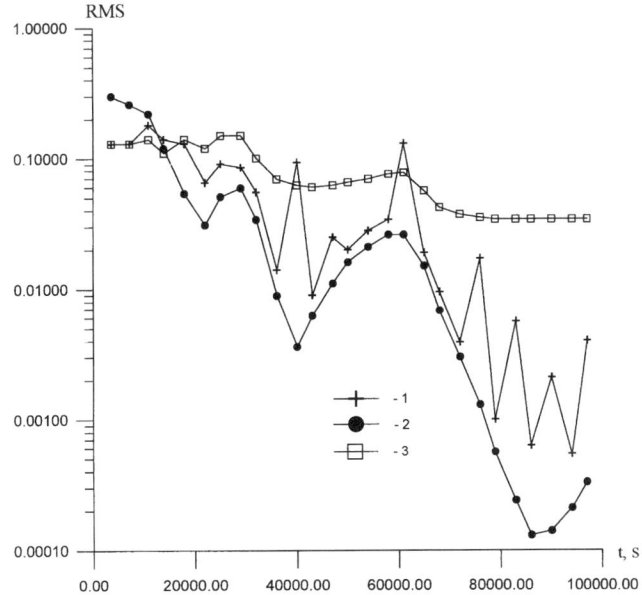

Fig. 3.6. Root mean square error calculations for different methods and integration time step $\Delta t = 1$ hour. 1 – semi-implicit method (3.47); 2 – implicit method (3.52); 3 – optimal order method (3.49)

schemes lose their stability with increasing integration time step. They become practically inapplicable for the simplest conditions of wave generation with time steps $\Delta t \geq 5$ minutes (for wind speed $U = 15 \, \text{m s}^{-1}$).

The numerical scheme using the formula (3.49) (the optimal order method) produces quite good results in comparison with other methods. This scheme is more stable and can be used with large time steps up to $\Delta t \approx 15$ minutes.

The semi-implicit numerical scheme proposed in the WAM model (3.47) is stable for sufficiently large time steps. However, it should be noted that the calculation accuracy is diminished with increasing integration time step.

The implicit scheme (3.52) is the most stable. It shows stable results not only for time steps $\Delta t \approx 60$ minutes, but for $\Delta t \approx 3$ hours and even for $\Delta t \approx 12$ hours. Though during the first integration steps the calculation error can be sufficiently appreciable, after passing some "threshold" the error is diminished, with its value being quite satisfactory. It is enough to make 4–6 numerical iterations to obtain a sufficiently accurate result.

Splitting method. Numerical solution of the wave energy balance equation was considered earlier without the non-linear interaction G_{nl}, which is discussed in the next chapter. It should be mentioned that numerical calculation of the non-linear interaction function G_{nl} is rather difficult. Besides it leads to an unstable solution in the high-frequency band of the wave spectrum. Additional solution limitations are introduced in its numerical implementation (The WAM model, 1988; Komen et al., 1994). As noted, such limitations are not physically well-grounded. The application of the implicit scheme is difficult for numerical integration of the wave energy balance equation due to the complicated expression of the non-linear interaction. In this case either the explicit schemes (3.40)–(3.41) or the semi-implicit scheme (3.47) can be used.

In this section an attempt is undertaken to consider a more efficient alternative scheme than those mentioned above. For this purpose the splitting method is introduced.

The total source function G is supposed to include the wind wave energy input, non-linear dissipation and non-linear energy transfer G_{nl}. It can be written as:

$$G(S) = BS(1 - cS^\alpha) + G_{nl} \ . \tag{3.55}$$

In order to solve the problem of the evolution of the spectrum in time, the energy balance equation is expressed in the form:

$$\frac{\partial S}{\partial t} = G(S) \ . \tag{3.56}$$

The numerical form of (3.56) is as follows:

$$\frac{S^{n+1} - S^n}{\Delta t} = BS^{n+\rho} - Bc(S^{\alpha+1})^{n+\rho} + G_{nl} \ . \tag{3.57}$$

This scheme can be used for any value ρ: $0 \leq \rho \leq 1$. It should be remembered that there is the Euler explicit scheme for $\rho = 0$, an implicit scheme for $\rho = 1$ and it is assumed that $\rho = 1/2$ for the rectangle method.

Now the scheme (3.57) with fractional steps is considered. At the first step the variation in the solution due to non-linear interactions can be calculated as:

$$\frac{S^{n+\rho} - S^n}{\Delta t} = G_{nl} \ . \tag{3.58}$$

At the second step the variation in the solution due to the remaining part of the source function can be determined:

$$\frac{S^{n+1} - S^{n+\rho}}{\Delta t} = BS^{n+\rho} - Bc(S^{\alpha+1})^{n+\rho} \ . \tag{3.59}$$

If $S^{n+\rho}$ is deduced from the first equation (3.58) and substituted into the left-hand part of the second equation (3.59) or the sum of these two equations by terms, the scheme with fractional steps is reduced to the scheme (3.57). The finite difference scheme (3.59) is actually a simple Euler one, requiring a small time step compared to other known schemes.

3.7 Numerical Integration of the Source Function

On the other hand, a reverse transformation can be done from a finite-difference equation to a differential equation, solving it at the time step $(1-\rho)\Delta t$. The equation (3.59) is rewritten in the form:

$$\frac{S^{n+1} - S^{n+\rho}}{(1-\rho)\Delta t} = \frac{1}{1-\rho}(BS^{n+\rho} - Bc(S^{\alpha+1})^{n+\rho}). \qquad (3.60)$$

The following differential equation corresponds to it:

$$\frac{\partial S}{\partial t} = \frac{1}{1-\rho}(BS - Bc\, S^{\alpha+1}) \qquad (3.61)$$

with the initial condition $S^{n+\rho}$ determined by (3.58). There is the analytical solution (3.53) of (3.61) with the value B replaced by $B/(1-\rho)$.

The scheme with fractional steps (3.58) and (3.59), where the second equation is replaced by the differential equation (3.61) at $\rho = 0$, is called a splitting or a semi-analytical method, the solution of (3.61) exists in the analytical form (3.53). It can be used at every time step, when it is necessary to solve the general equation (3.57).

Tests of the numerical scheme of the solution to the energy balance equation with non-linear energy transfer function. Unlike the previous tests, it is impossible to obtain an accurate analytical solution for the wave energy balance equation when it includes a non-linear energy transfer function (see Sect. 4.1). That is why the most accurate numerical solution should be used instead of the analytical one. The splitting method can be tested by comparing the numerical solutions with the results obtained with the help of the predictor–corrector method (3.41).

The function of the wind wave energy input is prescribed by (4.20), as used in the WAM model (Komen et al., 1994). The wind speed is taken as $18.45\,\mathrm{m\,s^{-1}}$. The dissipation is determined according to (3.50) with the varying parameter α. At first it is assumed to be equal to 2, corresponding to dissipation depending cubically on the spectrum. Then the calculations are made, with this parameter dependent on the frequency function: $\alpha = \nu(\omega/\omega_{\max})^\gamma$, where $\nu = 1$, $\gamma = 2$.

According to numerical experiments, if the integration time step is assumed to be not more than 3 minutes, stable results of integrating equation (3.56) are obtained with the help of the predictor–corrector method (3.41). However, a constraint has to be introduced to reduce the wave energy spectral density at the high-frequency band (at frequencies greater than or equal to double the frequency of the spectral maximum), so that its values do not exceed the equilibrium interval.

As shown by numerical calculation, the solution obtained by the splitting method is stable for time steps of 1, 3 and even 6 hours. That is why there is no longer any need to introduce any solution or source function limitations.

The results of numerical calculations for the frequency spectrum at a time moment $t = 30$ hours are shown in Fig. 3.7. The calculations are performed

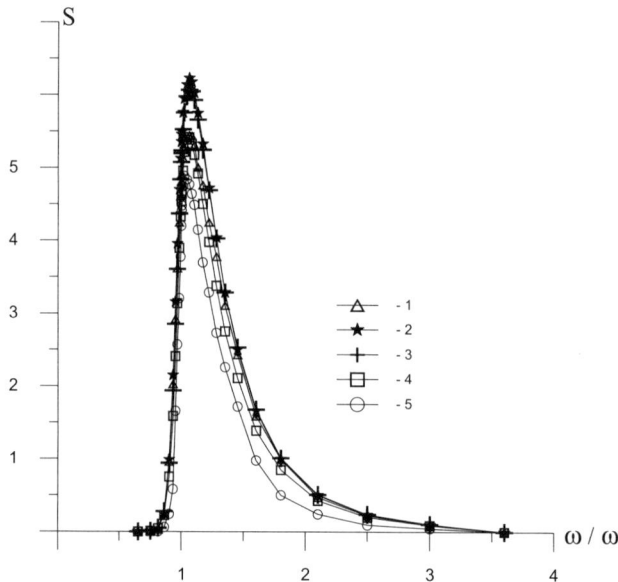

Fig. 3.7. Numerical spectrum solutions at time moment $t = 30$ hours, obtained using different methods: 1 – solution without non-linear energy transfer function; solutions with non-linear energy transfer function: 2 – predictor–corrector method with 3 minute time step, 3 – spliting method with 1 hour time step, 4 – spliting method with 3 hour time step, 5 – spliting method with 6 hour time step

with the help of the predictor–corrector and splitting methods with different integration time steps. A comparison shows that the calculation results obtained with the help of the predictor–corrector method (with the 3 minute integration time step) practically coincide precisely with those obtained by the splitting method with a 1 hour integration time step.

Moreover, with increasing integration time step up to 3 and 6 hours, the spectral value is close to the results obtained using the predictor–corrector method. There is only a 10 per cent difference towards smaller values for the integration time step of 3 hours and 20 per cent for a time step of 6 hours. In this case the spectral density value is closer to that calculated without taking into account the non-linear interaction. It is an indication that there is an underestimation of this mechanism with large numerical integration time steps.

The calculation results of the total energy change in time, made using the same methods, are presented in Fig. 3.8. The comparison results reveal a sufficiently high calculation accuracy, obtained by the splitting method within the entire time range (up to 2.0×10^5 s $= 55.6$ hours) with the 1 hour integration step. The wave energy growth is decreased as the integration step is increased up to 3 and 6 hours.

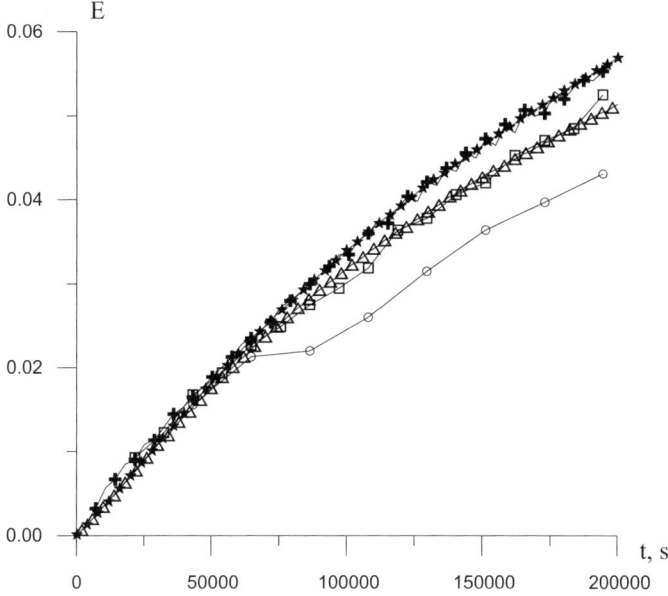

Fig. 3.8. Energy time variation obtained using different integration methods. Designations are the same as in Fig. 3.7

This difference is probably connected with the use of the Euler integration method (3.40), whose result is rough enough at the first fractional integration step (3.58). The numerical error can be significantly reduced if the more precise explicit (3.41) or semi-implicit (3.47) integration methods are used.

In conclusion it should be noted that the performed tests reveal the high efficiency of the splitting method for integrating the wave energy balance equation with the non-linear energy transfer function. An additional problem consists in substantiating the source function with energy wind input and dissipation, which allows a precise analytical solution (3.53). The problem of the optimal choice of the source function is discussed in the next chapter. It should be noted that the adopted dissipation function contains a number of free parameters. Their adjustment allows the model to be adapted to a number of functions describing the non-linear mechanism of wind wave energy dissipation.

Adjustment of integration of the source function with the wave energy propagation scheme. The aforementioned wave time evolution calculations were made for one spatial point and a uniform wind field. And, naturally, the question arises about the validity of the application method for a more complicated case of wave calculation in a specific area, with the wind field being non-uniform and non-stationary. It should also be noted that the wind field cannot be considered to be uniform and stationary for waves travelling distances in one large time step.

The use of the interpolation-ray method as well as the effective numerical method of source function integration facilitates the solution of the problem and essentially enlarges the numerical integration time step. Obviously, the time step can be limited by the solution accuracy as well as by the time scale of the wind field variation.

As the first step it is possible to determine the initial wave coordinates $(\vartheta_i^0, \varphi_j^0)$, which arrived at the regular grid point (ϑ_i, φ_j), using (3.2)–(3.4). These initial coordinates $(\vartheta_i^0, \varphi_j^0)$ may not coincide with a regular grid point. The polynomial interpolation (3.27) is used to determine the initial value of the spectrum S_{ij}^0 at the point $(\vartheta_i^0, \varphi_j^0)$. In case the source function G is not equal to zero, the spectral density is not constant along the trajectory of the wave packet propagation. In order to integrate numerically the source function along the characteristic, one of the above described predictor–corrector (3.41), semi-implicit (3.47) or splitting (3.58), (3.59) methods can be used.

It is important to take into account that at the initial moment of every time step at the point $(\vartheta_i^0, \varphi_j^0)$ the energy spectral density S_{ij}^0, the source function G_{ij}^0 and the wind speed $\boldsymbol{U}$ are determined by interpolation (3.27) at $t = t_n$. But at a finite step moment these values are taken in the point (ϑ_i, φ_j) at $t = t_{n+1}$.

In conclusion it should be noted that this method of solution can also be easily generalized for the general case (1.84), (1.86)–(1.90), i.e. in a spherical surface with non-uniform currents and a basin with an uneven bottom. For practical calculations, the sets of coordinates and the angles of harmonic propagation should be previously calculated for every regular grid point, taking into account wave refraction in the current and shallow water. Their typical scale variation should be much larger than the spacing between the numerical grid points. At the next step the numerical calculation of the wave energy balance equation can be made more quickly, as the solution is reduced to interpolation with the coefficients calculated beforehand.

3.8 Conclusions

The implementation of an optimal numerical scheme for the wave energy balance equation is considered in this chapter. A sufficiently effective algorithm for the solution of the problem has been elaborated using the fact that at every time step the solution can be divided into calculations of the wave energy propagation and the spectral density change due to the total source function. An attempt has been made to apply the accurate analytical solution in combination with the interpolation methods and the simplest finite-difference schemes.

The examples of numerical calculations of swell propagation reveal that the first-order numerical scheme of solving the wave energy balance equation used in the WAM model can lead to significant numerical errors in calculating the wave energy propagation. The interpolation-ray method is proposed as an alternative numerical one.

There is a major difference between the WAM model numerical scheme and the interpolation-ray method. The former is based on using the finite-difference approximation of the wave energy balance equation. The stability of this method is limited by the CFL condition. The integration time step is limited by the condition:

$$\Delta t \leq \{(R\Delta\vartheta \cos\varphi/C_g) \; ; \; (R\Delta\varphi/C_g)\} \; ,$$

where $\Delta\varphi$ and $\Delta\vartheta$ are special latitude and longitude steps. In the WAM model this restriction (especially for high latitudes) results in using a small time step: $\Delta t \leq 15-20$ minutes. On the other hand, the interpolation-ray method uses an accurate analytical solution describing the wave energy propagation. That is why it is absolutely stable for any time step and should not meet the CFL condition. The time step used in the interpolation-ray method is limited principally by the physical boundaries of the problem. The accuracy of the interpolation-ray method depends on the interpolation error in determining the initial spectrum value at every time step. Naturally, this error is less for a smaller spatial grid and for larger time steps, as the general number of steps, where errors are accumulated, are smaller for the total time of wave propagation.

There is certainly an advantage of the interpolation-ray method not only over the first-order numerical scheme, used in the WAM model, but also it can be considered as a general alternative for wind wave estimation. Its synthesis with the implicit or semi-analytical numerical method of integrating the right-hand part of the wave energy balance equation allows the construction of the optimal scheme to solve the problem in the most general case. This scheme is sufficiently stable for large time steps taking into account the source function in its generalized form, including non-linear interaction in the wind wave spectrum. It gives the possibility of using a minimum number of iterations to obtain the most accurate solution at the time moment coinciding with the synoptic term, i.e. at the time of output of the result.

4 Study of Physical Mechanisms Forming the Wind Wave Energy Spectrum in Deep Water

Introduction. As has been noted, the study of different physical mechanisms forming the wind wave spectrum is one of the central problems associated with wave simulation. The source function in the right-hand side of the wave energy balance equation reflects the formal representation of these mechanisms. At present there are many papers devoted to this problem. The most detailed description can be found, for example, in the latest monographs published in Russia (Problems of Research and Mathematical Modeling of Wind Sea, 1995) and in some other countries (Komen et al., 1994; Massel, 1996; Young, 1999). There is no need to present all these results in detail, but the most important aspects should be considered.

It is assumed in most wind wave models (Davidan et al., 1985; Problems of Research and Mathematical Modeling of Wind Sea, 1995; Komen et al., 1994; Ocean Wave Modeling, 1985) that the source function G in deep water includes additively three main components: G_{in} – wind wave energy input, G_{ds} – wave energy dissipation and G_{nl} – non-linear energy transfer within the wave spectrum, caused by a four-wave resonance interaction between the spectral components.

It is reasonable to consider the theoretical and practical importance of the problem once more.

4.1 Non-Linear Energy Transfer in Wind Wave Spectrum

4.1.1 Problem Review

The problem of non-linear energy transfer in the wind wave spectrum was formulated by K. Hasselmann (1960, 1962, 1963, 1965, 1966) and V. Zakharov (1968) in the 1960s. As a result of non-linear interaction, the wave spectrum evolution equation can be written as follows:

$$\frac{\partial N(\boldsymbol{k})}{\partial t} = \iiint T(\boldsymbol{k},\boldsymbol{k}_1,\boldsymbol{k}_2,\boldsymbol{k}_3)\, \delta(\boldsymbol{k}+\boldsymbol{k}_1-\boldsymbol{k}_2-\boldsymbol{k}_3)\, \delta(\sigma+\sigma_1-\sigma_2-\sigma_3)$$
$$\times \{N_2 N_3 (N+N_1) - N_1 N (N_2+N_3)\}\, \mathrm{d}\boldsymbol{k}_1\, \mathrm{d}\boldsymbol{k}_2\, \mathrm{d}\boldsymbol{k}_3\,, \qquad (4.1)$$

where $N_i = N(\boldsymbol{k}_i)$ is the spectral density of the wave action; $T(\boldsymbol{k}, \boldsymbol{k}_1, \boldsymbol{k}_2, \boldsymbol{k}_3)$ is the core function of the non-linear interaction between wave components; and $\delta(\boldsymbol{k})$ and $\delta(\sigma)$ are the Dirac delta-function describing the resonance interaction conditions between four wave components:

$$\boldsymbol{k} + \boldsymbol{k}_1 = \boldsymbol{k}_2 + \boldsymbol{k}_3 \ ; \tag{4.2a}$$

$$\sigma + \sigma_1 = \sigma_2 + \sigma_3 \ . \tag{4.2b}$$

The resonance condition is shown schematically in Fig. 4.1.

Hasselmann (1962) interpreted the integral (4.1) in the terms of a quadrupole interaction between three active wave components (defining the interaction intensity) and the fourth passive component, receiving energy without directly affecting the interaction.

The most important property of the kinetic equation (4.1) is the preservation of the following three integral values:

total wave action:

$$A = \int N(\boldsymbol{k}) \, \mathrm{d}\boldsymbol{k} \ ; \tag{4.3a}$$

total energy:

$$E = \int \sigma N(\boldsymbol{k}) \, \mathrm{d}\boldsymbol{k} \tag{4.3b}$$

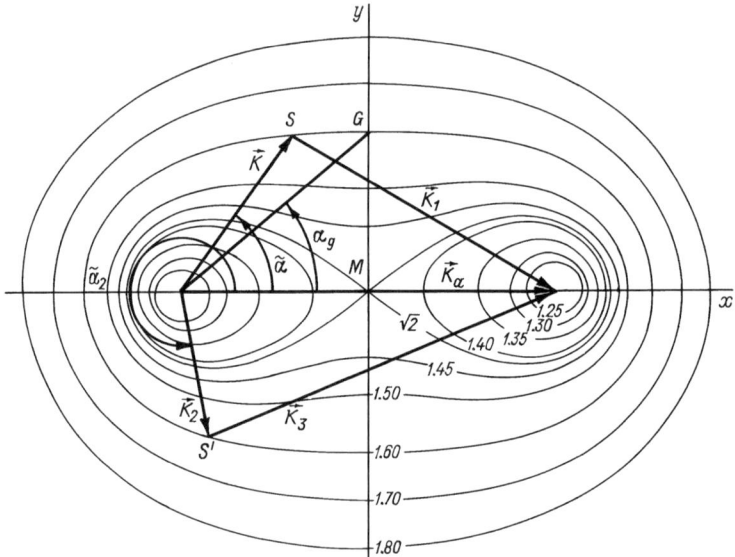

Fig. 4.1. Four-wave interaction diagram (Hasselmann, 1963)

and total momentum:

$$\boldsymbol{K} = \int \boldsymbol{k}\, N(\boldsymbol{k})\, \mathrm{d}\boldsymbol{k}\,. \qquad (4.3c)$$

It should be noted that these values remain constant in wave evolution.

In spite of the fact that Hasselmann derived the collision integral (4.1) for the first time in the early 1960s, for a long time it was practically impossible to obtain reliable estimations of this integral in its exact form. The numerical calculation is rather difficult and requires much computing time. This is, firstly, due to its six-fold form and, secondly, to the very complicated form of the core function $T(\boldsymbol{k}, \boldsymbol{k}_1, \boldsymbol{k}_2, \boldsymbol{k}_3)$. The six-fold integral was transformed to a three-fold form using the delta-functions (4.2), describing the resonance conditions (Hasselmann & Hasselmann, 1981; Dungey & Hui, 1985; Fox, 1976; Webb, 1978). However, these procedures resulted in singularities in the integrand, and this fact caused additional difficulties in computations. There were two ways of solving this problem: either by substituting the variables and using "stretched" coordinates (Hasselmann & Hasselmann, 1981), or by estimating the contribution of the singularity area (Masuda, 1981). However, accurate integral computation remained quite a difficult problem. As a result, some authors (Fox 1976; Longuet-Higgins, 1976; Resio, 1981; Webb, 1978) proposed simplified integral approximations for a narrow spectrum.

Webb (1978), Hasselmann and Hasselmann, (1981) and Masuda (1981) were among the first who overcame the numerical computation difficulties of the exact integral expression.

Hasselmann and Hasselmann (1981) proposed an integral calculation method using the symmetry of its expression. This significantly increased the calculation rate. In order to optimize the computation they proceeded from the asymmetric expression (4.1) describing the energy change of the wave component $\boldsymbol{k}$ (as an interaction result with other components $\boldsymbol{k}_1, \boldsymbol{k}_2$ and $\boldsymbol{k}_3$) to the detailed balance property description so that the maximum symmetry was used.

Subsequent studies (Davidan et al., 1985; Zakharov, 1968; Zakharov & Zaslavskii, 1983a,b; Hasselmann et al., 1973; Komen et al., 1994; Masuda, 1981; Ocean Wave Modeling, 1985) showed the importance of taking into account non-linear interaction in the wind wave spectrum and its role in the spectral maximum shift into the low frequency range with wave evolution.

The necessity for fast calculations of the non-linear interaction integral in operational wind wave models resulted in different simplified approximations. The most successful was the so-called "discrete interaction approximation" (DIA), suggested by Hasselmann, Hasselmann and Barnett (1985). This mainly used the symmetry of the integral expression. In spite of simplifications, it retained some principal properties of the initial integral. Later on, this approximation was used in the WAM model (Komen et al., 1994). The approximation was written in the following form:

$$\begin{pmatrix} \delta G_{\text{nl},1} \\ \delta G_{\text{nl},2} \\ \delta G_{\text{nl},3} \end{pmatrix} = D \begin{pmatrix} -2 \\ 1 \\ 1 \end{pmatrix} Cg^{-4} f^{11} \left[S_1^2 \left(\frac{S_3}{(1+\mu)^4} + \frac{S_4}{(1-\mu)^4} \right) - \frac{S_1 S_3 S_4}{(1-\mu^2)^4} \right],$$

where

$S_n = S(\sigma_n, \beta_n)$, $\mu = 0.25$, $\sigma_2 = \sigma_1$, $\sigma_3 = (1+\mu)\sigma_1$, $\sigma_4 = (1-\mu)\sigma_1$, $\beta_2 = \beta_1$, $\beta_3 = \beta_1 - 11.48°$, $\beta_4 = \beta_1 + 33.56°$, $f = \sigma/2\pi$, $C = 2.8 \cdot 10^7/2\pi$.

Unfortunately, this approximation takes into account a very limited number of wave components in the interaction. It leads to large discrepancies between these approximation results and the numerical estimations of the exact non-linear energy transfer.

A more developed approximation of non-linear energy transfer, the so-called diffusion approximation was proposed by Zakharov and Pushkarev (1999) and was more accurate than DIA. The diffusion approximation can be written as follows:

$$\frac{\partial N}{\partial t} = \frac{a}{\sigma^3} L N^3 \sigma^{24} = \frac{a}{\sigma^3} \left[\frac{1}{2} \frac{\partial^2}{\partial \sigma^2} + \frac{1}{\sigma^2} \frac{\partial^2}{\partial \beta^2} \right] N^3 \sigma^{24}$$

where a is an indefinite constant. The approximation preserves total energy, wave action and momentum. The disadvantage of this approximation is that the value a varies by two orders for different wind wave spectrum approximations. The diffusion approximation produces a non-linear energy transfer value with a narrower frequency form than the exact computation does.

Many qualitative properties of the integral were studied in further calculations (Hasselmann et al., 1985; Polnikov, 1988, 1989; Lavrenov & Ocampo-Torres, 1999, etc.).

There are two main extremes in the non-linear transfer function: one is a positive $G_{\text{nl}}^{(+)}$ and the other is a negative $G_{\text{nl}}^{(-)}$ for a typical wind wave spectrum. Their location and values are determined by the spectrum shape. The positive maximum $G_{\text{nl}}^{(+)}$ is usually located in the general spectral direction at the point where the non-dimensional frequency is $\tilde{\sigma}^{(+)} = \sigma/\sigma_{\max} \approx 0.94-1.00$ (where $\sigma_{\max}$ is the spectral maximum frequency). The negative minimum $G_{\text{nl}}^{(-)}$ is usually located in the general direction at the point $\tilde{\sigma}^{(-)} \approx 1.05-1.60$. There are two additional positive maximas symmetrical about the general direction. They are located at the point $\tilde{\sigma}_2^{(+)} \approx 1.5-3.0$, where the angle with general direction comprises $\beta \approx 25°-45°$.

However, in spite of these results, the problem of the accuracy of the calculation remained open in many papers. The estimations showed that a typical numerical error of collision integral calculations within the spectral maximum area was not less than 10–50 per cent and could be much higher in the other frequency-angular bands. Such accuracy might be sufficient for the usual practical wind wave evolution calculations taking into consideration the source function uncertainties and wind speed errors, but it did seem to be accurate for studying the finer effects of non-linear wave dynamics.

It should be noted that two advanced methods of calculating the collision integral are now known. The first one was suggested by Resio and Perrie (1991) with scaling and symmetry being used for calculating the integral. The second method was proposed by Snyder et al. (1993) using a hybrid integration scheme for the algorithm described by Hasselmann & Hasselmann (1981) and Hasselmann et al. (1985). This scheme uses the improvements of the earlier calculation method in combination with the advantages of the EXACT-NL model calculation (Ocean Wave Modeling, 1985) and allows increasing the calculation speed by an order of magnitude.

However, in spite of the obvious success, the problem of the calculation accuracy and its optimal algorithm is still of great interest. That is why an attempt to achieve some progress in this direction is made in this monograph. The results of the present study are of interest not only in terms of decreasing the computing time and providing guaranteed accuracy, but in obtaining more stable estimations of the non-linear energy transfer in the wind wave spectrum.

4.1.2 Optimal Algorithm of Computation of Non-linear Energy Transfer

The initial expression for the non-linear energy transfer integral G_{nl} is written in the right-hand part of (4.1). In further calculations the core function is used in the form (Webb, 1978):

$$T(\boldsymbol{k},\boldsymbol{k}_1,\boldsymbol{k}_2,\boldsymbol{k}_3) = \frac{\pi g^2 D^2(\boldsymbol{k},\boldsymbol{k}_1,\boldsymbol{k}_2,\boldsymbol{k}_3)}{4\rho^2 \sigma\, \sigma_1\, \sigma_2\, \sigma_3}, \qquad (4.4)$$

where:

$$\begin{aligned}
D(\boldsymbol{k},\boldsymbol{k}_1,\boldsymbol{k}_2,\boldsymbol{k}_3) = {}& 2\left[\frac{(\sigma+\sigma_1)^2(k k_1 - \boldsymbol{k}\boldsymbol{k}_1)(k_2 k_3 - \boldsymbol{k}_2\boldsymbol{k}_3)}{g\,|\boldsymbol{k}+\boldsymbol{k}_1| - (\sigma+\sigma_1)^2}\right.\\
& + \frac{(\sigma-\sigma_2)^2(k k_2 + \boldsymbol{k}\boldsymbol{k}_2)(k_1 k_3 + \boldsymbol{k}_1\boldsymbol{k}_3)}{g\,|\boldsymbol{k}-\boldsymbol{k}_2| - (\sigma-\sigma_2)^2}\\
& \left.+ \frac{(\sigma-\sigma_3)^2(k k_3 + \boldsymbol{k}\boldsymbol{k}_3)(k_1 k_2 + \boldsymbol{k}_1\boldsymbol{k}_2)}{g\,|\boldsymbol{k}-\boldsymbol{k}_3| - (\sigma-\sigma_3)^2}\right]\\
& + \frac{1}{2}\left[(\boldsymbol{k}\boldsymbol{k}_1)(\boldsymbol{k}_2\boldsymbol{k}_3) + (\boldsymbol{k}\boldsymbol{k}_2)(\boldsymbol{k}_1\boldsymbol{k}_3) + (\boldsymbol{k}\boldsymbol{k}_3)(\boldsymbol{k}_1\boldsymbol{k}_2)\right]\\
& - \frac{1}{4g^2}\left[(\boldsymbol{k}\boldsymbol{k}_1+\boldsymbol{k}_2\boldsymbol{k})(\sigma+\sigma_1)^4 - (\boldsymbol{k}\boldsymbol{k}_1+\boldsymbol{k}_1\boldsymbol{k}_3)(\sigma-\sigma_2)^4\right.\\
& \left. - (\boldsymbol{k}\boldsymbol{k}_3+\boldsymbol{k}_1\boldsymbol{k}_2)(\sigma-\sigma_3)^4\right]\\
& + \frac{1}{g^3}(\sigma+\sigma_1)^2(\sigma-\sigma_2)^2(\sigma-\sigma_3)^2(k+k_1+k_2+k_3) + \frac{5}{2}k k_1 k_2 k_3\,.
\end{aligned}$$

Using the symmetry of the variables k_2 and k_3 the integral (4.1) can be written as:

$$\int d\boldsymbol{k}_1 \iint d\boldsymbol{k}_2\, d\boldsymbol{k}_3 = 2 \int d\boldsymbol{k}_1 \iint_{|\boldsymbol{k}_2|\leq|\boldsymbol{k}_3|} d\boldsymbol{k}_2\, d\boldsymbol{k}_3, \qquad (4.5)$$

The first integration of (4.5) is carried out over the variable k_3. Transforming the variables from $\boldsymbol{k}_i = \{k_{xi}, k_{yi}\}$ to σ_i and to the angles $\beta_i = \arctan(k_{yi}/k_{xi})$ and from the wave action $N(\boldsymbol{k})$ to the wave energy $S(\omega,\theta)$, the expression $N(\boldsymbol{k}) = (\omega^2/2g^4)S(\omega,\theta)$ is obtained.

The integral (4.5) is written in the form:

$$G_{\mathrm{nl}}(\sigma,\beta) = 2 \iiiint T\{S\, S_1(S_2\sigma_3^4 + S_3\sigma_2^4) - S_2\, S_3(S\sigma_1^4 + S_1\sigma^4)\}$$

$$\times \frac{\delta(\sigma + \sigma_1 - \sigma_2 - \sigma_3)}{\sigma^4\, \sigma_1^4\, \sigma_2^4\, \sigma_3^4} \, d\sigma_2\, d\beta_2\, d\sigma_1\, d\beta_1 . \qquad (4.6)$$

Using the function $\delta(\sigma)$ the following expression is obtained:

$$G_{\mathrm{nl}}(\sigma,\beta) = 4 \sum_{\beta_2} \int_0^\infty \int_{-\pi}^{+\pi} \int_{\sigma_2} T\{S\, S_1(S_2\sigma_3^4 + S_3\sigma_2^4) - S_2\, S_3(S\sigma_1^4 + S_1\sigma^4)\}$$

$$\times \frac{\Theta(\sigma_1,\sigma_2,\beta_1)}{\sigma_1\sigma_2\sigma_3\, \sqrt{\sigma_a[(k_a + \sigma_3^2)^2 - \sigma_2^4]}\, \sqrt{B(\sigma_1,\sigma_2,\beta_1)}} \, d\sigma_2\, d\beta_1\, d\sigma_1, $$

$$(4.7)$$

where:

$$\sigma_a = \sigma_1 + \sigma; \quad \sigma_3 = \sigma_a - \sigma_2; \quad k_a^2 = \left[\sigma_1^4 + \sigma^4 + 2\sigma^2\sigma_1^2 \cos(\beta - \beta_1)\right];$$

$\Theta(\sigma_1,\sigma_2,\beta_1) = \Theta\left(2\,k_a - \sigma_a^2 \cos(\beta_2 - \beta_a)\right)$ is the Heaviside function:

$$\beta_2 = \beta_a \pm \arccos\left((k_a^2 + \sigma_2^4 - \sigma_3^4)/(2k_a\sigma_2^2)\right);$$

$$\beta_3 = \beta_a \mp \arccos\left((k_a^2 + \sigma_3^4 - w_2^4)/(2k_a\sigma_3^2)\right);$$

$$\beta_a = \arccos\left((\sigma^2 \cos\beta + \sigma_1^2 \cos\beta_1)/k_a\right)\, \mathrm{sign}(k_{1y} + k_y).$$

The function $B = B(\sigma_1,\sigma_2,\beta_1)$ is written as

$$B = \left[\sigma_2 - \sigma_a/2 + k_a/(2\sigma_a)\right] \cdot \left[(\sigma_2 - \sigma_a/2)^2 - (k_a/2 - \sigma_a^2/4)\right]. \qquad (4.8)$$

It should be noted that the function $B = B(\sigma_1,\sigma_2,\beta_1)$ becomes zero at some points (see Fig. 4.2). It produces a singularity and causes difficulties in integrating (4.7) numerically.

4.1 Non-Linear Energy Transfer in Wind Wave Spectrum

The most optimal integration algorithm, making the present method different from the others (Hasselmann & Hasselmann, 1981; Masuda, 1981; Komatsu & Masuda, 1996; Polnikov, 1989; Resio & Perrie, 1991) can be based on utilization of the Jacobi weight functions (Krylov & Shulgina, 1966).

The integration is carried out within the range $0.5\,\sigma_a(1 - \varepsilon_a/2) \leq \sigma_2 \leq 0.5\,\sigma_a\left(1 - \sqrt{\varepsilon_a - 1}\right)$ in case $\varepsilon_a > 1$ (where $\varepsilon_a = 2k_a/\sigma_a^2$). There are two singularities at both integration range boundary points (see Fig. 4.2). Using the Jacobi weight functions, the integration over σ_2 (in case $\varepsilon_a > 1$) can be approximated as:

$$\tilde{F}_1(\sigma_1, \beta_1) = \int_a^b \frac{f_1(\sigma_2, \sigma_1, \beta_1)}{\sqrt{(\sigma_2 - a)(b - \sigma_2)}} \, d\omega_2 = \frac{\pi}{n} \sum_{j=1}^n f_1(\sigma_{2j}, \sigma_1, \beta_1) , \quad (4.9a)$$

where $f_1(\sigma_2, \sigma_1, \beta_1)$ is a function without singularities; and

$$a = \sigma_a/2 - k_a/(2\sigma_a) ; \quad b = \sigma_a(1 - \sqrt{\varepsilon_a - 1})/2 ;$$

$$\sigma_{2j} = (b + a)/2 + (b - a)/2 \cos\left[(2j - 1)\pi/2/n\right] .$$

The integration range is $0.5\,\sigma_a(1 - \varepsilon_a/2) \leq \sigma_2 < 0.5\,\sigma_a$ in case $\varepsilon_a < 1$. The function $B = B(\sigma_1, \sigma_2, \beta_1)$ becomes equal to zero (see Fig. 4.2) at the integration range boundary (i.e. $\sigma_2 = 0.5\,(1 - \varepsilon_a/2)$). In this case the following formula can be applied:

$$\tilde{F}_2(\sigma_1, \beta_1) = \int_a^d \frac{f_2(\sigma_2, \sigma_1, \beta_1)}{\sqrt{\sigma_2 - a}} \, d\sigma_2 = \sqrt{d - a} \sum_{j=1}^n A_j f_2(\sigma_{2j}, \sigma_1, \beta_1) , \quad (4.9b)$$

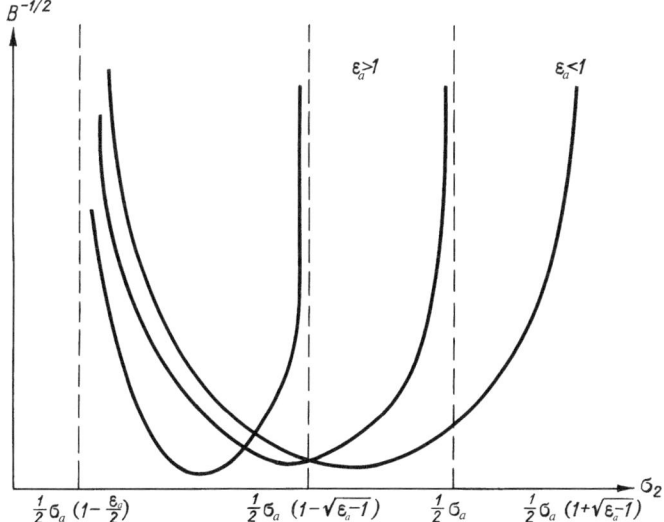

Fig. 4.2. The dependence of the function $B^{-1/2}(\sigma_2, \varepsilon_a)$ on its arguments

88   4 Physical Mechanisms Forming the Wave Spectrum in Deep Water

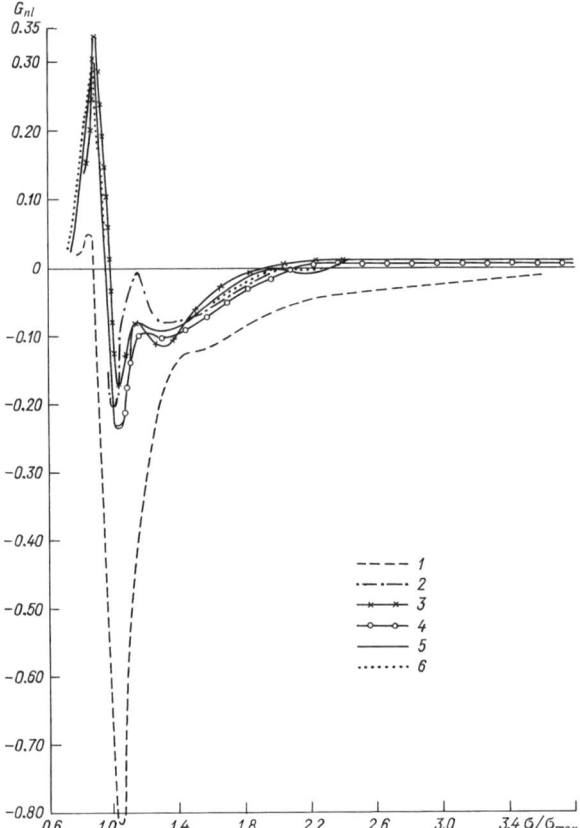

Fig. 4.3. Computations of non-linear energy transfer function with different number of points n in (4.9) and (4.10): $1 - n = 2$; $2 - n = 3$; $3 - n = 4$; $4 - n = 5$; $5 - n = 7$; $6 - n = 8$

where A_j are weight coefficients; σ_{2j} are function ordinates; and $f_2(\sigma_2, \sigma_1, \beta_1)$ is a function without singularity at the point $\sigma_2 = a$. As shown by the numerical results (see Fig. 4.3) it is enough to use $n = 7$ in order to obtain good results with a relative error less than 1–2 per cent. It is enough to use $n = 4$ for many practical computations.

The following integration is carried out with respect to β_1 assuming that the function $\tilde{F}_2(\sigma_1, \beta_1)$ is periodic. It is known (Krylov & Shulgina, 1966) that the numerical integration algorithm of the highest precision is the ordinary rectangular method:

$$\tilde{J}(\sigma_1) = \int_{-\pi}^{\pi} \tilde{F}(\sigma_1, \beta_1) \, d\beta_1 = \frac{2\pi}{m} \sum_{i=0}^{m-1} \tilde{F}(\sigma_1, \beta_{1i} + \xi), \qquad (4.10)$$

where $\beta_{1i} = \frac{2\pi}{m}i$. In case $\xi = \frac{\pi}{2m}$ or $\xi = \frac{3\pi}{2m}$, the expression (4.10) is valid for $C_m(\beta) = T_{m-1}(\beta) + a_m \cos(m\beta)$, where T_{m-1} is a trigonometrical polynomial with the power of $m-1$. In order to obtain results with an error less than 1–2 per cent, a sufficiently large number of ordinates, i.e. $m \geq 90$, should be taken.

The problem is that the latter integration over β_1 is not optimal. A large number of ordinates is to be used because the function $\tilde{F}(\sigma_1, \beta_1)$ includes singularities as well. The function $\tilde{F}(\sigma_1, \beta_1)$ becomes infinite when $\varepsilon_a = 2k_a/\sigma_a^2 = 1$. The most effective integration can be achieved by transforming the variables. In this case the Jacobi functions can be used to obtain the cubature formulas. Thus, the function (4.10) can be written as:

$$\tilde{J}(\sigma_1) = \int_{-\pi}^{\pi} \tilde{F}(\sigma_1, \beta_1) \, \mathrm{d}\beta_1 = \int_{-\pi}^{\pi} \tilde{\tilde{F}}(\sigma_1, \beta_1)/\sqrt{|\cos(\beta - \beta_1) - A|} \, \mathrm{d}\beta_1 , \quad (4.11a)$$

where

$$\tilde{\tilde{F}} = \tilde{F}\sqrt{|\cos(\beta - \beta_1) - A|} ,$$

$$A = \left((\sigma + \sigma_1)^4 - 4\left(\sigma_1^4 + \sigma^4\right)\right) \Big/ \left(8\sigma^2 \sigma_1^2\right) .$$

The function $A \leq 1$ takes its maximum value at the point $\sigma = \sigma_a$.

In order to introduce a new variable $x = \cos(\beta - \beta_1)$, the integral (4.11a) can be written in the following form:

$$\tilde{J}(\sigma_1) = \sum \int_{\beta_1^{\pm} - 1}^{1} \tilde{\tilde{F}}(\sigma_1, \beta_1^{\pm}) \frac{1}{\sqrt{1 - x^2}\sqrt{|A - x|}} \, \mathrm{d}x , \quad (4.11b)$$

where $\beta_1^{\pm} = \beta \pm \arccos(x)$.

The function $\tilde{\tilde{F}}(\sigma_1, x)$ is smooth enough. The integral (4.12) includes the same singularities as the first-order elliptical integral. Numerical results show that it is enough to use 6–8 ordinates to obtain rather good accuracy.

The last integration over σ_1 can be carried out effectively taking into account that the function $\tilde{J}(\sigma_1)$ is approximated as $\tilde{J}(\sigma_1) \sim \sigma_1^{-6}$ for large values of σ_1, and as $\tilde{J}(\sigma_1) \sim \sigma_1^{15 \div 25}$ for small values of σ_1. This allows the use of the traditional cubature method of integration. It should be noted that in order to speed up the computation, the part of the function in (4.7), not depending on the spectral value, is computed using the symmetric quantity (Hasselmann & Hasselmann, 1981).

It should be noted that the main advantage of the algorithm is that the integration is based on a relatively small number of grid points compared to the usual methods (Masuda 1981; Polnikov 1988). This speeds up the computation by at least two orders of magnitude (Lavrenov, 1998).

Tests of the collision integral calculation algorithm and assessment of the calculation accuracy. Before further calculations the experiments are made to test the aforementioned integration algorithm, using the quadrature formulas (4.9) described above. For this purpose the calculations of the non-linear transfer integral for the JONSWAP spectrum are made using a different number of integrand points. Thus, the parameter n is assumed to be equal to 2, 3, 4, 5, 7 and 8, respectively. The results of these calculations are presented in Fig. 4.3. They show that with increasing n, the calculation results converge quite rapidly to its accurate value. The numerical values become practically indiscernible for $n > 5$. For $n = 4$ the relative differences in the energy range are not more than 15 per cent (of the corresponding value at $n = 8$) sufficient for most practical calculations. The integration error is estimated as 1–2 per cent with $n = 7$.

The accuracy for the typical JONSWAP spectrum is estimated by repeated calculations with double the number of grid points until the difference of the calculation results for two successive cases reaches the prescribed value. By executing a series of successive calculations, an error is obtained not greater than 1–2 per cent in the area of the spectral maximum ($0.9 \leq \tilde{\sigma} \leq 1.5$, where $\tilde{\sigma} = \sigma/\sigma_{\max}$). Moreover, the calculation error is not greater than 3–5 per cent in the frequency ranges ($0.8 \leq \tilde{\sigma} \leq 0.9$) and ($1.5 < \tilde{\sigma} \leq 2.5$) and not greater than 5–10 per cent in the ranges ($0.7 \leq \tilde{\sigma} \leq 0.8$) and ($2.5 < \tilde{\sigma} \leq 3.5$). In addition, the assessment of the numerical accuracy of the conservation of the collision integral is performed. The expression $G_{\mathrm{nl}}(\boldsymbol{k})$ (4.1) is integrated over a wave vector $\boldsymbol{k}$. The error of the estimate of conserving the non-linear energy transfer is not higher than 1 per cent.

Then the calculations of the integral (4.1) are analysed, comparing with the results obtained by other investigators. As a criterion, the results of Hasselmann and Hasselmann (1981) are used. They present different functions for the non-linear energy transfer which are calculated for the JONSWAP spectrum at different determining parameters.

The calculation results for the one-dimensional function $G_{\mathrm{nl}}(\sigma)$, obtained by Hasselmann & Hasselmann (1981) for the peakness parameter $\gamma = 7$ and angular energy distribution $\sim \cos^2(\beta)$, are shown in Fig. 4.4a. This figure also shows calculations by the algorithm proposed in this monograph. As it is seen, the coincidence of results is quite good, taking into account that the frequencies used in the calculation are somewhat different.

Although the agreement between the calculation results for peakness $\gamma = 3.3$ is good enough, there are some differences at frequencies $\sigma > 1.2\sigma_{\max}$. Comparing the calculation results, it can be concluded that the calculations in the present study are of a more stable (smooth) character. For peakness $\gamma = 1.0$ this difference is already significant (see Fig. 4.4b). The proposed algorithm gives a noticeably smoother curve indicating greater stability of the result obtained. This conclusion becomes even more pronounced, if the calculations of a two-dimensional function $G_{\mathrm{nl}}(\sigma, \beta)$ are compared.

4.1 Non-Linear Energy Transfer in Wind Wave Spectrum

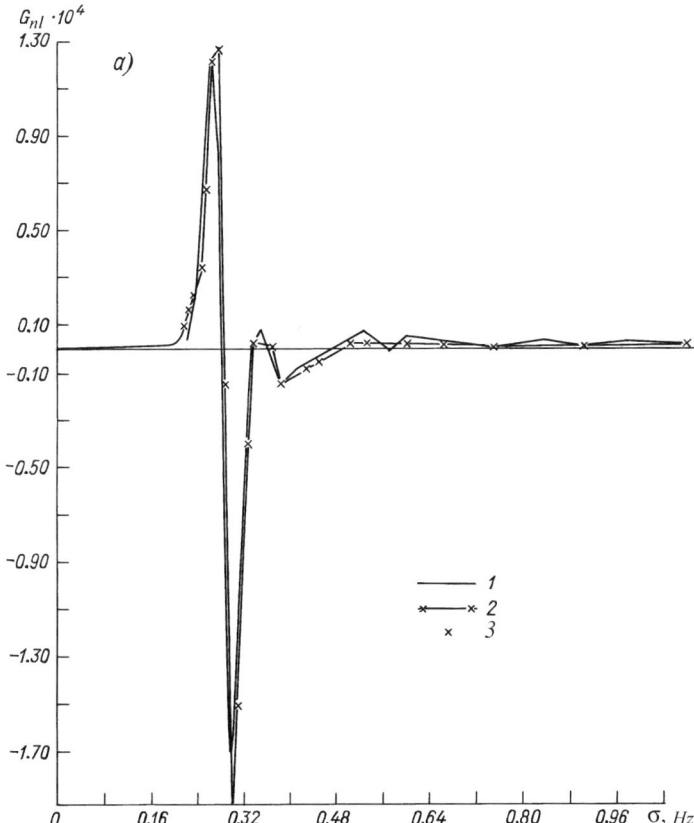

Fig. 4.4. (a) Non-linear transfer function for the JONSWAP spectrum with $\gamma = 7$: 1 – according to results (Hasselmann and Hasselmann, 1981); 2 – by the given algorithm; 3 – calculated points (Fig. 4.4(**b**) see next page)

Thus, it can be concluded that this algorithm obtains sufficiently stable results of calculating the non-linear energy transfer integral with limited computing time.

It should be noted that an explicitly analytical separation of the singularities of integrands (4.7), (4.8) in the form (4.9) and (4.11), as well as selection of the corresponding quadrature formulas and use of the most accurate numerical integration methods are a successful "finding" for the numerical integration algorithm. Probably, this is the main difference of this approach in comparison with those proposed by Polnikov (1988), Hasselmann and Hasselmann (1981), Masuda (1981), Komatsu and Masuda (1996). They preferred "struggling" against the integrand singularities and performed integration using less efficient methods.

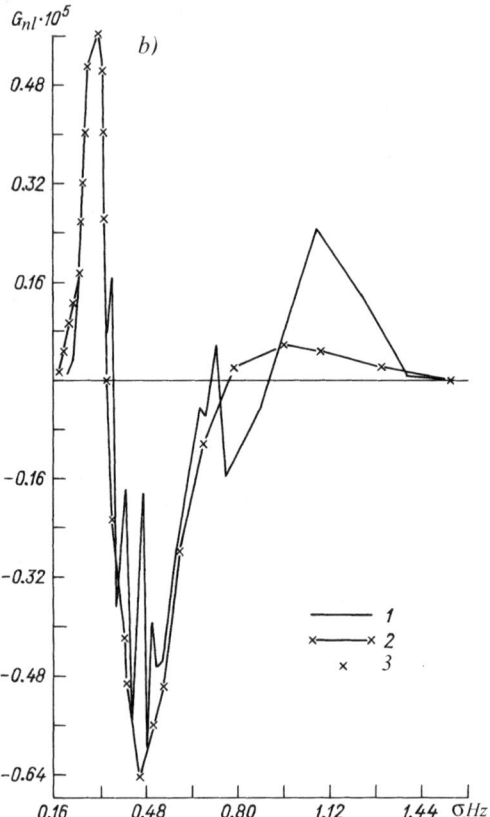

Fig. 4.4. (b) Non-linear transfer function for the JONSWAP spectrum with $\gamma = 1.0$: 1 – according to (Hasselmann and Hasselmann, 1981); 2 – according to the given algorithm; 3 – calculated points

4.1.3 Calculation Results of Non-linear Energy Transfer in Wind Wave Spectrum[1]

Calculations of non-linear energy transfer in the wind wave spectrum can be executed for its most typical frequency-angular approximations $S(\sigma, \theta)$ given in the following form:

$$S(\sigma, \beta) = S(\sigma) \; Q(\sigma, \beta), \qquad (4.12)$$

where $S(\sigma)$ is the frequency spectrum and $Q(\sigma, \beta)$ is the angular energy distribution.

The frequency approximation in the form of the JONSWAP spectrum (Hasselmann et al., 1973) is used as:

---
[1] These calculations were made with the assistance of Dr. F. J. Ocampo-Torres (CICESE, Mexico).

4.1 Non-Linear Energy Transfer in Wind Wave Spectrum

$$S(\sigma) = \alpha g^2 \sigma^{-5} e^{-\frac{5}{4}\left(\frac{\sigma_{\max}}{\sigma}\right)^4} \gamma^{\exp\left[-(\sigma-\sigma_{\max})^2/(2\sigma_J^2 \sigma_{\max}^2)\right]}, \qquad (4.13)$$

where

$$\sigma_J = \begin{cases} 0.07 & \text{at } \tilde{\sigma} \leq 1; \\ 0.09 & \text{at } \tilde{\sigma} > 1; \end{cases} \qquad \tilde{\sigma} = \frac{\sigma}{\sigma_{\max}}.$$

The angular energy distribution is used consecutively in the form of two approximations, one of them being an ordinary cosine energy distribution:

$$Q(\sigma,\beta) = \begin{cases} \left[\frac{\pi \Gamma(n_\beta+1)}{2^{n_\beta} \Gamma^2((n_\beta/2+1))}\right]^{-1} \cos^{n_\beta}(\beta-\bar{\beta}) & \text{at } |\beta-\bar{\beta}| \leq \pi/2; \\ 0 & \text{at } |\beta-\bar{\beta}| \geq \pi/2. \end{cases} \qquad (4.14)$$

The second angular distribution is used in the form obtained according to the JONSWAP experimental data (Hasselmann et al., 1980):

$$Q_J(\sigma,\beta) = \left[2^{2s-1}\pi \Gamma^2(s+1)/\Gamma(2s+1)\right] \cos^{2s}\left((\beta-\bar{\beta})/2\right), \qquad (4.15)$$

where $s = s_{\max}(\tilde{\sigma})^\mu$; $s_{\max} = 9.774$; $\mu = 4.06$ for $\tilde{\sigma} \leq 1$ and $\mu = -2.34$ in all other cases.

The wind wave spectrum approximation, proposed by Donelan et al. (1985), is used in the calculations as follows:

$$S(\sigma) = \alpha_D g^2 \sigma^{-5} e^{-\left(\frac{\sigma_{\max}}{\sigma}\right)^4} \gamma_D^{\exp\left[-(\sigma-\sigma_{\max})^2/(2\sigma_D^2 \sigma_{\max}^2)\right]}, \qquad (4.16)$$

where $\alpha_D = 0.006 \, (U/c_{\max})$ at $0.83 < U/c_{\max} < 5.0$;

$$\sigma_D = 0.08 \left[1 + 4 \, (U/c_{\max})^{-1}\right] \text{ at } 1.0 < U/c_{\max} < 5.0;$$

$$\gamma_D = \begin{cases} 1.7 & \text{at } 0.83 < U/c_{\max} < 1.0; \\ 1.7 + 0.6 \log(U/c_{\max}) & \text{at } 1.0 \leq U/c_{\max} < 5.0; \end{cases}$$

U is the wind velocity at a 10-m height; $c_{\max}$ is the phase velocity of waves, whose frequency coincides with the maximum spectral frequency.

The angular energy distribution is prescribed by the formula:

$$Q(\sigma,\beta) = \frac{1}{2} B \sec h^2 \left(\beta - \bar{\beta}(\sigma)\right), \qquad (4.17)$$

where

$$B = \begin{cases} 2.61 \, \tilde{\sigma}^{1.3} & \text{at } 0.56 < \tilde{\sigma} < 0.95; \\ 2.28 \, \tilde{\sigma}^{-1.3} & \text{at } 0.95 < \tilde{\sigma} < 1.6; \\ 1.24 & \text{at } \tilde{\sigma} > 1.6. \end{cases}$$

Banner (1990) specified the formula (4.17), using high-frequency stereo-photography data, and obtained a new approximation for the parameter B in the high-frequency band $\tilde{\sigma} \geq 1.60$:

$$B = 10^y, \qquad (4.18)$$

where $y = -0.4 + 0.8393 \, \exp\left[-0.567 \, \ln(\tilde{\sigma}^2)\right]$.

In the calculations below the aforementioned frequency-angular spectrum approximations are used.

First the results of the non-linear energy transfer calculation are given for the JONSWAP spectral approximation (4.13) with the cosine angular distribution (4.14). The results are presented on the plane $\{\tilde{\sigma}, \beta\}$ as isolines of the normalized non-linear energy transfer and frequency-angular spectrum (see Fig. 4.5a,b):

$$\tilde{G}_{\mathrm{nl}}(\tilde{\sigma}, \beta) = G_{\mathrm{nl}}(\sigma, \beta)/(\sigma_{\max}^{11} \, S_{\max}^3/g^4),$$

$$\tilde{S}(\tilde{\sigma}, \beta) = S(\sigma, \beta)/S_{\max}, \qquad (4.19)$$

where $S_{\max}$ is the frequency-angular spectral maximum.

The JONSWAP normalized spectrum for the peakness $\gamma = 3.3$ and the cosine angular energy distribution, with $n_\beta = 12$ are shown in Fig. 4.5a. The non-linear transfer values for the same case are shown in Fig. 4.5b. As can be seen, the non-linear energy transfer value, different from zero, is localized in the area along the horizontal axis and limited by the angles $\pm 45°$. The non-linear transfer function has two main extremes of the plus-minus type $G_{\mathrm{nl}}^{(\pm)}$. The maximum function $G_{\mathrm{nl}}^{(+)}$ is located in the general direction at the point $\tilde{\sigma}^{(+)} = \sigma/\sigma_{\max} \approx 0.95$. The negative extreme $G_{\mathrm{nl}}^{(-)}$ is located at the point $\tilde{\sigma}^{(-)} \approx 1.08$.

The local maxima and minima can be found in the area of the main extremes. Thus, for a sufficiently narrow angular distribution $\sim \cos^{n_\beta}(\beta)$, where $n_\beta \geq 10$, the main maximum is divided into two symmetrical directions relative to the general direction (the results are shown in Fig. 4.5c at a smaller scale). This is probably evidence of the stabilizing effect of the non-linear interaction on the angular energy distribution. A sufficiently narrow angular distribution becomes wider, and vice versa. Moreover, in a general direction the non-linear transfer becomes negative at frequencies less than the maximum of the function $G_{\mathrm{nl}}^{(+)}$ with respect to the frequency for the narrow angular distribution. Its extreme is located at the frequency $\tilde{\sigma}_3^{(-)} \approx 0.74$, and its value is two orders less than the main maximum function $G_{\mathrm{nl}}^{(+)}$ (see Fig. 4.5c).

There is a second extreme at the frequency $\tilde{\sigma}_2^{(-)} \approx 1.35$ in the main negative extreme area. It is about 52 per cent of the main negative extreme value. There are also two additional symmetrical maxima relative to the general direction. They are located at larger frequencies at the point $\tilde{\sigma}_2^{(+)} \approx 1.62$ at the angle $\beta \approx 20.5°$ comprising 35 per cent of the main maximum function.

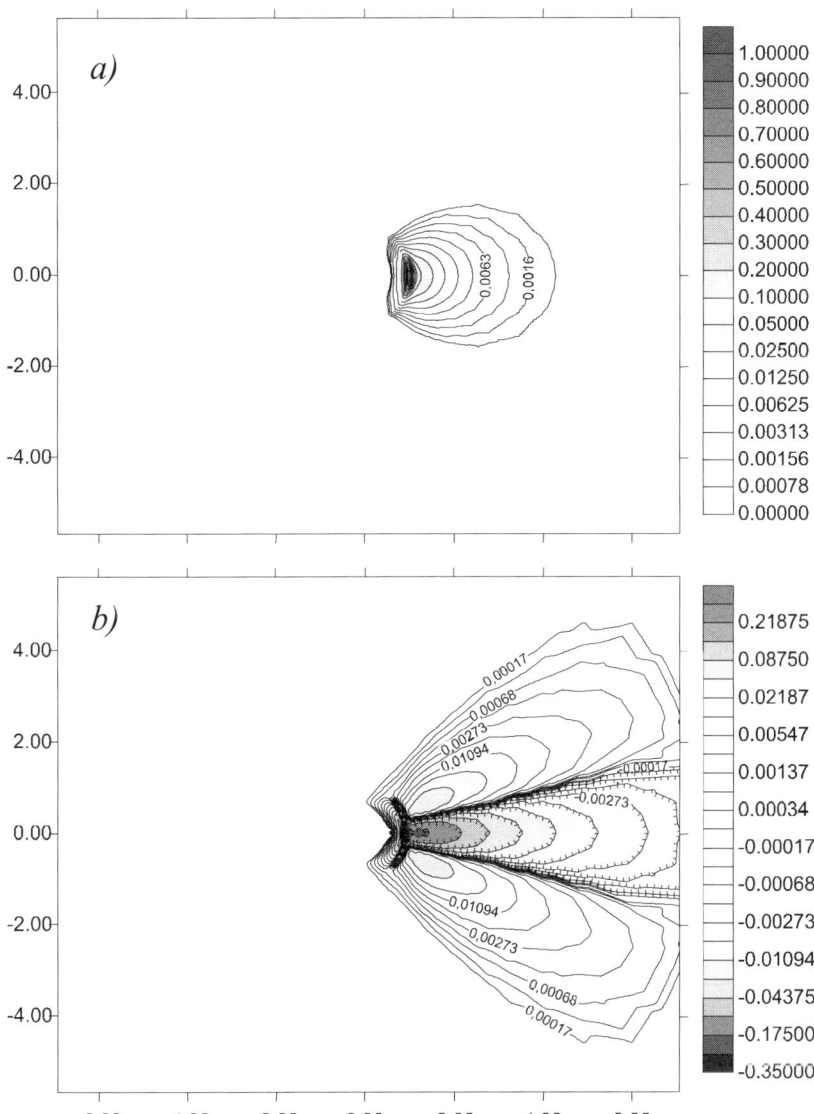

Fig. 4.5. JONSWAP normalized spectrum (**a**) and corresponding non-linear transfer function (**b**) for peakness $\gamma = 3.3$ and cosine angular energy distribution with $n_\beta = 12$. (Fig. 4.5(**c**) see next page)

A similar spectrum with the same peakness, but for the cosine angular distribution (4.14) with $n_\beta = 2$ and non-linear transfer values is shown in Fig. 4.6a,b.

The surface of the non-linear transfer function is smoother and wider. It covers almost the whole right-hand semi-plane $\{\tilde{\sigma}, \beta\}$. The non-linear transfer

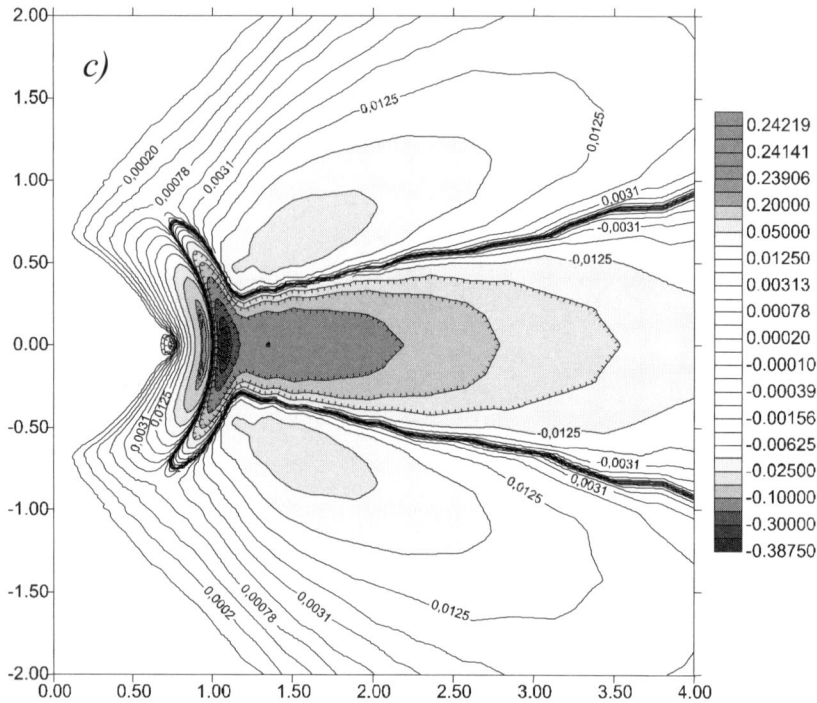

Fig. 4.5. (c) Non-linear transfer function for peakness $\gamma = 3.3$ and cosine angular energy distribution with $n_\beta = 12$ (at larger scale)

function is practically equal to zero in the left-hand semi-plane and the main extremes are approximately the same. Two additional positive extremes are located at slightly higher frequencies at the point $\tilde{\sigma}_2^{(+)} \approx 2.2$, where the angle is $\beta \approx \pm 42°$. The maximum of these values comprises about 11 per cent of the main maximum $G_{nl}^{(+)}$. The second negative extreme is located at the same place with approximately the same relative value. The local extremes observed in the previous case at the main maximum area have disappeared.

At the next stage, calculations of the non-linear energy transfer for the same frequency spectrum (4.12)–(4.13), but for the approximation of the angular energy distribution function, prescribed by the formula (4.15), are made.

The normalized spectral energy density and non-linear transfer function values for the spectrum peakness $\gamma = 3.3$ are shown in Fig. 4.7a,b. It should be noted that the frequency-angular spectrum form is nearly the same. But there are small non-zero spectral density values in the vicinity of the axis $\beta = \pm \pi/2$. The value is decreased quite rapidly from this axis to the left-hand semi-plane. As shown below, this difference of the spectrum shape from the previous one (see Fig. 4.6a) is of crucial importance.

4.1 Non-Linear Energy Transfer in Wind Wave Spectrum 97

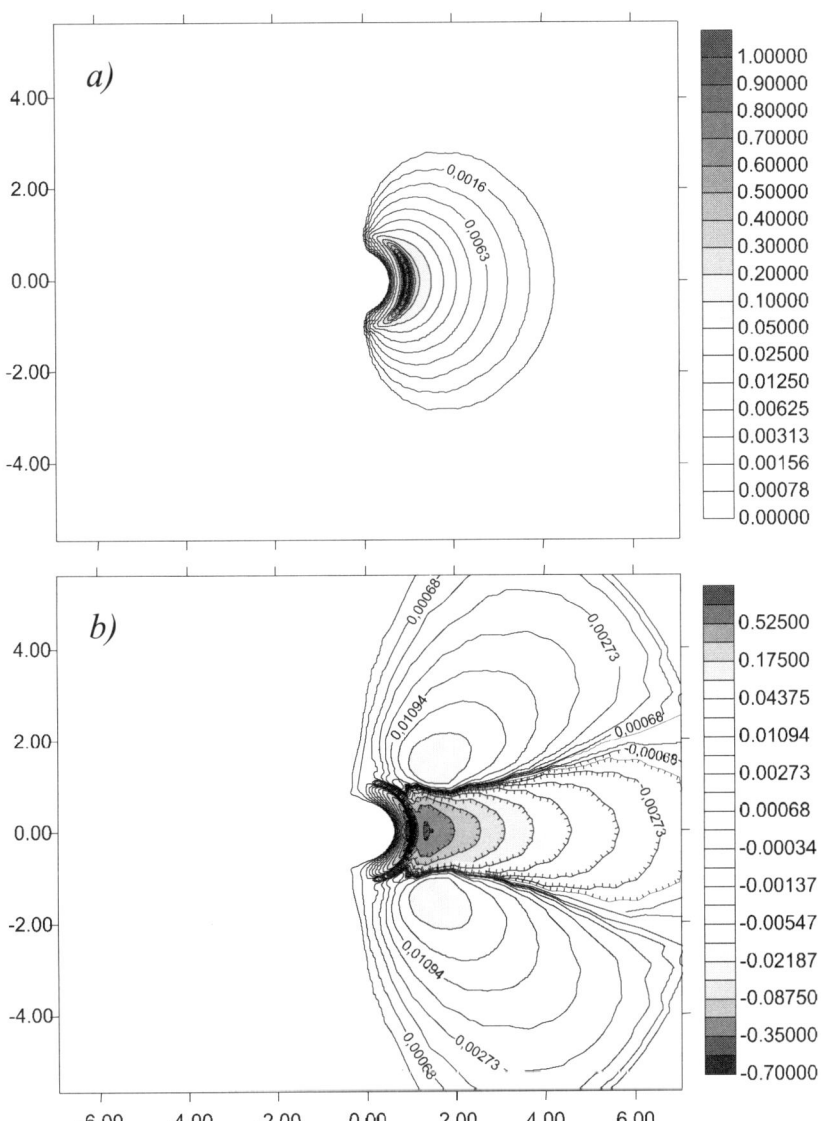

Fig. 4.6. JONSWAP normalized spectrum (a) and non-linear transfer function (b) for peakness $\gamma = 3.3$ and the cosine angular energy distribution with $n_\beta = 2$

The non-linear energy transfer function has been changed more significantly (see Fig. 4.7b). Now it occupies not only the left-hand, but almost the entire right-hand semi-plane $\{\tilde{\sigma}, \beta\}$. The shape of the isolines resembles a medusa, being turned vertically counterclockwise. The "medusa tentacles"

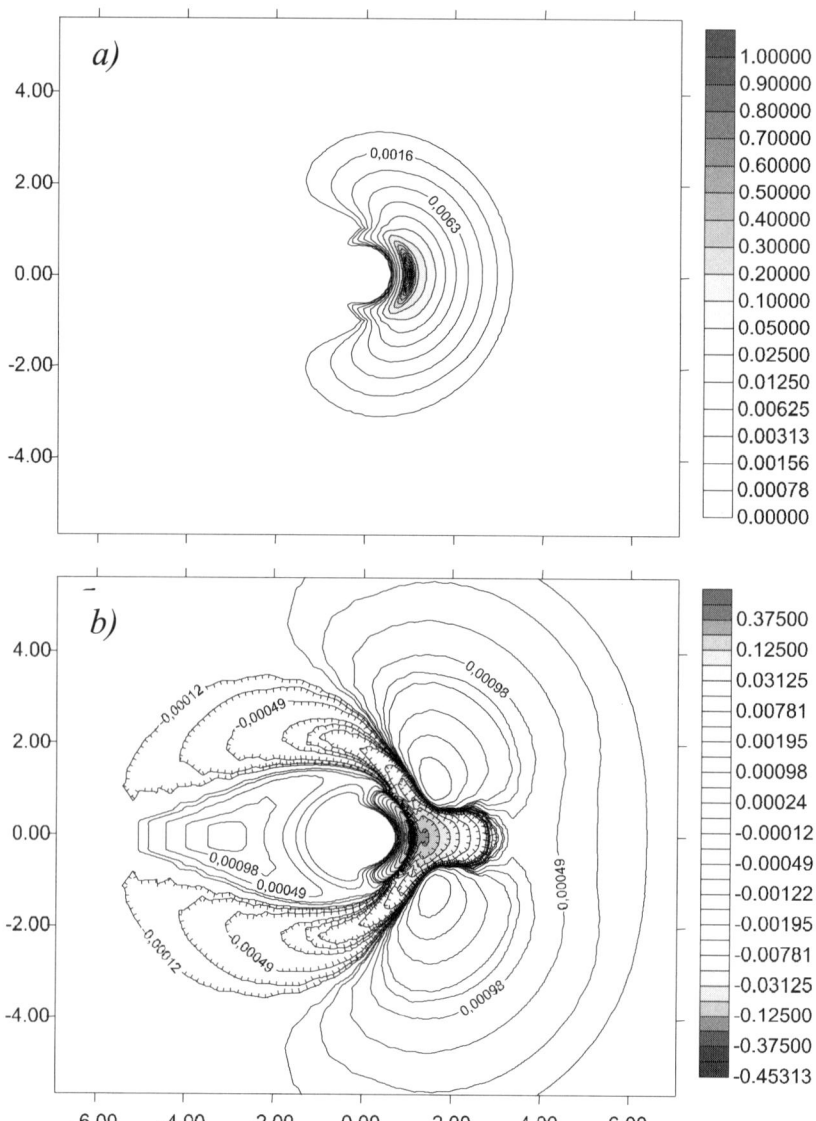

Fig. 4.7. Normalized spectral energy density (**a**) and corresponding non-linear transfer function (**b**) for peakness $\gamma = 3.3$ and angular distribution (4.15)

extend along the centre of the polar coordinate system $\{\tilde{\sigma}, \beta\}$ from the right-hand to the left-hand semi-plane.

On the right-hand semi-plane the presence of the same main details of the non-linear energy transfer function can be noted, although it has be-

4.1 Non-Linear Energy Transfer in Wind Wave Spectrum

come much wider in this case. The non-linear transfer has the same positive–negative structure of the extremes. The positive maximum $G_{\rm nl}^{(+)}$ is located in a general direction at the point $\tilde\sigma^{(+)} \approx 0.95$. Two additional maxima are located at the points $\tilde\sigma_2^{(+)} = 1.9$, where the angle $\beta \approx \pm 37.5°$. The main negative minimum $G_{\rm nl}^{(-)}$ is located at the point $\tilde\sigma^{(-)} \approx 1.06$. The area of the negative non-linear transfer function extends from the main minimum to the left semi-plane over a large distance. It passes round the centre of the polar coordinate system $\{\tilde\sigma, \beta\}$, decreasing with distance from the centre. In close proximity to the beginning of the polar coordinate system $\{\tilde\sigma, \beta\}$ the non-linear transfer is practically equal to zero. The non-linear transfer begins increasing at the opposite side of the zero area (at $\beta \approx \pm 180°$). A new area with positive non-linear energy transfer values is observed, being surrounded with negative values. The maximum area is located at the point $\tilde\sigma_3^{(+)} = 3.3$, $\beta \approx \pm 180°$ comprising about 0.5 per cent of the main maximum of the positive non-linear energy transfer.

Similar calculations, but for the JONSWAP spectrum with peakness $\gamma = 7.0$ have been made for studying a new positive maximum. The relative non-linear transfer value is less in the new maximum area. The new maximum comprises about 0.1 per cent of the main value.

Similar values for the spectrum peakness $\gamma = 1.0$ are shown in Fig. 4.8a,b. The main details are the same. A new positive maximum is even more explicitly defined. The relative maximum value has been increased and it comprises 3.6 per cent of the main positive maximum of the non-linear energy transfer.

Similar calculations have been performed for the spectrum approximation proposed by Donelan (4.17) with the spectrum peakness $\gamma = 3.3$, as well as for the specified version of this approximation (4.18), suggested by Banner (see Fig. 4.9a,b). The main details of the non-linear transfer function are the same, i.e. they have positive and negative extremes. The area of negative values extends to the high-frequency range along the general direction due to the high-frequency "tail" of the spectrum approximation decreasing more slowly as $\sim \sigma^{-4}$. This is contained in the angular sector $\beta \approx \pm 45°$, starting from the point $\tilde\sigma \approx 1.0$. In the left-hand semi-plane, a new positive area is located approximately in a similar symmetrical angular sector. Its maximum is located at the point $\tilde\sigma \approx 1.74$, $\beta = 180°$ comprising 0.5 per cent of the main positive extreme value.

The problem of high-frequency radar measurements of the sea surface is discussed by Crombie et al. (1978). The existence of spectral components, propagating against the wind, is shown. The angular energy distribution function (4.15) with power index $2s = 4$ is used to estimate the non-linear interaction. In this study calculations are repeated for the similar spectral approximation with peakness $\gamma = 3.3$. The obtained results are presented in Fig. 4.10a,b. The main calculation details of the non-linear interaction are similar to the previously obtained ones. Four clearly separate areas in the plane $\{\tilde\sigma, \beta\}$ can be identified. Three of them are characterized by positive

100 4 Physical Mechanisms Forming the Wave Spectrum in Deep Water

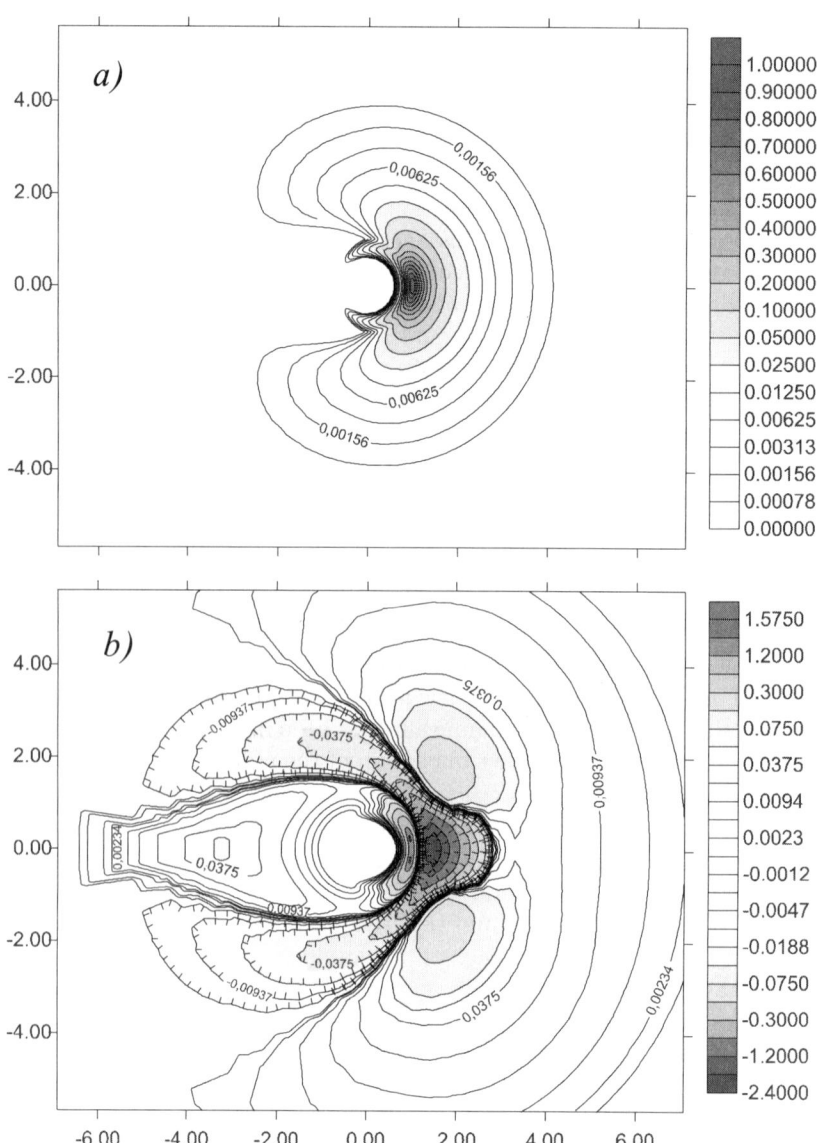

Fig. 4.8. Normalized spectral energy density (**a**) and non-linear transfer function (**b**) for peakness $\gamma = 1.0$ and angular distribution (4.15)

values, and another area contains negative ones. The maximum positive value is located at the point $\tilde{\sigma} \approx 1.06$, $\beta = 180°$ on the right-hand semi-plane. Its value comprises 2 per cent of the maximum. The same calculations made

4.1 Non-Linear Energy Transfer in Wind Wave Spectrum 101

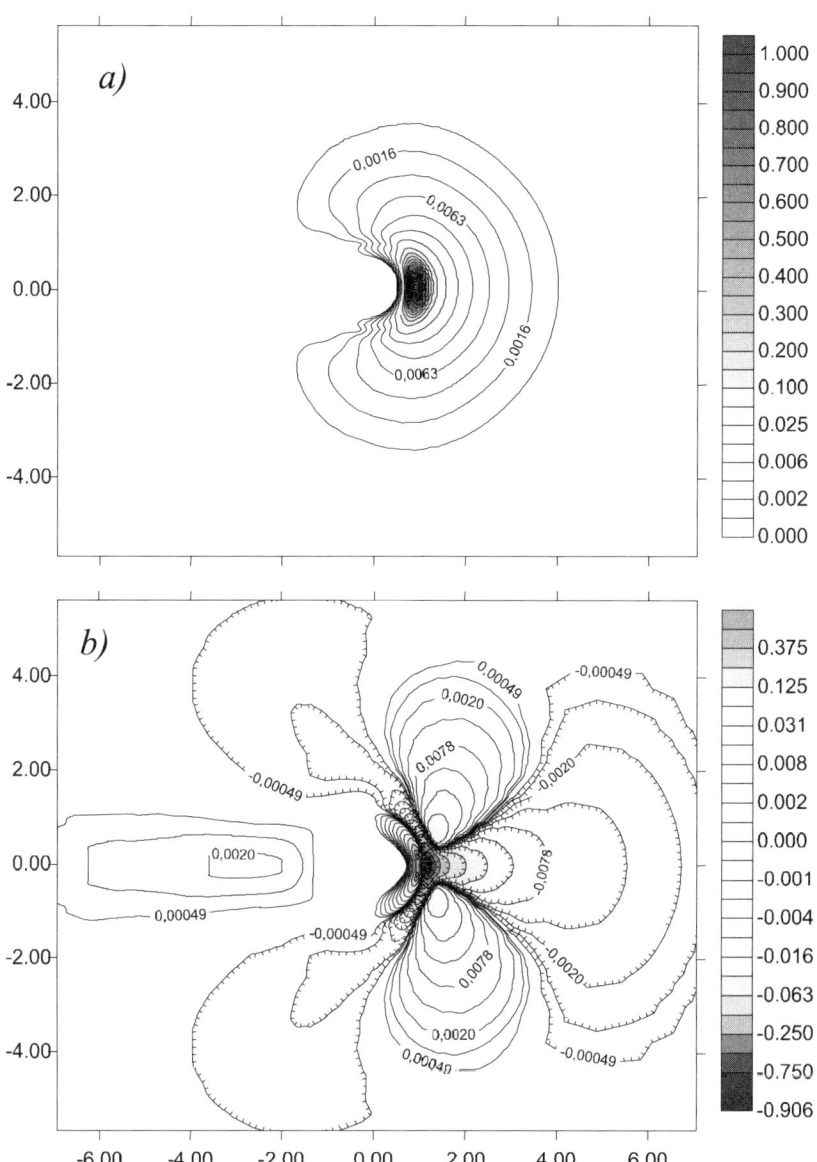

Fig. 4.9. Normalized spectral energy density (4.16) **(a)** and non-linear transfer function **(b)** for peakness $\gamma = 3.3$ and angular distributions (4.17), (4.18)

for the peakness $\gamma = 1.0$ show analogous details, but the non-linear transfer function is much more extended over the plane $\{\tilde{\sigma}, \beta\}$. The maximum of the new positive value comprises 4.5 per cent of the main maximum.

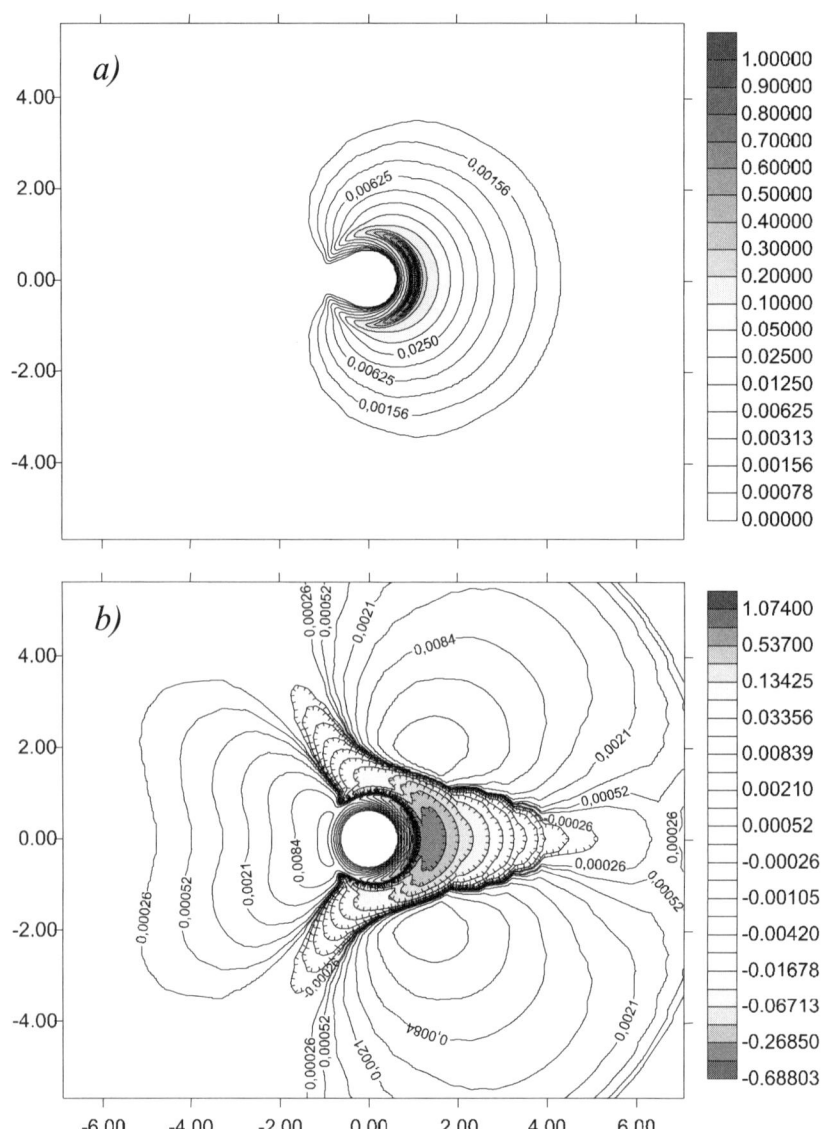

Fig. 4.10. Normalized energy spectral density (JONSWAP) (**a**) and non-linear transfer function with power index $2s = 4$ (**b**) for peakness $\gamma = 3.3$ and angular distribution (4.15)

Discussion of results. The estimates of non-linear interactions in the wind wave spectrum for its different approximations are made as a result of numerical calculations using the numerical integration algorithm of the highest accuracy. Some of the main details of the non-linear interaction function (such as the location and value of the main positive and negative extremes, etc.) are confirmed and specified by the numerical results. The strong influence of the frequency-angular spectrum form on the non-linear interaction function is revealed.

In particular, the non-linear energy transfer is found to be limited by approximately the same frequency-angular interval for the typical cosine spectral approximations (4.14). Its non-zero values are located in the semi-plane $|\beta| \leq \pi/2$. In the case of a wider angular distribution function, having small non-zero values at $|\beta| \geq \pi/2$, significant changes occur with the non-linear energy transfer function at $|\beta| \geq \pi/2$. Its value differs from zero over the entire frequency-angular plane $\{\tilde{\sigma}, \beta\}$. The calculation results show great sensitivity of the non-linear energy transfer towards the angular energy distribution function, especially for the left-hand semi-plane $\{\tilde{\sigma}, \beta\}$ ($|\beta| \geq \pi/2$).

The non-zero values of the non-linear transfer in the direction opposite to the general direction of the wave spectrum propagation (i.e. at $\beta = 180°$) are of special interest. In spite of the fact that the angular distribution function of the energy spectrum in this direction is practically equal to zero, the stable existence of the area of positive non-linear energy transfer values is observed. These values depend both on the angular distribution function of the energy spectrum and its frequency approximation. As for the same angular distribution function, the relative value of the non-linear energy transfer becomes much greater for a wider frequency spectrum. Thus, its value is changed by more than an order of magnitude with the spectrum peakness being varied from $\gamma = 7.0$ to $\gamma = 1.0$. This indicates an increase of the non-linear transfer in the direction opposite to the general one with wave development.

It should be noted that the non-linear energy transfer function becomes equal to zero near the origin of the polar coordinates $\{\tilde{\sigma}, \beta\}$, i.e. for the small frequency values. In other words, the non-linear energy transfer by-passes the origin of the coordinate system $\{\tilde{\sigma}, \beta\}$, as soon as it is "prohibited" by the resonance conditions (4.2).

The following interpretation of the spectral components generated in the opposite direction to the wind can be suggested with the help of the aforementioned facts. At the initial stage of offshore wave development, when they are formed under a uniform offshore wind, their spectrum is represented by a sufficiently narrow frequency-angular approximation. Its general direction coincides with the wind. The angular wave distribution becomes wider with its further development as a result of wind speed variations and non-linear transfer. At some moment, components whose direction is different from the average wind direction by an angle greater than $90°$ ($|\beta| \geq \pi/2$), can appear in the wave spectrum.

From this moment on, the generation of spectral components directed against the wind occurs due to non-linear transfer.

It should be noted that Hasselmann (1962) interpreted the integral (4.1) in terms of quadrupole interactions between three active wave components, which define the interaction intensity, and a passive fourth component, receiving energy, but not influencing the interaction directly. The interaction integral (4.1) can be rewritten in the form of the sum of two items. The first one does not depend directly on the wave action value $N = N(\boldsymbol{k})$, which is a function of the wave vector $\boldsymbol{k}$. But the second one depends explicitly on $N = N(\boldsymbol{k})$. This means that when $N(\boldsymbol{k}) = 0$ the first item of the non-linear interaction integral for this component $G_{\rm nl}(\boldsymbol{k})$ can be different from zero.

Taking into consideration the aforesaid, it is obvious that if three components are directed along the wind (or, at least, the angle between their direction and the wind speed is less than 90°), there may be one component with direction opposite to the wind and it can receive energy from the other three. It should be noted that the discrete interaction approximation (DIA) of the non-linear transfer used in the WAM model (Komen et al.,1994) does not produce this effect at all.

The spectral density of the given component starts growing linearly in accordance with the integral (4.1). After that its non-linear evolution starts. As shown above, a more intense development of waves propagating against the wind occurs with wave development. On the one hand the wind results in further wave development; on the other hand it creates an energy dissipation for the spectral components propagating against the wind.

While discussing a report by Phillips "Dynamics of random finite amplitude waves", Barber (1961) mentioned that he remembered standing on the shore of a lagoon, which was 10 m wide. The offshore wind was about $3 \, \rm m\, s^{-1}$, and wind waves with a period of about 1 s developed towards a distant shore. The water near him was calm and shallow, and he noticed very small waves with a period of about 1 s approaching him. These waves were propagating against the wind. He did not know what had caused them. There was no boat around. Probably, the waves were reflected from the other shore. But there was a beach there, rather than a wall. In connection with this presentation it could be suggested that the waves were generated by non-linear interaction of wind waves.

This is probably one of the first discussions of the physical mechanism of wave generation propagating against the wind. More precise measurements of this phenomenon are made later. Thus, the presence of spectral components propagating against the wind is detected in the high-frequency radar measurements of signals reflected from the sea surface. As shown by the BOMEX experiment (Crombie, 1972), the amplitudes of such components increase with distance from the shore.

The first explanation of this phenomenon was suggested by M.S. Longuet-Higgins in 1961 as a result of a weak wave non-linear interaction in the wind wave spectrum. In the paper by Crombie et al. (1978), K. Hasselmann showed that a non-linear interaction could transfer the energy in the direction opposite to the wind. However, the value of this non-linear interaction is two orders of magnitude less than its maximum. Unfortunately, it was practically impossible at that moment to obtain a more accurate estimation of the non-linear interaction in the wind wave spectrum due to the difficulty of calculating the collision integral. The problem was not only in the complexity of the traditional calculation of the non-linear interaction integral, but in the fact that its real contribution to this component was considerably less than the corresponding value of the wave components directed along the wind. The numerical methods for the calculation of the collision integral must be sufficiently accurate to distinguish this effect from the numerical error background, which could be quite considerable. Asshown in this section, such calculations have become possible nowadays only by using the accurate algorithm presented in this monograph.

4.1.4 Numerical Study of Non-Stationary Solution of the Hasselmann Equation

In spite of the fact that the main integral features have been investigated (Hasselmann & Hasselmann 1981, 1985; Komen et al., 1994; Komatsu & Masuda, 1996; Masuda, 1981; Lavrenov, 1991a,b, 1998, 2000; Lavrenov & Ocampo-Torres, 1999; Polnikov, 1989, 1990, 1993; Resio & Perrie, 1991; Snyder et al., 1993; Webb, 1978 etc.), not enough attention was paid to the problem of the analytical and numeric study of the kinetic equation (4.1). Among the stationary analytical solutions, the following one is known as "thermodynamic":

$$N = (a + b\omega + c\boldsymbol{k})^{-1}, \qquad (4.20)$$

where a, b and $\boldsymbol{c}$ are constants. The solution (4.20) reduces the integral value to zero. However, the solution (4.20) is of no great physical importance, because a small disturbance produces a divergence of the integral (4.1).

Investigation of the solution to (4.1) remains an important and difficult problem. The main progress in its analytical study was achieved by Zakharov et al. (1966, 1981, 1982, 1983a,b). The solution of the stationary analytical spectra of (4.1) is derived for the isotropic angle case and infinite frequency range $[0, \infty]$. Analytical methods of solution were developed further by Zaslavskii (1989a,b, 2000), based on "the narrow directional approximation". According to Zaslavskii (2000), the evolution equation (4.1) can be presented as self-similar solutions. The spectral frequency dependence is found to be equal to $S(\sigma) \sim \sigma^{-13/2}$ (for $\sigma > \sigma_p$, where σ_p is the peak spectrum frequency), the evolution of the spectral frequency maximum is estimated as $\sigma_p \sim t^{-1/11}$, the angular narrowness in the vicinity of the spectral maximum is $D_p \cong 1$. But, unfortunately, the confidence equations used by Zaslavskii (2000) do not provide a complete study of the whole evolution equation. His self-similar spectrum approximation does not satisfy the energy conservation law.

There are not so many papers devoted to the problem of numerical simulation of the evolution equation (4.1). In the papers by Polnikov (1990, 1999) it is shown that the spectral form does not depend on the initial spectrum and the integral spectral parameters vary within narrow bands for a large time scale of the non-linear evolution. The effect is considered to establish one of the self-similar spectral forms and is confirmed by the analytical estimations (Zaslaskii, 2000).

The problem of a self-similar solution seems to have been solved. But in the paper by Komatsu & Masuda (1996) the numerical solutions differ from the results of Polnikov (1990) and Zaslavskii (2000). The difference is not only in another spectral tail frequency dependence, but also in the integral parameters. Thus, the numerical results lead to the following approximation $S(\sigma) \sim \sigma^{-4}$ (for $\sigma > 1.5\sigma_p$). The problem remains unsolved and can be formulated this way: is there an exact self-similar form of the spectrum? And then: if the answer is positive, what are the values of its parameters?

An attempt is undertaken to solve the problem in this section.

Time scale of establishing a self-similar solution. The time scale T of the spectral non-linear evolution, on which the non-linear energy transfer exchanges fully a spectral form, is considered in this chapter. The spectrum should not depend on the initial details. It is possible to estimate the time scale with the help of the conservation law of the total wave action:

$$\frac{\partial N(\mathbf{k})}{\partial t} + \mathrm{div}_\mathbf{k} \mathbf{F}_n = 0 , \qquad (4.21)$$

where $\mathbf{F}_n$ is a vector of the action flux. According to Polnikov (1999) the value T can be estimated with the help of the ratio of the spectral maximum N_p to the maximum of the non-linear transfer $(\partial N/\partial t)_p$:

$$T \cong N_p/(\partial N/\partial t)_p . \qquad (4.22)$$

In order to get the estimation T one can transfer from the wave action spectrum $N(k)$ to the frequency-angular spectrum $S(\sigma, \beta)$, using the formula:

$$N(\mathbf{k}) \, \mathrm{d}\mathbf{k} \propto \frac{1}{\sigma} S(\mathbf{k}) \, \mathrm{d}\mathbf{k} \propto \frac{g^3}{2\sigma^4} S(\sigma, \beta) \, \mathrm{d}\sigma \, \mathrm{d}\beta , \qquad (4.23)$$

where $\sigma^2 = gk$ is the deep-water dispersion relation. Application of the spectrum $S(\sigma, \beta)$ gives a reliable estimation so long as its form is well known. For the typical values $\sigma_p = 1\,\mathrm{rad\,s^{-1}}$, $S_{\max} = 0.2\,\mathrm{m^2 s}$ it is possible to obtain:

$$T \cong 10^5 \tau , \qquad (4.24)$$

where $\tau = 2\pi/\sigma_p$ is the period of the initial spectral maximum.

Thus, a non-linear evolution establishing a self-similar spectral form happens at a time scale larger than the estimation (4.24) at least at one or two orders of magnitude: $\tilde{t} = t/\tau \gg T/\tau$.

Non-linear evolution of JONSWAP spectrum. The non-linear energy spectrum evolution is computed for the initial frequency-angular JONSWAP spectrum (4.12)–(4.16).

4.1 Non-Linear Energy Transfer in Wind Wave Spectrum

Table 4.1. Values of initial spectrum parameters

γ	$2s = 2$		$n_\beta = 2$		$n_\beta = 8$	
	B	D_p	B	D_p	B	D_p
1.0	0.69	0.32	0.69	0.64	0.69	1.16
3.3	0.34	0.32	0.34	0.64	0.34	1.16
7.0	0.25	0.32	0.25	0.64	0.25	1.16

In the process of computation the main parameters are estimated, including the frequency width B, defined as:

$$B = \int S(\sigma) \, d\sigma / S(\sigma_p)\sigma_p , \qquad (4.25)$$

where $S(\sigma) = \int\limits_{-\pi}^{\pi} S(\sigma, \beta) \, d\beta$, and σ_p is the frequency of the spectrum maximum of the time-evaluation spectrum and the directional width D:

$$D(\sigma) = S(\sigma, \beta_p)/S(\sigma) . \qquad (4.26)$$

A series of the initial spectrum parameters γ, n_β, $2s$ and the initial value B and D defined in the general direction: $D_p = D(\sigma_p) = S(\sigma_p, \beta_p)/S(\sigma_p)$ are shown in the Table 4.1.

Numerical algorithm. The establishment of a self-similar spectral form, controlled by non-linear energy transfer, demands computation for time scales larger than the estimation (4.24) at least at one or two orders of magnitude $\tilde{t} = t/\tau > 10^6 – 10^7$. The main problem is that the numerical computation of the integral (4.1) takes a lot of CPU time. This means that an algorithm for computations of non-linear energy transfer should be very fast. To solve the evolution equation the integral (4.1) is to be computed thousands of times to reach the estimated time scale of the spectrum evolution. That is why the most optimal algorithm is used in the present study, described in the Sect. 4.1.2.

A numerical solution of the evolution equation (4.1) is carried out using the semi-implicit method (The WAM Model, 1988; Lavrenov, 1998). It should be noted that the traditional method of numerical integration usually leads to numerical instabilities in the high-frequency spectral range. This requires using a small time step for the numerical integration and applying special limits on the numerical solution and on the value of the source function (Lavrenov, 1998; Tolman, 1992). Unfortunately, these methods can disturb energy conservation, leading to misinterpretation of the physical results. But, utilization of the semi-implicit numerical method does not produce any instabilities for relatively large time steps of numerical integration (Lavrenov, 1998) and there is no need to use any limits.

The numerical accuracy and the preservation of the main integrals (4.3) are controlled in the process of computation. So, the numerical error of the total energy conservation does not exceed 10 per cent and the error of the total momentum estimation does not exceed 25 per cent in comparison to its initial values.

Numerical results. Numerical results for the initial condition (4.12)–(4.14) with $\gamma = 3.3$ and $n = 2$ for four different time moments: $\tilde{t} \cong 0$, 10^3, 10^5, 10^7, respectively, are presented in Figs. 4.11–4.14. Numerical simulation results of the relative frequency-angular spectrum values in a polar coordinate system, where the radius vector is the relative frequency value $\tilde{\sigma} = \sigma/\sigma_p$, are presented in Fig. 4.11a–d. The spectrum time evolution, resulting in a moon-like form, is shown in the figures. The spectrum becomes narrower in the vicinity of the spectral maximum and wider in the high and low frequency range.

Two-dimensional non-linear transfer values for the same steps of wave evolution are presented in Fig. 4.12a–d. The non-linear evolution changes the spectrum form and shifts the frequency of the spectrum maximum to lower frequencies. The non-linear energy transfer function is changed to a larger extent in comparison with the spectrum. It becomes narrower with the concentration intensity in the vicinity of the spectral maximum.

The frequency spectrum for the same evolution time steps is presented in Fig. 4.13. The spectrum is proportional to $\sim \sigma^{16.3}$ in the low-frequency range $\sigma_p > \sigma$. It is proportional to $\sim \sigma^{-6.1}$ within the range $\sigma_p < \sigma < 1.5\sigma_p$. The spectrum decreases as $\sim \sigma^{-2.6}$ at the larger frequencies $\sigma > 1.5\sigma_p$.

The evolution of the parameter D (4.26) for different time moments is presented in Fig. 4.14. The value of the parameter D becomes larger and equals 1.33 in the vicinity of the spectral maximum, and it gets smaller in the low and high frequency ranges. It is evidence of spectral isotropization within these frequency ranges.

Similar results for the initial frequency-angular spectrum (4.13) and (4.15) with $\gamma = 3.3$, $2s = 2$ for the following evolution steps $\tilde{t} \cong 0$, 10^4, 10^6, 10^7 are presented in Figs. 4.15–4.18. It is interesting to note that at times $\tilde{t} \leq 10^4$ the frequency-angular spectrum (see Fig. 4.15) varies in such a way that it becomes more isotropic as mentioned by Lavrenov & Ocampo-Torres (1999). It also becomes smoother and rounder. But later on (at $\tilde{t} \geq 10^5$) it is transformed into a more directional form. The two-dimensional spectrum could not become isotropic, as follows from the total momentum conservation law (4.3c). So, the previous suggestion about the spectrum becoming isotropic with the initial angular approximation (4.14) is not proved. The spectrum becomes narrower in the vicinity of its maximum and it is wider at low and high frequencies.

An important feature of the non-linear energy transfer (see Fig. 4.16) is the presence of an area with positive values in the direction opposite to the main direction of the wave propagation. Due to the action of the non-

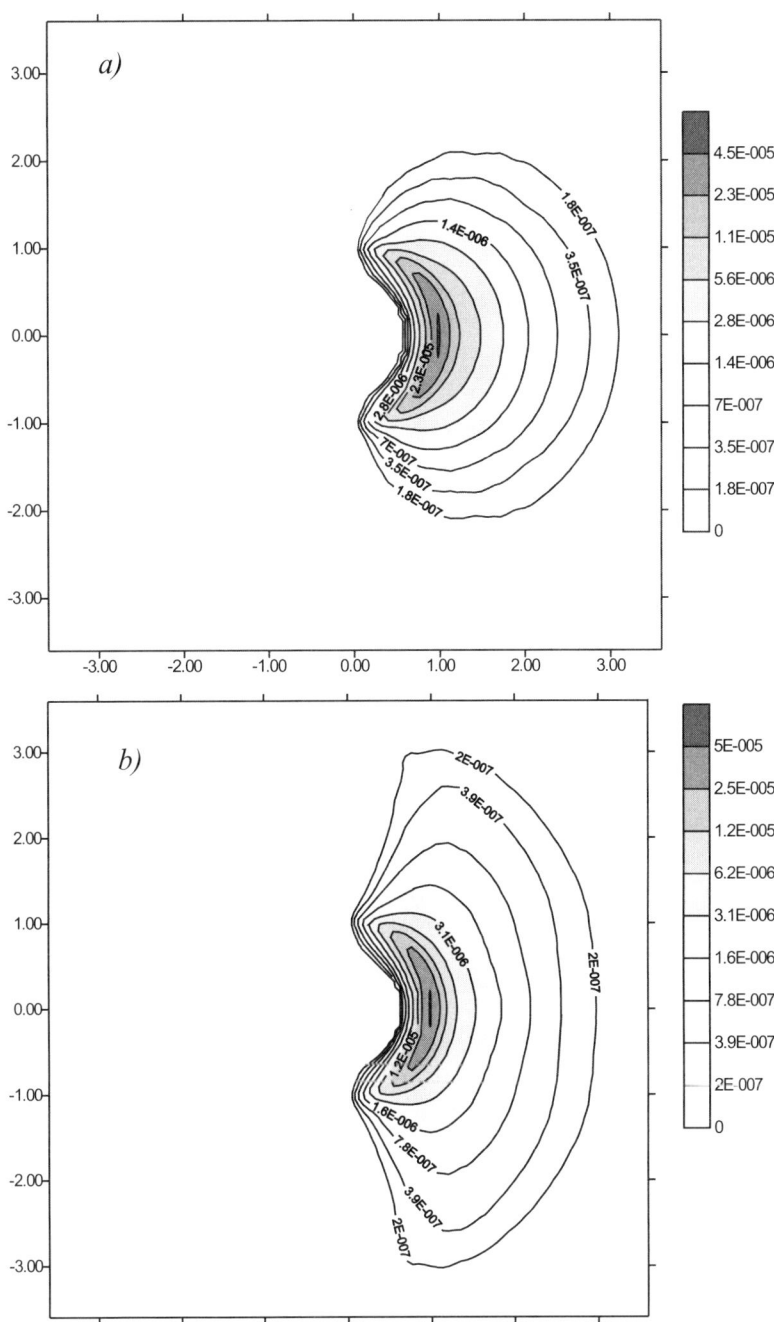

Fig. 4.11. Non-linear two-dimensional spectrum evolution for the initial spectrum value (4.12)–(4.14) and $\gamma = 3.3$, $n = 2$ at different time moments: (**a**) $\tilde{t} = 0$ and (**b**) $\tilde{t} = 10^3$ (Fig. 4.11(**c,d**) see next page.)

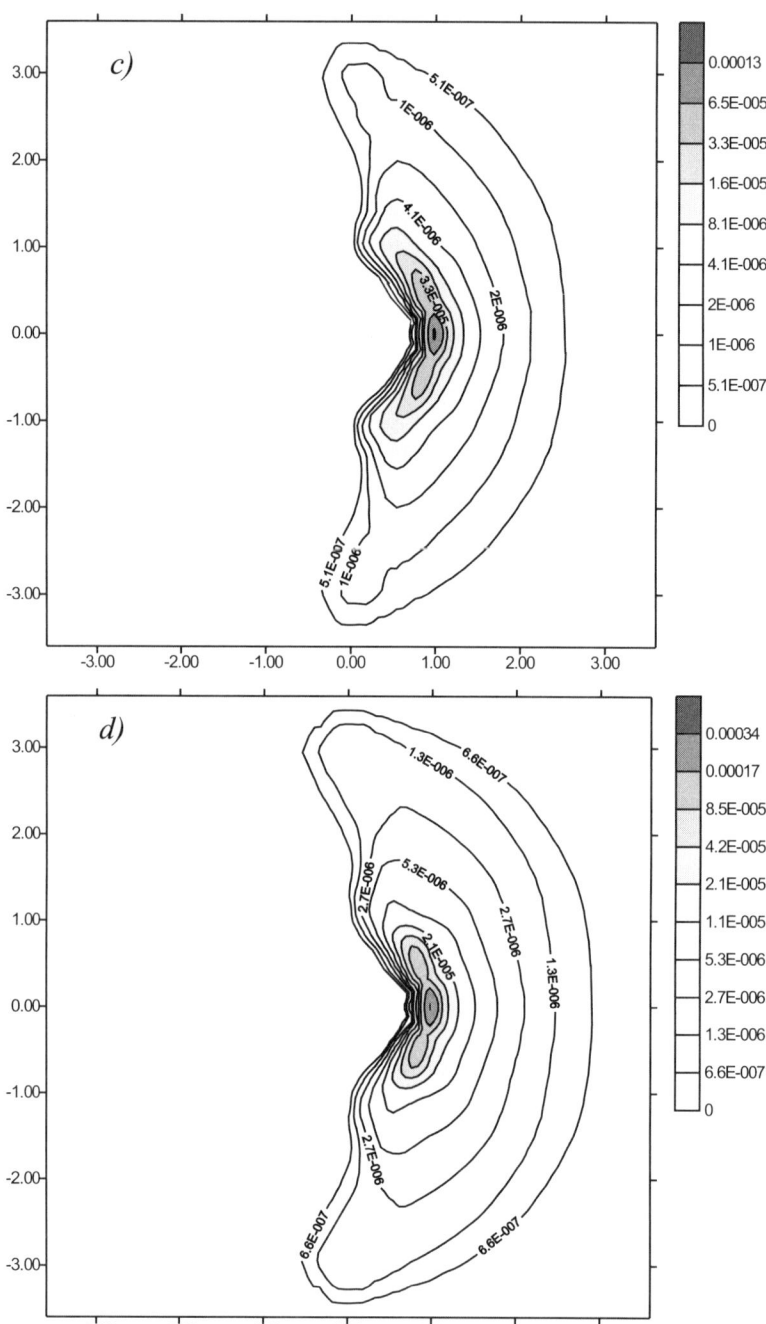

Fig. 4.11. Non-linear two-dimensional spectrum evolution for the initial spectrum value (4.12)–(4.14) and $\gamma = 3.3$, $n = 2$ at different time moments: **(c)** $\tilde{t} = 10^5$ and **(d)** $\tilde{t} = 10^7$

4.1 Non-Linear Energy Transfer in Wind Wave Spectrum 111

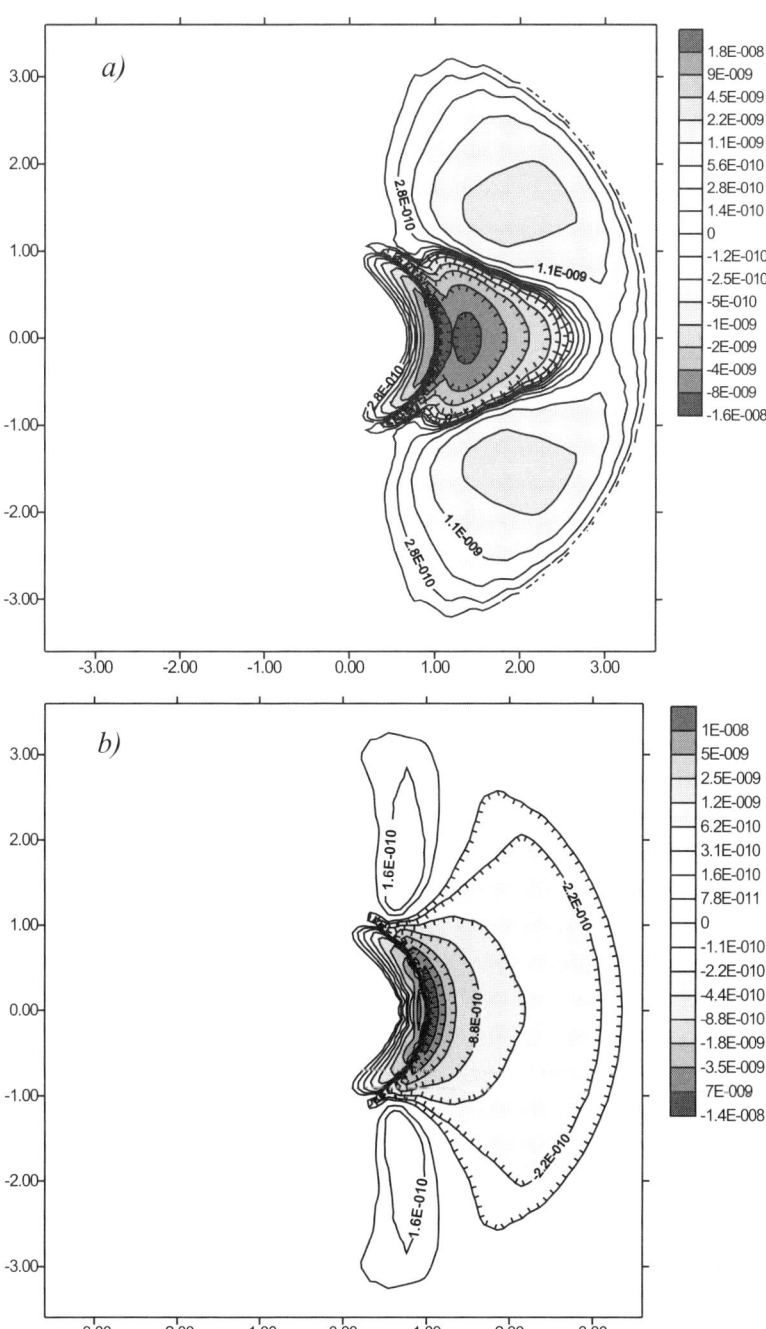

Fig. 4.12. Two-dimensional non-linear energy transfer function for the initial value (4.12)–(4.14) and $\gamma = 3.3$, $n = 2$ at different time moments: **(a)** $\tilde{t} = 0$ and **(b)** $\tilde{t} = 10^3$ (Fig. 4.12(**c,d**) see next page.)

112 4 Physical Mechanisms Forming the Wave Spectrum in Deep Water

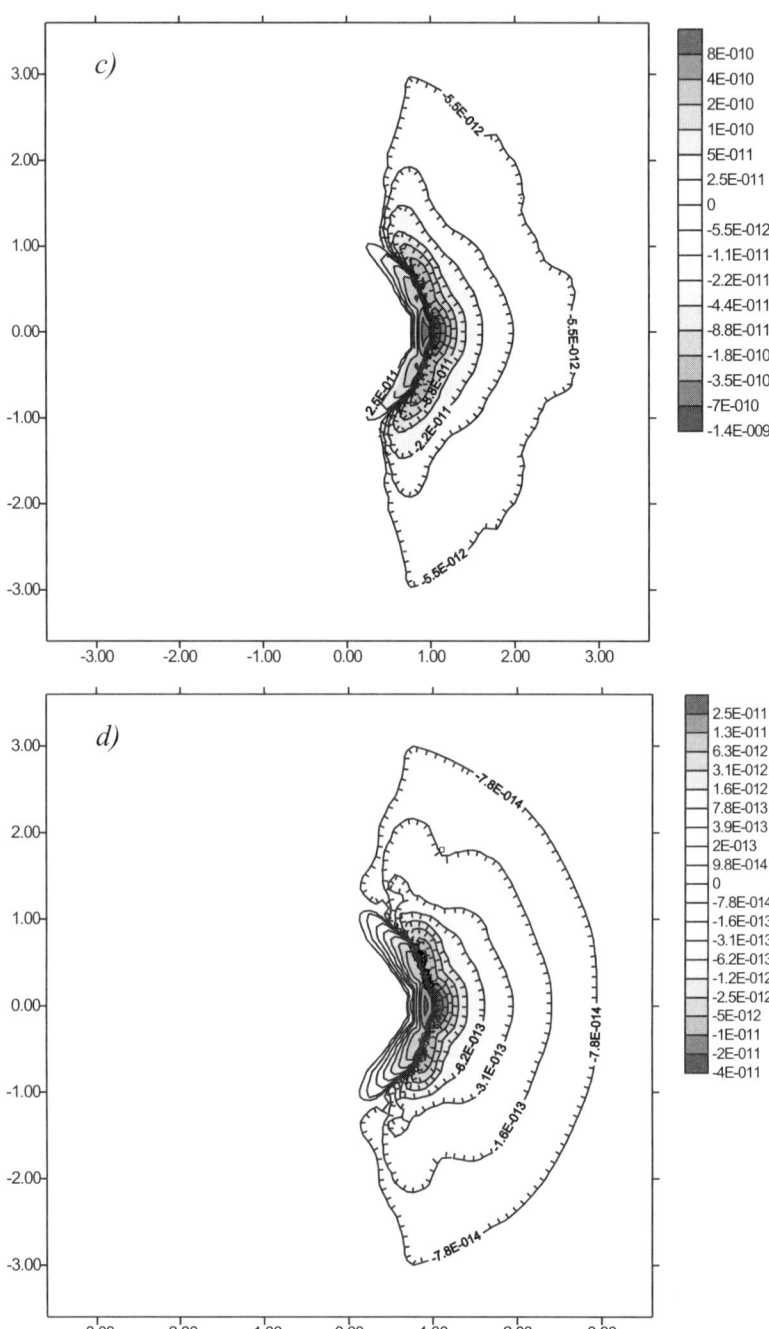

Fig. 4.12. Two-dimensional non-linear energy transfer function for the initial value (4.12)–(4.14) and $\gamma = 3.3$, $n = 2$ at different time moments: (**c**) $\tilde{t} = 10^5$ and (**d**) $\tilde{t} = 10^7$

4.1 Non-Linear Energy Transfer in Wind Wave Spectrum

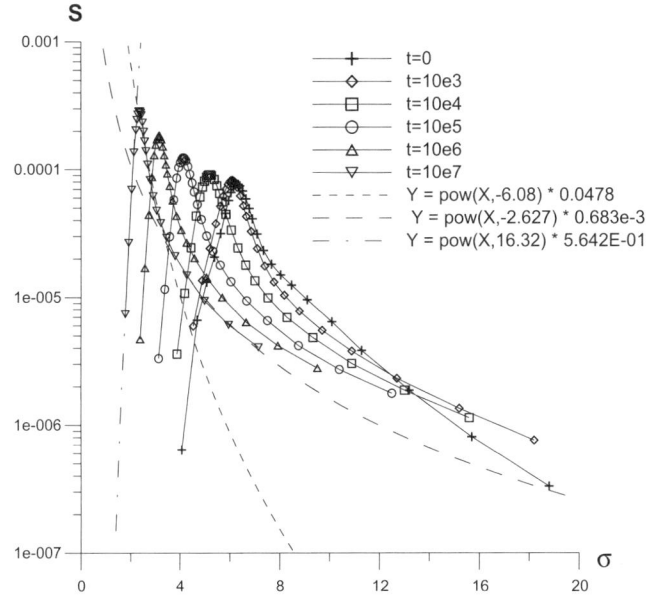

Fig. 4.13. Frequency spectrum evolution for the initial value (4.12)–(4.14) and $\gamma = 3.3$, $n = 2$ at different time moments

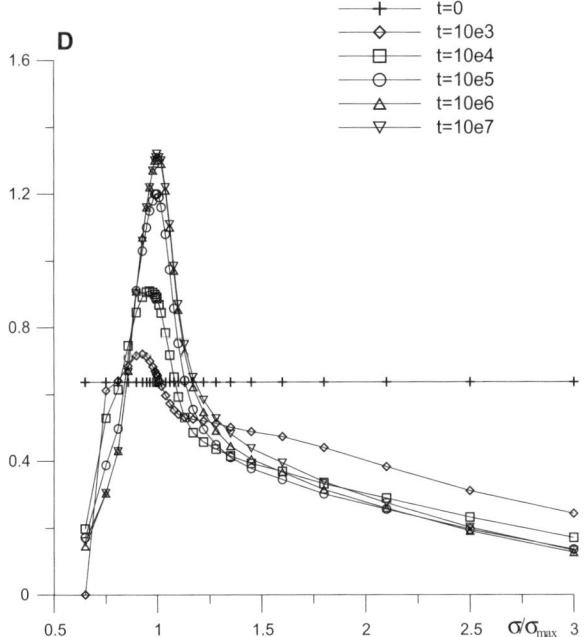

Fig. 4.14. Parameter D for the initial value (4.12)–(4.14) and $\gamma = 3.3$, $n = 2$ at different time moments

114 4 Physical Mechanisms Forming the Wave Spectrum in Deep Water

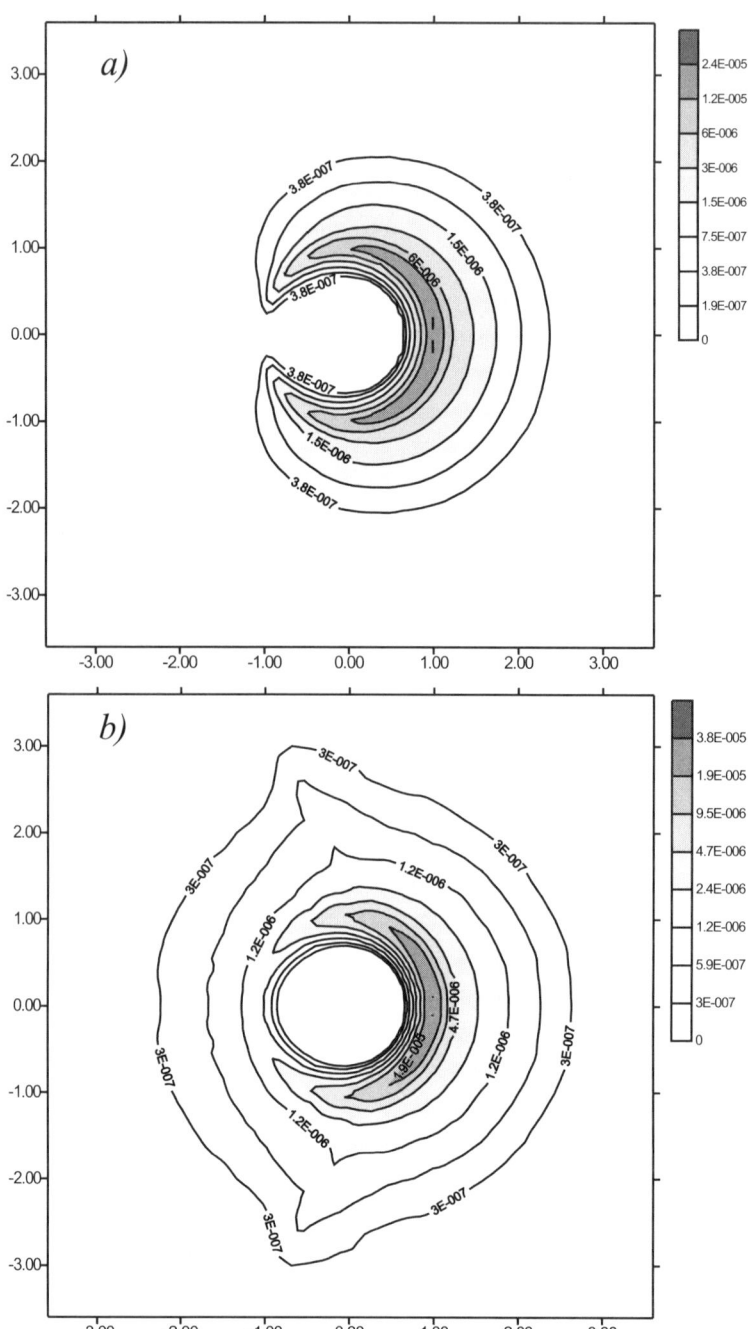

Fig. 4.15. Non-linear two-dimensional spectrum evolution for the initial value (4.13) and (4.15) and $\gamma = 3.3$, $2s = 2$ at different time moments: **(a)** $\tilde{t} = 0$ and **(b)** $\tilde{t} = 10^3$ (Fig. 4.15(**c,d**) see next page.)

4.1 Non-Linear Energy Transfer in Wind Wave Spectrum 115

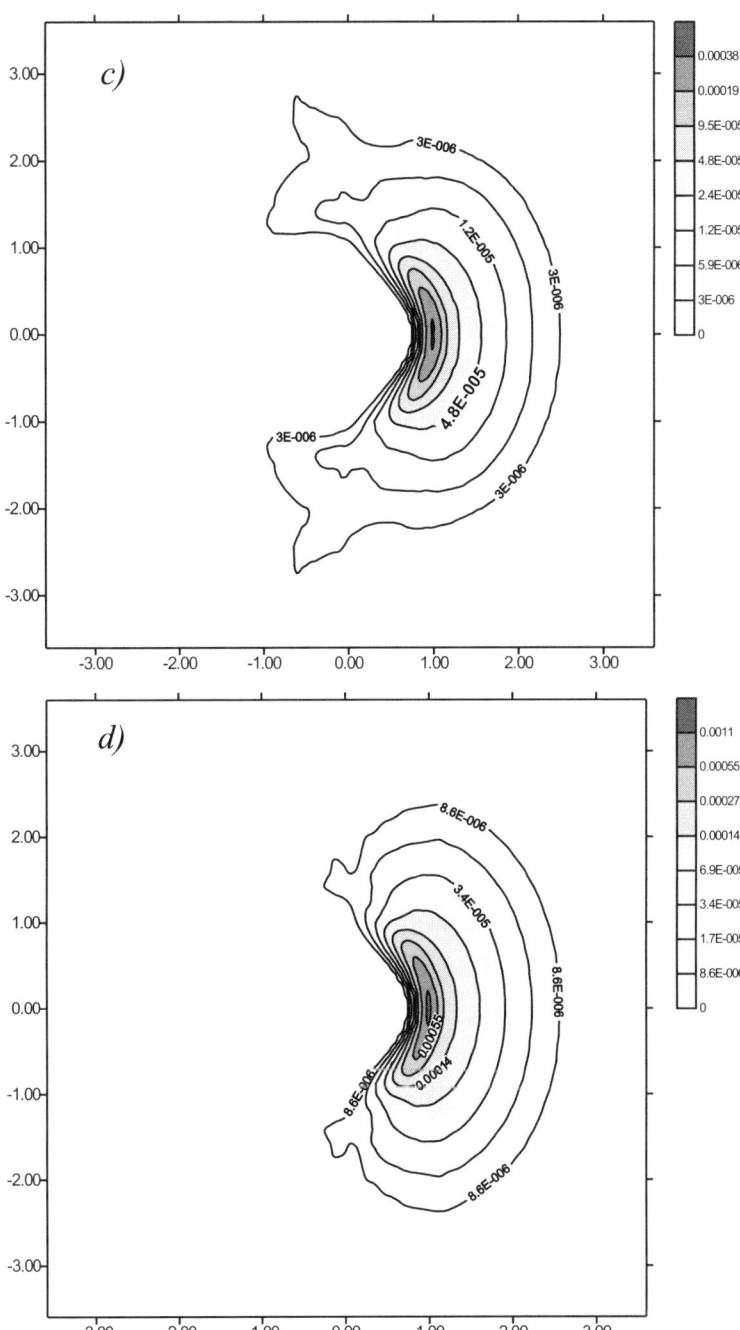

Fig. 4.15. Non-linear two-dimensional spectrum evolution for the initial value (4.13) and (4.15) and $\gamma = 3.3$, $2s = 2$ at different time moments: **(c)** $\tilde{t} = 10^6$ and **(d)** $\tilde{t} = 10^7$

116   4 Physical Mechanisms Forming the Wave Spectrum in Deep Water

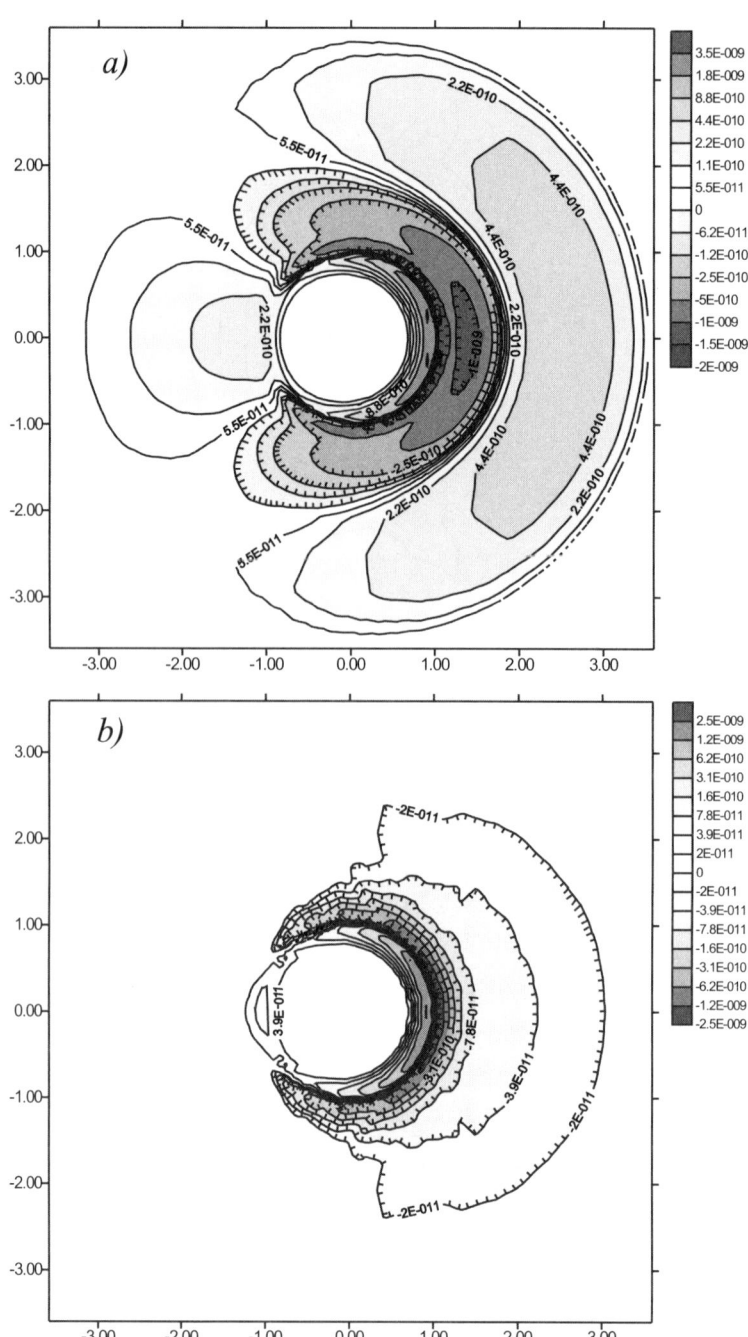

Fig. 4.16. Two-dimensional non-linear energy transfer function for the initial value (4.13) and (4.15) and $\gamma = 3.3$, $2s = 2$ at different time moments: (**a**) $\tilde{t} = 0$ and (**b**) $\tilde{t} = 10^3$ (Fig. 4.16(**c**,**d**) see next page.)

4.1 Non-Linear Energy Transfer in Wind Wave Spectrum 117

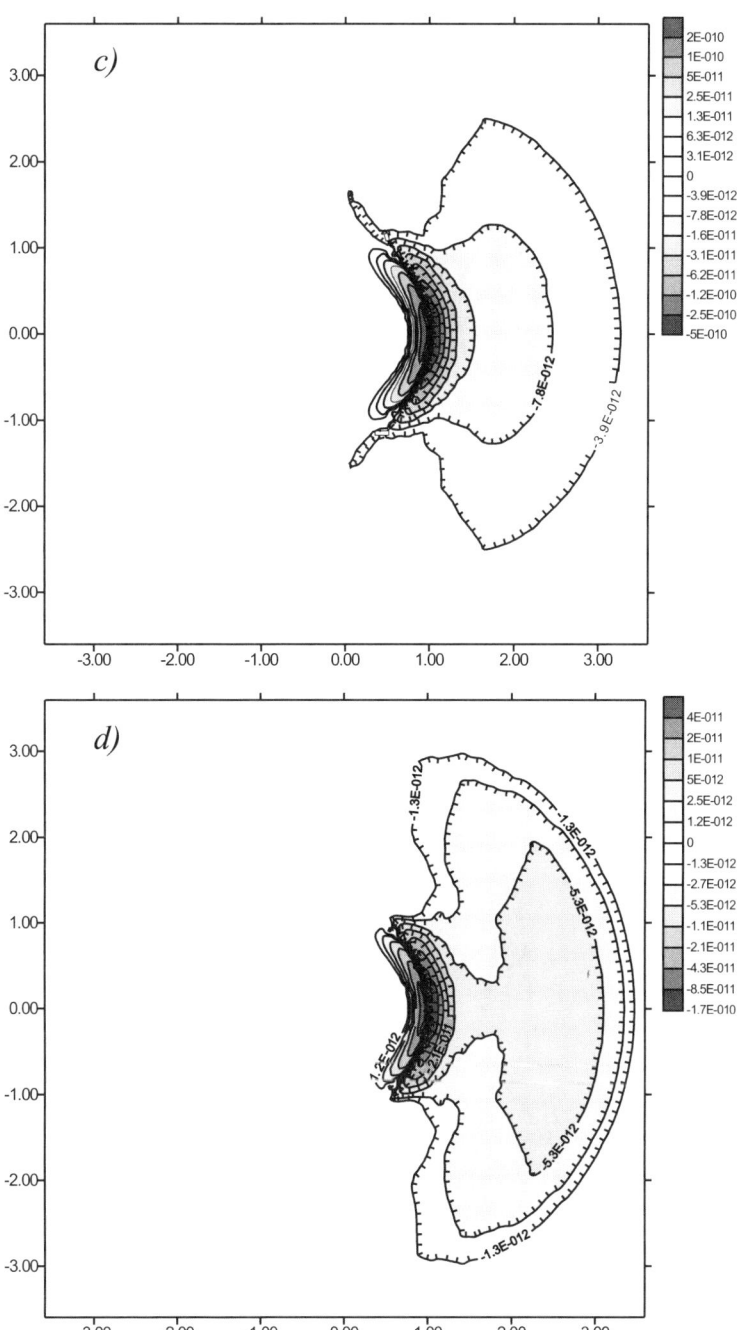

Fig. 4.16. Two-dimensional non-linear energy transfer function for the initial value (4.13) and (4.15) and $\gamma = 3.3$, $2s = 2$ at different time moments: **(c)** $\tilde{t} = 10^6$ and **(d)** $\tilde{t} = 10^7$

118   4 Physical Mechanisms Forming the Wave Spectrum in Deep Water

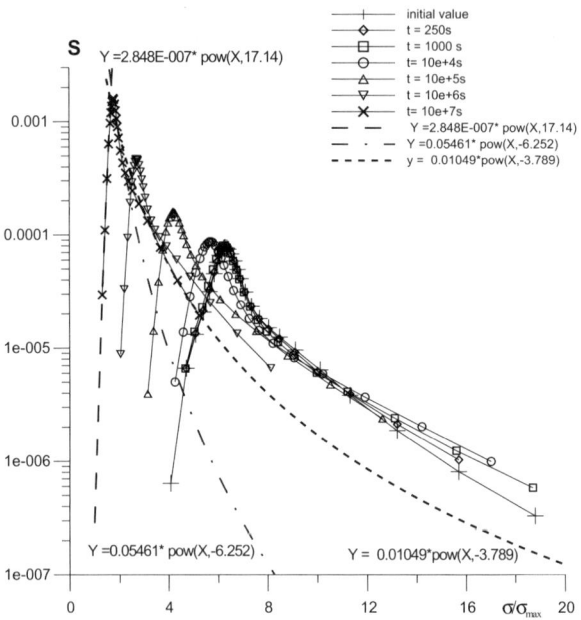

Fig. 4.17. Frequency spectrum evolution for the initial value (4.13) and (4.15) and $\gamma = 3.3$, $2s = 2$ at different time moments

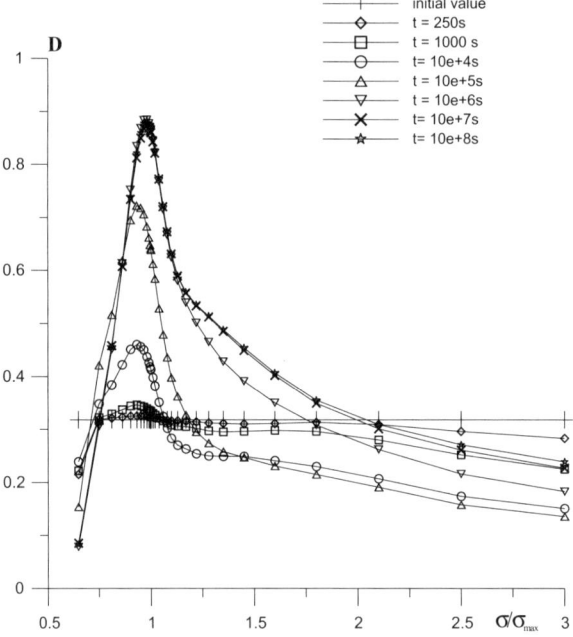

Fig. 4.18. Parameter D for the initial value (4.13) and (4.15) and $\gamma = 3.3$, $n = 2$ at different time moments

4.1 Non-Linear Energy Transfer in Wind Wave Spectrum

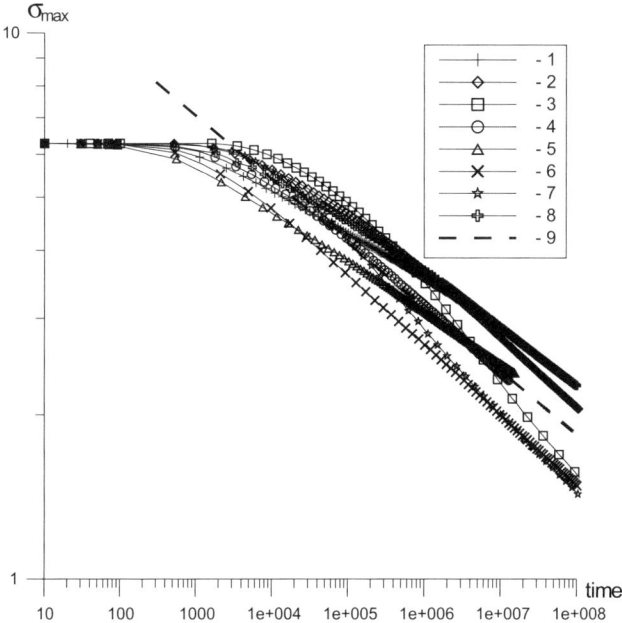

Fig. 4.19. Evolution of the maximum frequency of the spectrum in time for different initial spectra parameters (logarithmic scale): $1 - \gamma = 3.3$, angular distribution $\cos^8 \beta$; $2 - \gamma = 1.0$, angular distribution $\cos^2 \beta$; $3 - \gamma = 1.0$, angular distribution $\cos^2(\beta/2)$; $4 - \gamma = 3.3$, angular distribution $\cos^2 \beta$; $5 - \gamma = 7.0$, angular distribution $\cos^8 \beta$; $6 - \gamma = 7.0$, angular distribution $\cos^2 \beta$; $7 - \gamma = 3.3$, angular distribution $\cos^2(\beta/2)$; $8 - \gamma = 1.0$, angular distribution $\cos^8 \beta$; 9 – approximation $\sigma_{\max} = 15.85 t^{-0.1168}$

linear energy transfer the spectrum becomes more isotropic in the first stages of its evolution. Later on, this positive non-linear energy transfer area disappears, with the spectrum becoming more directional and moving to the low-frequency range.

The numerical solution of all initial spectra (Table 4.1) for time durations up to $\tilde{t} = 10^7$–10^8 are obtained. The frequency of the spectrum maximum evolution $\sigma_p = \sigma_p(t)$ for all these cases is presented in Fig. 4.19. The frequency of the spectrum maximum shifts to the lower frequency range as $\sigma_p(t) \sim t^{-0.11}$.

The time evolution of the parameter $B = B(t)$ for all initial conditions is presented in Fig. 4.20. The most important feature of the parameter $B = B(t)$ is that it approaches the same similar value for all initial conditions. It is approximately equal to $B \approx 1/3$ (the "law of 1/3").

As shown by the numerical results, not only the parameter B, but also the parameter D_p approaches some constant value for all initial spectra. The final numerical values of the parameters B and D_p for all initial spectra conditions are presented in Table 4.2.

120   4 Physical Mechanisms Forming the Wave Spectrum in Deep Water

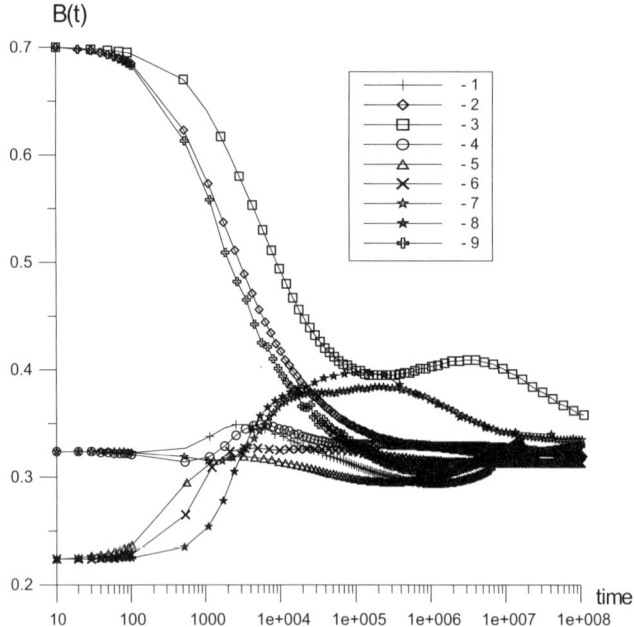

Fig. 4.20. Time evolution of the parameter B for different initial spectra: $1 - \gamma = 3.3$, angular distribution $\cos^8 \beta$; $2 - \gamma = 1.0$, angular distribution $\cos^2 \beta$; $3 - \gamma = 1.0$, angular distribution $\cos^2(\beta/2)$; $4 - \gamma = 3.3$, angular distribution $\cos^2 \beta$; $5 - \gamma = 7.0$, angular distribution $\cos^8 \beta$; $6 - \gamma = 7.0$, angular distribution $\cos^2 \beta$; $7 - \gamma = 3.3$, angular distribution $\cos^2(\beta/2)$; $8 - \gamma = 7.0$, angular distribution $\cos^2(\beta/2)$; $9 - \gamma = 1.0$, angular distribution $\cos^8 \beta$

Table 4.2. Final values of the parameters B and D_p for all initial spectrum conditions

γ	$2s = 2$		$n_\beta = 2$		$n_\beta = 8$	
	B	$D_{\max}$	B	$D_{\max}$	B	$D_{\max}$
1.0	0.358	0.988	0.328	1.31	0.330	1.29
3.3	0.342	0.899	0.320	1.32	0.321	1.12
7.0	0.342	0.963	0.313	1.31	0.326	1.09

Intermediate asymptotic approximation of the self-similar frequency spectrum. It is shown by the numerical result that the spectrum reveals itself as a self-similar solution at large times of non-linear evolution. It is possible to assume that the frequency spectrum does not depend on the features of the initial spectrum, but it should depend on the frequency of the spectrum maximum σ_p, the time t, the total energy E and the wave action A. Thus the frequency spectrum S can be written as the following function:

4.1 Non-Linear Energy Transfer in Wind Wave Spectrum

$$S = S(\sigma, \sigma_p, \sigma_\pi, t, E, A) \tag{4.27}$$

where σ_π is the frequency of transition from the main frequency domain to the high-frequency range.

Using the Π theorem (Barenblatt, 1984) it is possible to define a self-similar approximation as a function of non-dimensional parameters in the following form:

$$\frac{\sigma S(\sigma)}{E} = F\left(\frac{\sigma}{\sigma_p}, \frac{\sigma}{\sigma_\pi}, t\sigma_p, \frac{A\sigma_p}{E}\right), \tag{4.28}$$

where E is the total energy, and F is a function to be defined.

The number of variables in (4.28) can be reduced using the dependence between the frequency of the spectrum maximum and the time $\sigma_p = \sigma_p(t)$. In order to satisfy the numerical results the function F, describing the main features of the numerical solution, can be defined in the following way:

$$F = \begin{cases} F_1 = (n+1)\left(\dfrac{\sigma_p}{\sigma}\right)^n \exp\left(-\dfrac{n+1}{n}\left(\dfrac{\sigma_p}{\sigma}\right)^n\right) & \text{for } \sigma \leq \sigma_\pi \\[2ex] F_2 = (n+1)\left(\dfrac{\sigma_p}{\sigma}\right)^n \left(\dfrac{\sigma_\pi}{\sigma}\right)^{n_\pi - n} \exp\left(-\dfrac{n+1}{n}\left(\dfrac{\sigma_p}{\sigma_\pi}\right)^n\right) & \text{for } \sigma > \sigma_\pi, \end{cases} \tag{4.29}$$

where n and n_π are new non-dimensional parameters, which are functions of the non-dimensional value $A\sigma_p/E$. As is shown by the numerical results, the value n is estimated as $n = 5\text{--}7$ and $n_\pi \approx n/2$. It should be noted that integrating the spectrum (4.27) with (4.29) within the frequency range results in the total energy value E, which is equal to

$$E = m_0\left(1 + \frac{n+1}{n}\left(\frac{\sigma_p}{\sigma_\pi}\right)^n \left[\frac{n}{n_\pi} - 1\right]\exp\left(-\frac{n+1}{n}\left(\frac{\sigma_p}{\sigma_\pi}\right)^n\right)\right) \tag{4.30}$$

where $m_0 = \int\limits_0^\infty \sigma^{-1} F_1(\sigma)\, d\sigma$.

In order to define the value n, the total wave action A can be used as the moment of the -1 power:

$$A = m_{-1} = \int\limits_0^\infty \sigma^{-1} S(\sigma)\, d\sigma = m_0\, \sigma_p^{-1} \left[\left[\frac{n+1}{n}\right]^{-1/n} \Gamma\left(1 + \frac{1}{n}\right)\right.$$

$$\left. + \left(\frac{\sigma_p}{\sigma_\pi}\right)^{n+1}\left(\frac{n - n_\pi}{n_\pi + 1}\right) \exp\left(-\frac{n+1}{n}\left(\frac{\sigma_p}{\sigma_\pi}\right)^n\right)\right] \tag{4.31}$$

where $\Gamma(n)$ is the gamma function.

Thus, the non-dimensional value $A\sigma_p/E$ can be written as follows:

$$\frac{A\sigma_p}{E} = \frac{\left[\left[\frac{n+1}{n}\right]^{-1/n}\Gamma\left(1+\frac{1}{n}\right) + \left(\frac{\sigma_p}{\sigma_\pi}\right)^{n+1}\left(\frac{n-n_\pi}{n_\pi+1}\right)\exp\left(-\frac{n+1}{n}\left(\frac{\sigma_p}{\sigma_\pi}\right)^n\right)\right]}{\left[1+\frac{n+1}{n}\left(\frac{\sigma_p}{\sigma_\pi}\right)^n\left(\frac{n}{n_\pi}-1\right)\exp\left(-\frac{n+1}{n}\left(\frac{\sigma_p}{\sigma_\pi}\right)^n\right)\right]}. \tag{4.32}$$

Using the approximation (4.27) with (4.28), the frequency narrowness parameter (4.25) can be defined as:

$$B = E/S(\sigma_p)\sigma_p = \exp\left(\frac{n+1}{n}\right)/(n+1). \tag{4.33}$$

Due to the change of the n parameter from 4 to 8, the value B is varied within the range from 0.69 to 0.34. This estimation corresponds approximately to the numerical results obtained in this chapter.

Thus, it can be assumed that the expression (4.28) with (4.29)–(4.32) describes the intermediate self-similar approximation of the frequency spectrum. It preserves the total energy and the wave action.

Conclusions. The problem of non-linear spectral evolution based on Hasselman's equation is solved for large times. In order to solve the problem the most optimal numerical algorithm for non-linear energy transfer computation (Lavrenov, 1991b, 1998a) is used. It is shown that the non-linear spectrum evolution can be described by variation of different features. The numerical results reveal some general features common to all cases of spectrum evolution. They can be defined as:

– the non-linear energy transfer results in approaching, in time, all frequency–angular spectra to some general form depending on the initial conditions;
– the most definite evolution features are revealed for the frequency spectrum $S(\sigma)$ and the function of the angular narrowness $D(\sigma)$;
– the sharp peak maximum with steep enough low and high frequency ranges is shown in the frequency spectra $S(\sigma)$, with the low-frequency range $\sigma < \sigma_p$ approximated by the dependence $S(\sigma) \sim \sigma^{16.0\pm1.0}$, whereas within the range $\sigma_p < \sigma < 1.5\sigma_p$ the spectrum is approximated as $S(\sigma) \sim \sigma^{-6.5\pm0.5}$;
– the frequency spectrum decreases as $S(\sigma) \sim \sigma^{-3.1\pm0.6}$ in the high-frequency range $\sigma > 1.5\sigma_p$, with the frequency power depending on the angular energy distribution of the initial spectrum; a wider initial angular distribution gives a larger negative value of the power, so the isotropic initial spectrum provides the frequency dependence $S(\sigma) \sim \sigma^{-3.67}$;
– the spectrum maximum shifts in time to a low-frequency range as $\sigma_p \sim t^{-0.11\pm0.01}$;

- the frequency narrowness of the value B is established as 0.33 ± 0.02 ("the law of $1/3$") and is practically not dependent on the initial spectral form;
- the function of the angular narrowness $D(\sigma)$ reveals a sharp maximum in the vicinity of the frequency spectral maximum, with its value becoming smaller in the range of low and high frequencies;
- the maximum of the parameter D_p is established within the range $D_p = 0.9\text{--}1.3$, depending on the initial angular narrowness $D_p(t=0)$.

As shown by the numerical computation results, a self-similar spectrum is established in the process of non-linear evolution. This is revealed as the establishment of the parameter values B and D_p, and the total form of the spectrum. It has two main domains: a sharply defined peak and a slowly decreasing high-frequency tail. It should be noted that Kolmogorov's spectrum $S(\sigma) \sim \sigma^{-4.0}$, numerically found by Komatsu & Masuda (1996), is not seen in all initial conditions. Kolmogorov's spectrum is known to appear in the presence of an energy source located in the low-frequency range, with the energy sink at the high-frequency range (Zakharov & Zaslavskii, 1982). But, there are no sources and sinks in our case. Using these numerical results a simplified intermediate self-similar frequency spectrum approximation is offered (4.28). This approximation describes the main features of the non-linear spectrum evolution and provides conservation of the total energy and wave action.

4.2 Wind Wave Energy Input

Miles' model of wind wave energy input. The component G_{in} of the wind wave energy input is usually determined with the help of the relation based on the model of averaged air flow interaction with waves proposed by Miles (1960). In spite of the fact that this model was proposed in 1957, it describes accurately enough the mechanism of wind wave energy input. It is still used nowadays. This mechanism which is specified by using full-scale observational data (Snyder et al., 1981) can be described as follows:

$$G_{in}(\sigma,\beta) = \max\left\{0;\ 0.25 a_1 \frac{\rho_a}{\rho_W}\sigma\left(a_2\frac{U_{10}}{c}\cos(\beta-\beta_U)-1\right)S(\sigma,\beta)\right\}, \tag{4.34}$$

where U_{10} is the wind speed at the 10 m level; $\beta - \beta_U$ is the angle between the wind speed and the direction of spectral wave component propagation; and a_1 and a_2 are parameters, assumed to be about 1. It follows from (4.34) that the wind energy is supplied to the wave spectrum range with $a_2(U_{10}/c)\cos(\beta-\beta_U) > 1$. It should be noted that at lower frequencies, wave energy transfer takes place only due to the non-linear wave interaction G_{nl}.

The relation (4.34) can also be expressed with the help of the dynamic velocity (or friction velocity) U_* instead of the wind speed. It is believed to be more universal with the wind speed U_{10} in (4.34) being replaced by the value $28 U_*$. However, the problem of determining U_* turns out to be rather complicated. In the original version of The WAM model (1988) it is assumed that the dynamical velocity can be determined as:

$$U_* = \sqrt{C_D} U_{10} \; ;$$

$$C_D = \begin{cases} 1.2873 \cdot 10^{-3} & \text{at} \quad U_{10} \leq 7.5 \, \text{m s}^{-1}; \\ 0.8 \cdot 10^{-3} + 0.65 \cdot 10^{-4} & \text{at} \quad U_{10} > 7.5 \, \text{m s}^{-1}. \end{cases} \qquad (4.35)$$

The drawback of this ratio is that it does not take into account the actual state of the near-water atmospheric boundary layer determined by the sea surface roughness and atmospheric stratification. In this connection the model wind wave energy input, proposed by P. Janssen, was used in the next version of the WAM model (Komen et al., 1994).

Relation of the friction velocity with the wave surface resistance coefficient and the wave development stage. The experimental results of investigating the dependence of the friction velocity and the wave surface resistance coefficient on the wave development stage are in Problems of Research and Mathematical Modeling of Wind Sea (1995); Huang et al., (1986); Komen et al., (1994). Thus, using full-scale data analysis (Problems of Research and Mathematical Modeling of Wind Sea, 1995) it is shown that the following dependence between the sea roughness z_0 and the non-dimensional frequency of the wind wave spectral maximum for the deep-water case, excluding the first stage of wind wave development, can be written:

$$z_0^* = 0.4 \, \sigma_{\max}^* \qquad (4.36)$$

The parameters are written in non-dimensional form normalized by the friction velocity in this relation. The relation (4.36) indicates a decrease of sea surface roughness depending on the wave development stage. Thus, at the initial stage of wave development the waves are very steep and their speed is relatively low, creating large resistance to airflow and significant surface roughness. An intense energy and momentum transfer from the wind flow to waves appears. The reverse influence of waves on the air flow is significant. It is revealed in the increase in dynamical velocity and in violation of the self-similar regime. The waves become flatter, their phase speed is increased while the effective roughness is decreased with wave development. Thus, the resistance coefficient is decreased. Correspondingly, the energy and momentum transfer rates to waves are decreased. The wave increase becomes slower.

If the wind speed profile is assumed to be described by a logarithmic law with known roughness parameter z_0, it can be written as:

$$U(z) = \frac{U_*}{\kappa} \ln\left(\frac{z}{z_0}\right), \qquad (4.37)$$

where $\kappa = 0.4$ is the Karman constant.

Now the following equation connecting the resistance coefficient C_{10} with the wind mean speed U_{10} and the non-dimensional frequency of the spectral maximum $\tilde{\sigma}_{\max}$ normalized by the value U_{10} can be obtained for the air flow case of neutral stratification:

$$\frac{3}{2}\ln C_z + \kappa\, C_z^{-1/2} = \ln\left(\frac{z\,g}{\kappa\, U_z^2}\right) - \ln \tilde{\sigma}_{\max}, \qquad (4.38)$$

where $z = 10$ m; and $C_z = U_*^2/U^2(z)$ is the resistance coefficient.

Chalikov–Makin model of wind wave energy input. The most accurate numerical model of the statistical structure of the near-water wave boundary layer based on the numerical solution of the Reynolds two-dimensional equations is described by Chalikov (1986); Burgers & Makin (1992); Chalikov & Belevich (1995); Belevich & Neelov (1998,2000). It is shown that the wind wave energy input term $G_{\text{in}}(\sigma,\beta)$ can be expressed in the following form:

$$G_{\text{in}}(\sigma,\beta) = \mathrm{B}_U \sigma\, S(\sigma,\beta), \qquad (4.39\text{a})$$

where B_U is a non-dimensional wind wave interaction parameter. Its value can be approximated as:

$$10^4 \mathrm{B}_U = \begin{cases} -a_1 \tilde{\sigma}_a^2 - a_2, & \tilde{\sigma}_a \leq -1; \\ a_3 \tilde{\sigma}_a (a_4 \tilde{\sigma}_a - a_5) - a_6, & -1 < \tilde{\sigma}_a < \Omega_1/2; \\ (a_4 \tilde{\sigma}_a - a_5)\tilde{\sigma}_a, & \Omega_1/2 < \tilde{\sigma}_a < \Omega_1; \\ a_7 \tilde{\sigma}_a - a_8, & \Omega_1 < \tilde{\sigma}_a < \Omega_2; \\ a_9(\tilde{\sigma}_a - 1)^2 + a_{10}, & \Omega_2 < \tilde{\sigma}_a, \end{cases} \qquad (4.39\text{b})$$

where $\tilde{\sigma}_a = (\sigma U_\lambda/g)\cos(\beta - \beta_U)$ is the non-dimensional "apparent" frequency of the wave propagating at the angle β; β_U is the wind direction; and U_λ is the wind speed at the height equal to the "apparent" wavelength $\lambda_a = 2\pi g/\sigma^2\,|\cos(\beta - \beta_U)|$. The parameters $a_1 - a_{10}$ and Ω_1, Ω_2 are dependent on the resistance coefficient C_λ at the level $z = \lambda_a$ as follows:

$\Omega_1 = 1.075 + 75\, C_\lambda;$ $\qquad \Omega_2 - 1.2 + 300\, C_\lambda;$

$a_1 = 0.25 + 395\, C_\lambda;$ $\qquad a_2 = 0.35 + 150\, C_\lambda;$

$a_3 = (a_0 - a_2 - a_1)/(a_0 - a_4 + a_5);$ $\qquad a_4 = 0.30 + 300\, C_\lambda;$

$a_5 = a_4\, \Omega_1;$ $\qquad a_6 = a_0(1 - a_3);$ $\qquad (4.40)$

$a_7 = \bigl(a_9(\Omega_2 - 1)^2 + a_{10}\bigr)/(\Omega_2 - \Omega_1);$ $\qquad a_8 = a_7\, \Omega_1;$

$a_9 = 0.35 + 240\, C_\lambda;$ $\qquad a_{10} = -0.05 + 470\, C_\lambda;$

$a_0 = 0.25\, a_5^2/a_4.$

The parameter B_U has been thoroughly studied by Chalikov & Belevich (1995). There are three main differences of this parameterization from Snyder's empirical relation. Firstly, the value of the function (4.39) becomes negative for waves propagating faster than the wind speed. In this case the wave phase speed projection onto the wind direction is compared with the wind speed. If the dynamical wind pressure on the frontal wave surface is higher in comparison with the rear surface pressure, this leads to the appearance of energy flow directed from the waves to the wind. Secondly, the integral energy flow to the waves becomes 2–3 times lower in case of fully developed wind sea. It is determined by the energy outflux from low-frequency components propagating faster than the wind speed and a relatively small influx to the waves with velocity being close to the wind speed. Thirdly, in the high-frequency range, a greater energy flux compared to Snyder's formula (4.34) is estimated with the help of the approximation (4.39a) since B_U in (4.39b) is proportional to $\tilde{\sigma}_a^2$ at $\tilde{\sigma}_a > 2$. The difference in the integral value of this wind wave energy input and Snyder's ratio becomes smaller for the initial stage of wind sea development. It should be noted that the value of the wind wave energy input function used in the WAM model (Komen et al., 1994) is also much smaller in comparison with Snyder's value.

Since there is not only an energy flux from the wind to the waves, but also a flux from the waves to the wind in the wave/wind interaction mechanism (4.39), it becomes possible to use this mechanism for achieving a more rapid spectrum shape stabilization at the developed wave stage.

The parameter approximation B_U is compared with observational data. Their consistency within the confidence interval of the measurement results is shown by Chalikov & Belevich (1995). Additionally a quadratic dependence of the parameter B_U on the frequency $\tilde{\sigma}_a$ for large values is confirmed.

The values U_λ and C_λ are defined at the level of the "apparent" wavelength in the approximation of the wind wave energy input function (4.39) As the wind speed and the resistance coefficient are changed with height, the introduction of the parameters U_λ and C_λ eliminates the ambiguity of choosing the readout level, typically existing in every calculation scheme. It reduces the number of determining parameters. Also it corresponds to the physics of the process, since the layer of the air flow/wind interaction becomes thinner with increasing frequency. In other words, what happens beyond this layer is not essential for the given wave.

The prevalent part of the momentum flux, connected with waves, is determined to be formed within the high-frequency spectrum range. The surface resistance can be taken into account with the help of the wind sea model using the roughness parameter z_0. Its value is dependent on the energy of the high-frequency components. This can be related to the Phillips parameter α for the JONSWAP spectrum approximation by the ratio:

$$z_0 = \chi \alpha^{1/2} \Lambda, \qquad (4.41)$$

where $\Lambda = U_*^2/g$, χ is a parameter in the range 0.15–0.25 and U_* is the frequency velocity. The relation (4.41) can be considered as a generalization of the well-known relation of Charnock (1955), taking into account the sea surface state.

In order to implement the aforementioned scheme in spectral calculations of the energy and momentum fluxes to waves in the first approximation, it can be suggested that the wind profile is described with the help of the logarithmic law using the known roughness parameter (4.37).

The relation (4.37) can be rewritten in terms of the resistance coefficient:

$$C_z = U_*^2/U^2(z) = \kappa^2 \left(R - \ln(C_z) \right), \tag{4.42}$$

where $R = \ln \left(\frac{z g}{\chi \sqrt{\alpha U^2}} \right)$ is a non-dimensional parameter.

It is necessary to estimate the parameter α in order to complete the parameterization of the wind wave energy input function G_{in}. An estimation of the resistance coefficient C_z can be obtained on the basis of the phase speed of components of the spectral maximum c_p in the following form:

$$C_z = \exp\left(0.07\, R_1 + 0.2345\, \ln(R_1) - 6.783 \right), \tag{4.43}$$

where $R_1 = \frac{U^2}{g z} \left(\frac{U}{c_p} \right)^{3/4}$.

This approximation completes the parameterization of the wind wave energy input function proposed by Chalikov and Belevich.

4.3 Wave Energy Dissipation in Deep Water

Statement of the problem. The dissipation mechanism of wave energy in deep water remains the least studied. The absence of a definite physical basis is, probably, connected with difficulties in the theoretical description of wind sea dissipation in the framework of the existing concepts of hydrodynamics.

The mechanism of wave dissipation in deep water is thought to be mainly associated with wave crest breaking. However, there is no scientifically recognized opinion whether its dependence on the energy spectral density is linear or non-linear.

There are some empirical approximations for wave dissipation in wind wave modelling (Abuzyarov, 1981; Davidan et al., 1985; Ocean Wave Modeling, 1985). A generalized review concerning this problem has been put forward by M. Donelan and R. Young (Komen et al., 1994) and M. Banner et al. (2002).

Hasselmann (1974) suggested wave energy dissipation parameterization, connected with wave breaking. In his opinion it can be considered as a random distribution of perturbing forces, making up pressure pulsations with small scales in space and time in comparison with the proper wavelength and period. All the weak processes are shown to be locally non-linear on the average, producing a source function, which is quasi-linear relative to interactions of the lowest order. In this case the source function dissipation is presented in the form of linear dependence on the spectrum. It is multiplied by the value depending on the integral spectrum parameters of the whole spectrum. The wave dissipation used in the WAM model (The WAM model, 1988; Komen et al., 1994) connected with wave breaking is accepted in the form of the quasi-linear approximation, as suggested by Komen et al. (1984) on the basis of the Hasselmann model:

$$G_{\mathrm{ds}}(\sigma,\beta) = -c_1 \bar{\sigma} \left(\frac{\sigma}{\bar{\sigma}}\right)^n \left(\frac{\bar{\alpha}}{\alpha_{\mathrm{PM}}}\right)^m S(\sigma,\beta) , \qquad (4.44)$$

where c, n and m are the model parameters; $\bar{\sigma}$ is the mean frequency of the wave spectrum; α_{PM} is the constant of the Pierson–Moskovits spectrum; and $\bar{\alpha} = m_0 \bar{\sigma}^4 / g^2$. The following parameters are accepted: $c_1 = 3.33 \cdot 10^{-5}$, $n = 2$, $m = 2$. The dissipation function (4.44) depends linearly on the spectrum as well as on its integral parameters.

A peculiarity of the dissipation parameterization is in permitting one (totally with other items of the source function such as the Pierson–Moskovits spectrum approximation) to obtain spectra of fully developed sea in the form of the Pierson–Moskovits approximations. It should be noted that the existence of such spectra has not yet been proved. Due to the initial supposition that wave breaking is considered to be a random distribution of disturbing forces with small scales, use of the relation (4.44) is limited for the following reasons:

- The initial supposition is disturbed in the high-frequency spectral area. As a consequence, the relation (4.44) does not guarantee the stable convergence of the solution to the wave energy balance equation to values of the equilibrium interval spectrum in the area of high frequencies $\sigma \gg \sigma_{\mathrm{max}}$. This happens because the wind energy input and dissipation are linear in the spectrum intensity $S(\sigma,\beta)$. That is why an unjustified small time step and additional limitations of spectral value and source function are used in the WAM model (The WAM model, 1988; Komen et. al., 1994).
- According to the relation (4.44), the value $G_{\mathrm{ds}}(\sigma,\beta)$ is not dependent on wind speed in its explicit form. It does not agree with observations of the wind wave whitecapping process.
- The value $G_{\mathrm{ds}}(\sigma,\beta)$ does not take into account the increasing energy dissipation due to wave breaking in the case of wind being contrary to wave propagation.

There are some other dissipation parameterizations, depending non-linearly on the spectral density function. They were investigated more than 40 years ago in the first semi-empirical wind wave models (Abuzyarov, 1981; Davidan et al., 1985; Ocean Wave Modeling, 1985). Phillips (1985) suggested a theoretical basis for wave energy dissipation depending non-linearly on the spectral density. In contrast to Hasselmann, he put forward the supposition that wave breaking was a local character, i.e. energy losses due to dissipation of concrete spectral components depend on its spectral density of energy and were not determined by the integral parameters of the whole spectrum. He came to the conclusion that wind wave energy input, dissipation and non-linear interaction were of the same size in the spectrum equilibrium range. Using the wave energy balance equation he found that the dissipation should depend cubically on the spectrum:

$$G_{\mathrm{ds}}(\boldsymbol{k}) = -C_{\mathrm{ds}}\sigma\, k^8 S^3(\boldsymbol{k})\ , \qquad (4.45)$$

where C_{ds} is a constant.

It should be noted that wave energy dissipation in the form (4.45) gives more stable convergence to the limiting value with $\sigma \gg \sigma_{\max}$. However, it is rather doubtful to use it for low frequencies of wind waves and swell (Komen et al., 1994).

Wind wave frequency-angular spectrum obtained at the Leningrad Branch of the State Oceanographic Institute (LB SOIN). A great number of empirical frequency and frequency-angular spectra has been obtained with the help of wave records. Some known approximations are presented above (4.13)–(4.18). Their reliability is determined by measurement accuracy, the chosen spectral analysis methods, selective variability of the spectral density and statistic estimations specified by the limits of wave records used. However, in spite of some differences in the results of generalizing natural data, the main features of the wave frequency-angular spectrum are well known. The simplest shape of the frequency spectrum is revealed in case of wind wave development under stable wind speed and in the absence of swell. The characteristic features of this spectrum are: a sharp increase of spectral density from low frequencies to spectral maximum; slow spectral decrease with transition to high frequencies and displacement of the spectral maximum to lower frequencies with wind sea development.

The frequency spectrum of a wind wave can be divided into three intervals: spectral maximum including the spectral increasing branch and the main maximum of spectral density, transitional and equilibrium spectral intervals (see Fig. 4.21).

Spectra boundaries between different gravitational areas are not always expressed clearly, depending, probably, on the reliability of the spectral density statistic estimations. In the case of spectral density graphs, obtained by extended wave records with wind waves developing under stable uniform wind field, the boundaries between the main maximum interval and the transitional one are determined by the minimum; and between the transitional area and the equilibrium one, by the second maximum of the spectral density function. In some cases these second minima and maxima in the boundaries of the transitional interval are not definitely expressed. Sometimes instead are only disturbance of the smooth course of the spectral density noted in the spectrum.

It should be noted that 70–80 per cent of total wave energy is contained in the main maximum range, some per centage in the transitional interval and

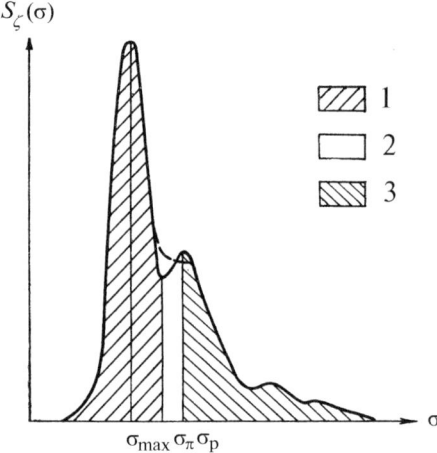

Fig. 4.21. Frequency spectrum ranges: 1 – main maximum range; 2 – transitional range; 3 – equilibrium range

20–30 per cent in the equilibrium interval. In spite of the fact that there is less energy in the latter interval than in the main maximum interval, the equilibrium interval plays an essential role in forming wave surface roughness. It determines the reflection and dispersion conditions of electromagnetic waves.

The only approximation of the frequency spectrum, taking into account the division of its area in three ranges, was obtained in the LB SOIN earlier and published by Davidan (1969); and Davidan et al. (1978).

The approximation can be presented in the form:

$$S(\sigma, \beta) = S(\sigma)\, Q(\sigma, \beta), \qquad (4.46)$$

where the frequency spectrum $S(\sigma)$ is written as follows:

$$S(\sigma) = \begin{cases} S_1 = (n+1)m_0(\sigma_\pi)\left(\frac{\sigma_{\max}}{\sigma}\right)^n \sigma \exp\left\{-\frac{n+1}{n}\left[\left(\frac{\sigma_{\max}}{\sigma}\right)^n - \left(\frac{\sigma_{\max}}{\sigma_\pi}\right)^n\right]\right\} \\ \quad \text{when } \sigma_\pi > \sigma\,; \\ S_2 = S_1(\sigma_\pi) + \frac{S_3(\sigma_p) - S(\sigma_\pi)}{\sigma_p - \sigma_\pi}(\sigma - \sigma_\pi) \\ \quad \text{when } \sigma_\pi < \sigma < \sigma_p\,; \\ S_3 = 7.8 \times 10^{-2} g^2 \sigma^{-5} \\ \quad \text{when } \sigma > \sigma_p\,, \end{cases} \qquad (4.47)$$

The function $Q(\sigma, \beta)$ of the angular distribution is accepted according to (4.14), where n_β is the narrowness parameter of the angular energy distribution function (Davidan et al., 1978).

There are five parameters in the approximation of the frequency spectrum (4.46): $m_0(\sigma_\pi)$ is the zero moment of the spectrum truncated at the frequency σ_π (the boundary of the spectrum transitional area); the frequency of the spectral maximum σ_{max}; the parameter n, characterizing the spectrum sharpness, and σ_p is the low-frequency limitation of the spectral equilibrium area. All these parameters were obtained earlier using field data as functions of the stage of wind sea development. The value $\tilde{\sigma}_{max}$ is the dimensionless frequency of the spectral maximum (Davidan, 1969; Davidan et al., 1978). At each stage of wind sea development these parameters are connected with the non-dimensional frequency of the main spectral maximum $\tilde{\sigma}_{max} = U_{10}\sigma_{max}/g$ as follows:

$$n = 5\exp\left(0.5\tilde{\sigma}_{max} - 0.4\right)^2$$

$$\sigma_{p1} = 2\sigma_{max}\tilde{\sigma}_{max}^{-0.7} \quad \text{when} \quad \tilde{\sigma}_{max} < 2.4 \quad (4.48)$$

$$\sigma_\pi = 1.8\sigma_{max}\tilde{\sigma}_{max}^{-0.7} \quad \text{when} \quad \tilde{\sigma}_{max} < 2.4$$

$$m_0(\sigma_\pi) = 1.27 \cdot 10^{-3} g^{-2} U^4 \tilde{\sigma}_{max}^{-3.19}$$

As can be seen from adduced relations, actually it is enough to have a spectral maximum frequency and wind velocity in order to calculate the wave frequency spectrum. The peculiarity of the frequency spectrum (4.47) is the second maximum with its decreasing relative value, its location displacement relative to the spectral maximum is shifted to the higher frequency area.

Davidan passed from normalized spectral parameters with respect to U_{10} (wind velocity at 10 m height) to their normalization by the dynamic velocity U_* in order to make the approximation (4.48) more universal. On the basis of the data (Davidan et al., 1985), the other parameters of the wind wave spectrum are defined by transition from U_{10} to U_*:

$$m_0^*(\sigma_\pi) = m_0^* - 2.84 \cdot 10^{-3}\sigma_p^{*-4};$$

$$\sigma_{max}^* = 0.5\, m_0^{*-0.39}\, \text{th}\left(0.52 m_0^{*0.25}\right);$$

$$\sigma_\pi^* = \sigma_{max}^* \text{cth}^{0.28}\left(3.72 \cdot 10^3 \sigma_{max}^{*3.21}\right); \quad (4.49)$$

$$\sigma_p = 1.1\,\sigma_{max};$$

$$n = 8.0 + 3.0\,\text{th}^{0.75}\left\{\left|\left[1.2 \cdot 10^4\left(\sigma_{max}^* - 0.09\right)\right]\right|\right\}\text{sign}\left(\sigma^* - 0.09\right).$$

The spectral approximation (4.46) allows the analysis of the formation of wind wave spectral structure in different frequency bands.

Correlation of physical mechanisms with spectrum approximation.
One of the most interesting features of the LB SOIN spectrum is its frequency structure based on detailed analysis of field observations. In particularly, the second maximum at frequencies greater than the peak spectrum frequency as well as the transition to the equilibrium state approximation is evidence of

the different physical mechanisms that form the spectral structure in various frequency bands.

As concluded by Davidan et al. (1978), if the Miles model is accepted, the wind wave frequency spectrum can be divided into three ranges. These are determined by the different physical mechanisms on wind wave spectrum formation. That is why the total frequency range can be separated in those determined mainly by non-linear energy transfer G_{nl}; the range of wind energy input, with the source function determined by the sum of three components: G_{in}, G_{nl}, G_{ds} and the equilibrium range, with the source function determined by the sum of two components G_{in} and G_{ds}. It is also found that the peculiarities of the energy dissipation in the spectrum equilibrium range differ essentially from the those in other spectrum frequency ranges.

As for the first frequency range with the source function determined mainly by non-linear energy transfer, the papers by Zakharov & Zaslavskii (1982, 1983a,b) should be mentioned. In their opinion the spectral evolution of developed waves within the spectral maximum range could be described within the supposition of a "transparency window". This is a frequency spectrum area without any wind energy input, and dissipation can be neglected. The ideas of the Kolmogorov weak turbulence theory are used in this case.

The turbulence is known to appear as a result of laminar current instability. This is characterized by the large number of degrees of freedom. In media with dispersion, such as the sea surface, separate wave packages overlap during a short period, and their interaction is weak enough (this state is called the weak turbulent state). The smallest interaction energy between wave packages in comparison with the total wave energy allows the use of disturbance theory. The turbulence is described by a closed equation set, which yields analytical results in some cases.

Two physically sensible frequency spectra have been determined (Zakharov & Zaslavskii, 1982, 1983a,b):

$$S_1(\sigma) = C_1 \sqrt[3]{pg^4/\sigma^{11}}\,; \quad S_2(\sigma) = C_2 \sqrt[3]{qg^4}/\sigma^4\,,$$

where p is the wave action flow; and q is the wave energy flow. The first solution is interpreted as a model with energy input located at $\sigma = \infty$, with the spectrum being determined by the wave action flow directed to the long-wave area $\sigma = 0$. The second solution describes the wave energy input at $\sigma = 0$, forming an energy flow into the dissipation area $\sigma = \infty$. Both solutions are obtained in accordance with rather strict mathematical notions. They are justified in the sense of physical hypotheses accepted by the authors: the weak turbulence approximation in the presence of the transparency interval, with the wave energy input and dissipation being inessential.

If the wind energy is estimated according to (4.34) and the non-linear energy transfer is calculated using the four-wave interaction integral (4.1), the "transparency window" in its pure form is not observed in the wind wave spectrum even neglecting the energy dissipation (Davidan & Lavrenov, 1991;

Davidan et al., 1985; Theoretical bases and methods for wind sea calculation, 1988). It should be noted that Komen, Hasselmann and Hasselmann (1984) came to the same conclusion, estimating the wave energy dissipation as the difference between the empirical source function obtained by spectral density variations and the sum of the functions of the non-linear transfer and wind energy input. The dissipation values were estimated with sign opposite to the physical meaning of the mechanism.

As it has been noted, the results of Zakharov & Zaslavskii (1982, 1983a,b) were obtained in the form of the approximation of weak turbulence and within the transparency interval with the wave energy input and dissipation being inessential. This statement seems to be rather doubtful in actual situations. Probably it explains the difference between theoretical estimations and field spectra data. At the same time these solutions may be useful for qualitative research of wind wave dynamics, when they allow extracting the most significant physical mechanisms forming the wave spectrum in this frequency band.

It is interesting to note that Tolman and Chalikov (1996) came to the same conclusions as Davidan. They investigated the source function in the third-generation wind wave models and found principal difference of the physical mechanism effect in various bands of the frequency spectrum. The whole frequency range is divided into three sections: low, transitional and high-frequency parts, the dissipative mechanism being determined separately for every section. The wave energy dissipation is expressed in the form of a superposition of two components $G_{\text{ds}}^{(1)}(\sigma, \beta)$ and $G_{\text{ds}}^{(2)}(\sigma, \beta)$ as follows:

$$G_{\text{ds}}(\sigma, \beta) = A G_{\text{ds}}^{(1)}(\sigma, \beta) + (1 - A) G_{\text{ds}}^{(2)}(\sigma, \beta), \qquad (4.50)$$

where

$$A = \begin{cases} 1 & \text{when } \sigma < \sigma_1; \\ (\sigma - \sigma_2)/(\sigma_1 - \sigma_2) & \text{when } \sigma_1 \leq \sigma < \sigma_2; \\ 0 & \text{when } \sigma_2 \leq \sigma; \end{cases}$$

σ_1 and σ_2 are the boundaries of the transitional ranges depending on the characteristic frequency, where the main energy input is $\sigma_{\max,i}$:

$$\sigma_1 = 1.75\, \sigma_{\max,i}, \qquad \sigma_2 = 2.5\, \sigma_{\max,i}. \qquad (4.51)$$

The frequency $\sigma_{\max,i}$ is determined by the function of the wind wave energy input as follows:

$$\sigma_{\max,i} = \iint \sigma^{-3} \max[0, G_{\text{in}}(\sigma, \beta)]\, d\sigma\, d\beta \Big/ \iint \sigma^{-4} \max[0, G_{\text{in}}(\sigma, \beta)]\, d\sigma\, d\beta. \qquad (4.52)$$

Thus, the qualitative analogy can be seen between the interpretation of spectral frequency separation made by Davidan on the basis of detailed analyses of field measurements and the theoretical conclusions by Tolman and

Chalikov. It proves expedient to use different approaches for the description of the physical mechanism in different wind wave frequency bands.

In the dissipation mechanisms, offered by Tolman and Chalikov, the low-frequency dissipation connected with turbulence in the near-surface oceanic layer is described by the first function $G_{ds}^{(1)}(\sigma, \beta)$. On the basis of the Navier–Stockes equations and separation of movements into waves and turbulence the following formula for wave energy turbulent dissipation is proposed:

$$G_{ds}^{(1)}(\sigma, \beta) = -2 U_* \tilde{h} k^2 \phi(\xi) S(\sigma, \beta) , \qquad (4.53)$$

where $\tilde{h}$ is the turbulent mixing scale estimated as the wave height in the high-frequency spectral band, and $\phi(\xi)$ is a non-dimensional function.

The high-frequency dissipation $G_{ds}^{(2)}(\sigma, \beta)$ is written in the following form:

$$G_{ds}^{(2)} = -a_0 \left(\frac{U_*}{g}\right)^2 \sigma^3 \alpha_n^B(\sigma) S(\sigma, \beta) , \qquad (4.54)$$

where $B = a_1 (\sigma U_*/g)^{-a_2}$, and a_1, a_2, a_3 are non-dimensional adjusting parameters.

An important conformity between the empirical frequency spectrum (4.47) and theoretical estimations of the source function should be noted. The frequency spectrum (4.47) is characterized by the second maximum with decreasing relative value. Its location relatively to the spectral maximum is shifted to the higher frequency area. The function of wind energy input is similar and its maximum value is located around the second extreme of the frequency spectrum.

In conclusion it should be noted that theoretical and experimental results show the different dissipative mechanisms in various frequency spectrum ranges. At least, one thing can be said for sure: in the low-frequency spectral band with practically no wind wave energy, the dissipative value is so small that it can be neglected. At the same time the dissipation is sharply increased in the high frequency band where the wind energy input is important. In other words, the dissipation is dependent on wind speed. The greatest dissipation value is in the vicinity of the point, where the wind energy input wind is maximal, and slowly decreasing in the equilibrium interval.

As shown by field observations (Theoretical Bases and Methods for Wind Sea Calculation, 1988) the wave frequency spectrum is constantly undergoing quasi-oscillations relative to its mean value. The presence of the second spectral maximum is also quasi-periodical. Numerical calculations of non-linear interaction made by Resio and Perrie (1991) and Lavrenov (Davidan & Lavrenov, 1991) show that the appearance of a local maximum in the high-frequency area results in a sharp increase of the non-linear energy transfer intensity in the area of the local maximum point. As a consequence this maximum disappears, but later on it appears due to intensive wind energy input. It witnesses the necessity of taking into account non-linear energy transfer alongside other mechanisms within the whole frequency range.

The wave energy dissipation function is significantly different in various frequency bands. Thus, it can be presented as a linear or quasi-linear dependence on the spectral density in the case of wave interaction with turbulent movement of water masses in the low-frequency band. As for the high-frequency band with dissipation defined by crest wave breaking (being a strongly non-linear mechanism), its spectral density dependence should also be non-linear.

The high-frequency spectral area determines, mainly, sea surface roughness, with the main impulse and energy flows from wind to waves. The high-frequency spectral components reach their maximum development within a short period of time. As shown by numerical experiments, the presence of an equilibrium interval determines the essential weak non-linear energy flows to the low-frequency spectral area. For example, in order to create the low-frequency spectrum development it is enough to solve the problem of non-linear spectrum evolution with a constant equilibrium interval (4.47) as a condition and without taking into consideration the influence of the wind. The Phillips parameter, equal to $\alpha = 7.8 \cdot 10^{-3}$, provides more than 30 per cent of energy supply, necessary for typical development of wind waves. The same value, equal to $\alpha = 20.0 \cdot 10^{-3}$, provides the usual wave development without wind, due only to the action of non-linear energy transfer from the equilibrium interval to the low spectral area.

Parametrization of high-frequency dissipation. Since there is no theory of wave energy dissipation at present, an attempt is made to obtain its value, proceeding from general considerations. In order to do so, estimations of the known components of the source function at large frequency values can be used, i.e. when $\sigma \gg \sigma_{\max}$. It can also be assumed that the frequency spectrum value should asymptotically approach the value of the isotropic equilibrium interval: $S_\infty(\sigma) = \alpha g^2 \sigma^{-5}$.

The existence of the asymptotic stationary spectrum range means that physical mechanisms should be balanced properly. The sum of source functions should be equal to zero within the stationary spectrum interval. Integrating the source function (3.1) over directions one can obtain:

$$\int G_{\mathrm{ds}}(\sigma,\beta)\,\mathrm{d}\beta \approx -\int G_{\mathrm{in}}(\sigma,\beta)\,\mathrm{d}\beta - \int G_{\mathrm{nl}}(\sigma,\beta)\,\mathrm{d}\beta\,. \qquad (4.55)$$

The estimation of the non-dimensional relaxation time (normalized by dynamic velocity) of the wave process connected with energy dissipation is as follows:

$$\tilde{T}_{\mathrm{ds}}^{-1} \approx -\tilde{T}_{\mathrm{in}}^{-1} - \tilde{T}_{\mathrm{nl}}^{-1}\,, \qquad (4.56)$$

where the non-dimensional relaxation time is determined as:

$$\tilde{T}^{-1} \equiv \frac{U_*}{g} \int G(\sigma,\beta)\,\mathrm{d}\beta \bigg/ \int S(\sigma,\beta)\,\mathrm{d}\beta\,. \qquad (4.57)$$

Proceeding from the function of the wind wave energy input (4.39), the value $\tilde{T}_{\text{in}}^{-1}$ for large frequencies $\sigma \gg \sigma_{\max}$ can be estimated as:

$$\tilde{T}_{\text{in}}^{-1} = a_0 \tilde{\sigma}_*^3, \qquad (4.58)$$

where $\tilde{\sigma}_* = \sigma U_*/g$.

The estimation of the value $\tilde{T}_{\text{nl}}^{-1}$ can be obtained from (4.1) in the following form:

$$\tilde{T}_{\text{nl}}^{-1} = b_0 S^2 \sigma^{11} U_*/g^5. \qquad (4.59)$$

The values $\tilde{T}_{\text{in}}^{-1}$ and $\tilde{T}_{\text{nl}}^{-1}$ are of the same order within the considered frequency range $2\sigma_{\max} \leq \sigma \leq 4\sigma_{\max}$. That is why, to estimate the value $\tilde{T}_{\text{ds}}^{-1}$ any of them can be used. It should be noted that the value $\tilde{T}_{\text{in}}^{-1}$ does not depend on the spectrum, but the wave process relaxation time due to wave breaking should depend on (4.60) below. The dissipation mechanisms should be non-linear within this frequency interval. In order to estimate the value $\tilde{T}_{\text{ds}}^{-1}$ more accurately, the second value can be taken with some correction coefficients, which take into consideration the first value indirectly. Thus, the dissipation mechanism can be written as follows:

$$G_{\text{ds}}(\sigma) = B_0 \frac{\sigma^{11} S^3(\sigma, \beta)}{g^4 \alpha^2} \tilde{\sigma}_*^{B_1}, \qquad (4.60)$$

where B_0 and B_1 are adjusting coefficients, with the latter being $B_1 \approx 2$.

It should be noted that the obtained formula is a compromise between the Phillips approximation (4.45), as it depends cubically on the spectral density, and the approximation of Tolman and Chalikov (4.54), as it depends quadratically on wind speed. It decreases as $\sim \sigma^{-2}$ if the equilibrium interval is supposed to exist at large frequencies.

4.4 Influence of Mesoscale Effects on Wind Wave Evolution

Spatial–temporal scale variation of wind wave field. Wind waves represent a non-stationary probabilistic process. As shown by experimental data, the evolution of the wind wave field occurs within a wide range of spatial–temporal scales.

The period of wave movement fluctuations or simply the wave period is the shortest temporal scale. The wind wave period varies from several to ten seconds. Its value depends on a number of circumstances, such as the wave development stage, wind speed, presence of currents, swell, etc. For the first estimation, a typical scale of sea wind waves can be considered as $\tau_1 \approx 1\text{--}10$ s.

Wave movements connected with a wave group represent the second temporary scale (Davidan et al., 1978; Davidan et al., 1985). The periods of wave group fluctuations range from tens to hundreds of seconds. The wave groups repeat themselves approximately every 10–15 average wave periods ($\tau_2 \approx 10$–$15\,\tau_1$), consisting of 5–9 waves. As a first approximation, the temporal scale of their frequency of occurrence can be assumed as $\tau_2 \approx 10^2$ s.

The so-called "quasi-oscillations" can be referred to the third temporal fluctuation scale of the sea surface. They are fairly well traced in changes of the form of the frequency spectrum and in its parameters even under stable wave development conditions (Andreyev 1988; Zaslavskii & Krasitskii, 1993; Bitner-Gregersen & Gran, 1983). The quasi-oscillation periods are approximately equal to 5–20 min. The scale of these periods can be estimated as $\tau_3 \approx 10^3$ s. A description of this phenomenon will be given later this chapter.

The wave field changes occurring within a period of 3–6 hours represent the fourth temporal scale. This temporal scale is estimated as $\tau_3 \approx 10^4$ s. Significant changes in the wind wave spectrum and in all its parameters usually occur within this time interval. As far as the temporal scale is concerned the surface wind field data with a similar temporal step are input to the mathematical models for wind wave numerical simulations. The mathematical models used for practical application are based on the numerical solution of the wave energy balance equation (Davidan et al., 1985; Komen et al., 1994). The field changes occurring within smaller temporal and spatial scales are not taken into account. In other words, the numerical results represent estimations of the wave field calculated on the basis of the wind field averaged over this time interval.

The problem is that these results are usually compared with the full-scale measurements made over a shorter time interval, namely, within a quasi-stationary range. A question appears whether it is correct to compare the random process parameter estimations averaged over different time intervals.

The next temporal scale of the wave wind field evolution is a synoptic range order of $\tau_5 \approx 10^5$ (Davidan et al., 1978). Numerical simulations of wind waves in forecasting are usually carried out by means of mathematical models for these time intervals. The scale variation of the wind wave field (including seasonal, interannual variability, etc.) could be enlarged. But their detailed consideration is beyond the scope of this study.

It is important to point out that all the aforementioned temporal scales of wind wave field changes differ one from another by an order of magnitude. This allows considering the wave field evolution at one scale irrespective of the others. However, the problem of the influence of one scale of the spatial–temporal wave field evolution on the other still remains unstudied.

The problem of mesoscale effects in wind sea numerical simulations. The wave energy balance equation (3.1) is used in modern methods of wind wave numerical simulations. A grid with a specific spatial step is used to solve the problem numerically. The wind speed data is introduced in the model with a definite time interval. As a rule, the spatial step comprises tens or hundreds of miles. The near-surface wind values are introduced with a time interval of three or more hours. This traditional approach does not take into account "sub-grid" effects, i.e. describing finer details of the wind wave field evolution. That is why the solution is an averaging over these spatial–temporal steps.

The wave energy balance equation is assumed to describe sufficiently slow wave field variations, since uniform and steady state wave processes are suggested in the derivation of the kinetic equation. In reality, it is necessary to use a more complex equation relative to a fourth-order corrector for the non-uniform wave field (Yuen & Lake, 1987). Only in the case of a spatially uniform field can the traditional kinetic equation be applied.

In order to simplify the solution of the problem concerning wind wave modelling, the wave field non-uniformity is taken into account only in the advective term of the kinetic equation. The physical mechanisms forming the wave energy spectrum are considered to be local and appear in the right-hand side of the equation.

In this respect a question arises as to what extent such a solution is correct. Two aspects of the problem can be considered. The first one is an estimation of the error due to the discrete spatial and temporal wind field description; the second is the problem of the correct description of physical effects in the wind field under the given conditions.

It should be noted that the problem of obtaining sufficiently detailed spatial and temporal wind field data is connected with general problems of meteorology. This depends both on the state of this science and on the quality of the current synoptic data. The information is collected by satellites, vessels and synoptic stations. Nowadays meteorological centres produce wind field calculations and forecasting, using numerical models of global atmospheric circulation. The observation data is used as initial conditions to solve this problem.

There are a number of specific questions in this respect. For example, how to take into account the mesoscale wind speed changes such as wind gusts and squalls? What is the influence of atmospheric boundary layer stratification? How can these circumstances influence the accuracy of surface wind wave estimations? How can mesoscale effects be parameterized in wind wave models?

Two of these problems will be considered in this section. An attempt is undertaken to calculate the gust influence on the wind wave energy input and to estimate mesoscale wave parameter oscillations on long-term spectrum evolution.

Influence of gusts on the wind wave energy input. In order to estimate the influence of gusts on wind wave development, the parameterization of the wind wave energy supply mechanism (Miles 1957, 1960; Snyder et al., 1981) can be used. This theory can also be extended to the case when the mean wind speed profile is a slowly changing function of time (Komen et al., 1994). We consider how the wind field fluctuations can influence the wind wave energy input.

4.4 Influence of Mesoscale Effects on Wind Wave Evolution

The equation of spectral energy density evolution can be written in its simplest form. The linear term connected with the wind wave energy input according to the Miles theory is taken only into account on the right-hand side of the equation:

$$\frac{\partial}{\partial t} S(\sigma) = \gamma S(\sigma), \qquad (4.61)$$

where γ is the wave energy increment. This can be written in the simplified form as:

$$\frac{\gamma}{\sigma} = \max\left(0.2 \frac{\rho_a}{\rho_w}\left(28\frac{U_*}{c} - 1\right), 0\right). \qquad (4.62)$$

It should be noted that the value γ cannot be negative: $\gamma > 0$ with $28 U_*/c > 1.0$.

The friction velocity U_* can be written in the form of its mean value $\bar{U}_*$ and the fluctuations δU_*:

$$U_* = \bar{U}_* + \delta U_*. \qquad (4.63)$$

After substituting (4.63) into (4.62), the wave energy increment can be presented as its mean value and fluctuations:

$$\gamma = \bar{\gamma} + b\gamma_1, \qquad (4.64)$$

where $b \ll 1$.

Since γ is always positive, the mean value of $\langle b\gamma_1\rangle$ is not equal to zero. Thus, the mechanism of wind wave energy input within the approximation (4.62) acts as a "rectifier" in an electric circuit. The wind gusts increase the intensity of wave energy input for $28U_*/c > 1.0$, whereas in the opposite case, it converts to zero.

Using previous equations the following stochastic equation can be considered:

$$\frac{\partial}{\partial t} S(\sigma) = (\gamma + b\gamma_1) S(\sigma). \qquad (4.65)$$

When the correlation time of a random process is small, the equation of wave spectrum evolution $\langle S \rangle$, averaged over an ensemble of states, can be written in the following form:

$$\frac{\partial \langle S \rangle}{\partial t} = \langle \bar{\gamma} + b\gamma_1 \rangle \langle S \rangle. \qquad (4.66)$$

Assuming that the statistics of friction velocity fluctuations is described by the Gaussian distribution:

$$p_{U_*}(x) = \frac{1}{\sigma_U \sqrt{2\pi}} \exp\left[-\frac{(x - U_*)^2}{2\sigma_U^2}\right] \qquad (4.67)$$

the mean increment of the spectral energy can be written in the form:

$$\frac{\langle \bar{\gamma} + b\gamma_1 \rangle}{\varepsilon \sigma} = 5.6 \int_{X_0}^{\infty} \mathrm{d}x \, p_{U_*}(x) \left[x/c - 0.036 \right], \tag{4.68}$$

where $X_0 = 0.036\,c$. The integral (4.68) can be estimated as:

$$\frac{\langle \bar{\gamma} + b\gamma_1 \rangle}{\varepsilon \sigma} = 5.6 \left[\frac{\sigma_U}{c\sqrt{2\pi}} \exp\left\{ -\frac{1}{2\sigma_U^2} Y^2 \right\} + \frac{1}{2} \left(\frac{Y}{c} \right) \left(1 - \mathrm{erf}\left(\frac{Y}{\sigma_U \sqrt{2}} \right) \right) \right], \tag{4.69}$$

where $Y = 0.036c - U_*$, and $\mathrm{erf}(z) = \frac{2}{\sqrt{\pi}} \int_0^z e^{-t^2} \mathrm{d}t$ is the error function.

It follows from the relation (4.69) that the wind gusts result in the wave energy increment increasing proportionally to the friction velocity dispersion σ_U. The greatest effect is achieved at small frequencies with $U_*/c \approx 0.036$, which is important for wind wave energy input, especially for developed waves.

Cavaleri and Burgers (Komen et al., 1994) performed wave evolution calculations using the WAM model. The non-linear energy transfer, dissipation connected with wave breaking and wind wave energy input taking into account the wind gusts, were considered in this model. The Monte Carlo method was used to integrate the wave energy balance equation. The random value of the wind speed U_{10} (with a mean value of $\bar{U}_{10}$) was specified according to the Gaussian distribution. The one-minute time integrating step was used. It was shown that the gust effect did not significantly influence the wave development with $\tilde{\sigma}_U = \sigma_U/U = 0.1$, whereas the relative wave height increased by more than 30 per cent for $\tilde{\sigma}_U = 0.3$.

It was concluded that the gust intensity was connected with stratification of the atmospheric boundary layer. The parameter $\tilde{\sigma}_U = \sigma_U/U$ was increased for unstable stratification. Theoretical calculations were proved by full-scale observational data made on a platform in the Adriatic Sea (Cavaleri, 1999).

The extension of the Miles theory to wind gust cases is based on the atmospheric boundary layer model taking into account turbulent exchange. The wind gusts can be considered as the low-frequency spectrum of turbulent pulses in the atmospheric boundary layer. The problem is whether it is justified to use the Gaussian distribution (4.67) to describe the wind speed fluctuations. It can be applied for estimating wind gust speed. But its use for lower-frequency fluctuations such as squalls is disputable. That is why it is more difficult to solve the problem of taking into account the squalls in wave development. Squalls result in a considerable local wind increase. They are characterized by larger spatial and temporal scales than the wind gusts, and depend on a number of additional meteorological factors (for example, on cloudiness). The final solution for these factor parameterizations in wind wave models remains open.

Effect of wind wave parameter quasi-oscillation on the non-linear spectrum evolution. For several decades temporal wave records with about 20-minute duration were obtained in most wind wave measurements under field conditions. Records of duration were believed to be sufficient to obtain a representative estimation of the wave energy spectral density. The process was considered to be ergodic in the sense that the probabilistic characteristics of some records were assigned a known degree of similarity with the probabilistic characteristics of the hypothetical (obtained under analogous conditions) ensemble of selected functions. Such temporal intervals were called "quasi-stationary intervals" (Davidan et al., 1985). It was assumed that the local wave energy, defined as $E = \langle \eta^2 \rangle = \int S(\boldsymbol{k}) \, \mathrm{d}\boldsymbol{k}$, remained constant, i.e. $E = \langle \eta^2 \rangle = \mathrm{const}$ (where $\eta = \eta(\boldsymbol{r}, t)$ was the sea surface displacement, which was a function of the spatial coordinate $\boldsymbol{r}$ and time t, S was the spatial wind wave spectrum). Deviation from this condition was explained as sampling variability of the random process.

However, the assumption of the existence of "quasi-stationary intervals" became doubtful due to studies made in the last decade. The latest data reveal that the condition of constant process dispersion is not valid for the quasi-stationary interval. The local dispersion creates a quasi-periodic fluctuation (or simply quasi-oscillations) in both the temporal (τ_3) and the corresponding spatial scale range. This phenomenon, well known as SMIWEH (Smoothed Instantaneous Wave Energy History), was described by Zaslavskii & Krasitskii (1993), Bitner-Gregersen & Gran (1983), Mase (1989), and Sand (1982).

Measurements of the local wave dispersion $E = \langle \eta^2 \rangle$ obtained by Zaslavskii and Krasitskii (1993) in the North Atlantic served as experimental evidence of the existence of these effects. The experiment was performed under stationary conditions of wave development and in the absence of swell. The variation of the local dispersion $E(t)$, obtained by a running averaging method with a 60 s interval, is shown in Fig. 4.22a.

Fluctuations of the wave dispersion $E(t) = \bar{E} + E^{\mathrm{I}}(t)$ with an approximately 5 minute temporal scale can be seen. The amplitude $E^{\mathrm{I}}(t)$ comparable with $\bar{E}$ dispersion is estimated for the total record. These fluctuations are not dependent on the group structure of the wind waves estimated by a narrow band of their spectrum. The average wave period is equal to 6 s, whereas the temporal wave group scale is less than 60 s. It is important to note that the sampling variability estimates are smaller than the wave dispersion calculations.

Some other papers contain similar data. An example of time variation of wave dispersion (Andreyev, 1988) is presented in Fig. 4.22b. The dispersion is estimated for 150 s intervals with a fluctuation period of 7–15 minutes. Similar data for the variation $E(t)$ with 300 s averaging were obtained by Yefimov & Soloviev (1984). They estimated a 15–20 min fluctuation period.

The local wind wave dispersion fluctuation is connected with variations of the corresponding wave spectra. Such spectral density fluctuations of the

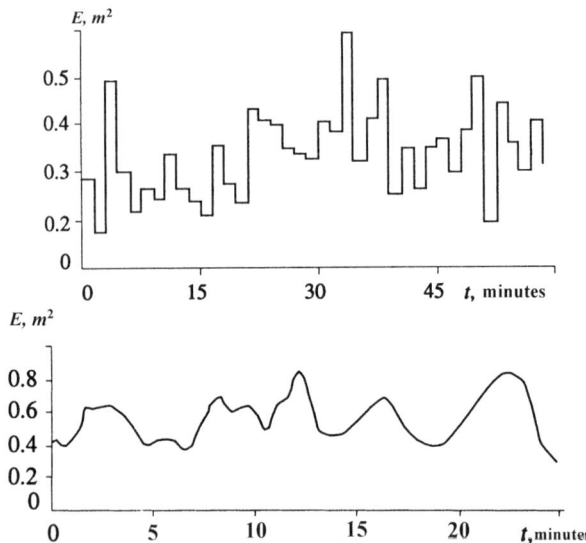

Fig. 4.22. Wave dispersion $E(t)$ obtained by the running averaging $\eta^2(t)$: (a) for the 60 s averaging interval (Zaslavskii & Krasitskii, 1993); (b) for the 150 s averaging interval (Andreev, 1988)

wave development with 3.5 min wave records (Andreev, 1988) is shown in Fig. 4.23.

The obtained spectra reveal the following features:

- the measured wave values are characterized by various spectral densities and wave dispersions;
- the most conservative parameter is the spectral maximum frequency ω_{max}, which is approximately of the same value in different wave records;
- the greatest fluctuations manifest themselves in the vicinity of the spectral maximum density.

The sea surface is assumed to respond to low-frequency wind gusts and squall. This phenomenon can be considered as "quasi-oscillations", clearly revealed in the changes of the frequency spectrum and wave dispersion even under the stable conditions of wave development (Andreev, 1988; Zaslavskii & Krasitskii, 1993).

It is possible to simulate these wave fluctuations of wind wave spectrum parameters with the help of a source function of the wave energy balance equation, including wind wave energy input. The role of gusts in the wind wave generation is described above.

There exists a micro-meteorological spectral maximum possessing a minute period of the atmospheric turbulence spectra (Monin & Yaglom, 1992). Thus, the fluctuations of wind wave parameters appear at these temporal scales. The mechanism of wind dispersion generation by atmospheric turbu-

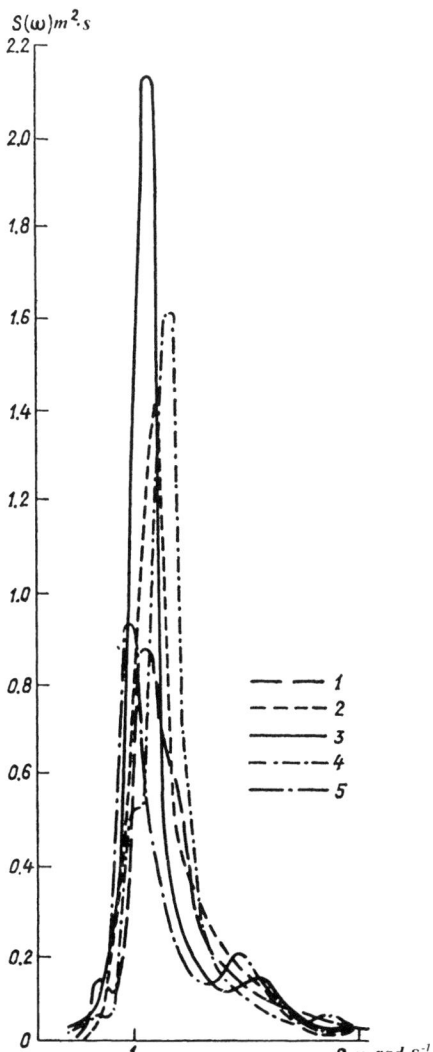

Fig. 4.23. Spectral density fluctuations of developing waves (1–5) estimated using a wave record of 3.5 min long intervals (Andreev, 1988)

lence may possibly be connected with both pressure and wind speed fluctuations. It should be noted that the temporal scales of the micro-meteorological spectral maximum in these atmospheric characteristics differ significantly. It is the order of several tens of minutes for the pressure spectra and the order of a minute for the wind speed. This quasi-oscillation period of wave spectrum parameters, as shown above, is correlated with the pressure fluctuation of atmospheric turbulence.

The presence of local variant fluctuations within the quasi-stationary state time intervals (τ_3) follows from the spatial–temporal wave field non-uniformity. It can be described by a kinetic equation for the wind wave spectrum evolution (Zaslavskii & Krasitskii, 1993). But in that paper the non-linear energy transfer is not taken into consideration due to energy conservation. The wave dispersion fluctuations are determined by the wind variations connected with "micro-meteorological gusts". The dispersion fluctuations of the sea surface are shown to be quasi-periodic, due to changes of the local wave spectrum and, in particular, to its peakness.

Wave generation by wind and wave dissipation are not considered here. The effect of wind wave parameter fluctuation on non-linear spectrum evolution is investigated by introducing a periodically changing term to the energy balance equation. The variations of the spectrum form including its peakness cause a local variation of the non-linear energy transfer in the wave spectrum. Thus it makes sense to estimate the influence of these variations at large temporal scales (τ_4–τ_5), taking into account the cubic proportion between the non-linear energy transfer and the spectrum value.

Problem formulation. In order to obtain quantitative estimates of the physical processes related to quasi-oscillations, it is necessary to define their parameters. A number of unsolved problems remain, namely the dependence of the oscillation period on the parameters of the spectrum itself, its peakness, wave stage development, etc. Using the results of Zaslavskii & Krasitskii (1993), the elementary approximation of spectral density in the quasi-stationary state interval with quasi-oscillations can be presented as an approximation of the JONSWAP spectrum (Hasselman et al., 1973) with changes in the periodic spectral density maximum (or spectrum peakness). This spectrum variation can be written as:

$$S(\omega, \varphi, \omega_{\max}, t) = S_{\text{JONSWAP}}(\omega, \varphi)\, \gamma^{\mu\, \sin(2\pi/\tau)\, f(\omega)}, \qquad (4.70)$$

where $S(\omega, \varphi, \omega_{\max}, t)$ is the frequency-angular spectrum of wind waves; $S_{\text{JONSWAP}}(\omega, \varphi)$ is the JONSWAP spectrum with the frequency dependence determined as:

$$S_{\text{JONSWAP}}(\omega) = \alpha g^2 \omega^{-5} \exp\left(-\frac{5}{4}\left(\frac{\omega_{\max}}{\omega}\right)^4\right) \gamma^{f(\omega)},$$

$$f(\omega) = \exp\left[-\frac{(\omega - \omega_{\max})^2}{2\sigma\, \omega_{\max}^2}\right],$$

where γ is the spectrum peakness parameter, μ is the relative oscillation amplitude of spectral maximum and τ is the period of oscillations of the order $\tau \sim \tau_3$. It should be noted that the temporal dependence of spectrum enhancement can be defined in this case as $\tilde{\gamma} = \gamma^{(1-\mu \sin(2\pi/\tau))}$.

4.4 Influence of Mesoscale Effects on Wind Wave Evolution

According to the experimental data (Andreev, 1988; Zaslavskii & Krasitskii, 1993) it is assumed that the spectrum maximum frequency $\omega_{\max}$ is changed significantly less at a given time interval than the spectrum value itself.

According to the experimental data the spectrum peakness parameter γ shows a decreasing tendency as waves are developing (Mitsuyasu et al., 1980; Donelan et al., 1985; Babanin & Soloviev, 1998). The JONSWAP spectrum transforms to the Pierson–Moskowitz spectrum with $\gamma = 1$, in case the fetch X is sufficiently large. High variability of the spectrum peakness parameter γ is well known according to measurement data. It is so high that a systematic dependence $\gamma(\tilde{x})$ cannot be determined for a wide range of dimensionless fetch range $\tilde{x} = Xg/U^2$. That is why the dependence $\gamma(\tilde{x})$ was not found in the JONSWAP experiment (Hasselmann et al., 1973).

Now the non-linear spectrum evolution depending on quasi-oscillations in the spatially homogenous case has to be calculated numerically. Consider the following equation:

$$\frac{\partial S}{\partial t} = G_{\mathrm{nl}}(S) - S f_2(\omega) \cos(2\pi t/\tau), \qquad (4.71)$$

where $f_2(\omega) = \mu f(\omega) \ln(\gamma) \, 2\pi/\tau$, and $G_{\mathrm{nl}}(S)$ is the non-linear interaction integral in the wind wave spectrum relative to the spectral energy density $S(\omega, \varphi)$.

It should be noted that in case $G_{\mathrm{nl}}(S)$ is equal to zero, the expression (4.70) is the solution to equation (4.71). On the other hand if oscillations are excluded ($f_2(\omega) = 0$) the traditional non-linear spectrum evolution can be described by (4.71).

Starting with fundamental papers by Hasselman (1962) and Zakharov (1969), where the right-hand term of (4.71) contains only the non-linear interaction term $G_{\mathrm{nl}}(S)$, (4.71) is traditionally applied in studying the non-linear wave spectrum evolution. But, unlike the aforementioned papers, the right-hand term of (4.71) includes an additional component, periodic in time, not used in the wind wave models. The second component of the right-hand term (4.71) may be assumed to be an approximation of the wind wave energy input and energy dissipation. But the problems of wave development under the wind effect and dissipation are not discussed here. An attempt to describe the spectrum evolution due to non-linear energy transfer and spectrum quasi-oscilation is undertaken below.

In order to solve the problem correctly the quasi-oscillation period should be sufficiently large for the waves "to have time" to interact, the latter being determined by the characteristic time of phase intermixing (Yuen & Lake, 1987):

$$\frac{1}{\tau_{\mathrm{ph}}} \approx \omega''(k) \, (\Delta k)^2, \qquad (4.72)$$

where Δk is the wave spectrum width.

The value τ_{ph} appears to be equal to several spectrum maximum periods. The condition of applicability of the method can be written as:

$$\tau_{ph} \ll \tau_3. \tag{4.73}$$

It should be noted that in this case the quasi-oscillation period can be smaller or of the same order with the characteristic time of non-linear wave evolution $\tau_{nl} \sim \tau$. Using (4.1), the value τ_{nl} can be estimated as (Yuen & Lake 1987):

$$\frac{1}{\tau_{nl}} \approx \frac{T N^2}{\omega''(k) (\Delta k)^2}, \tag{4.74}$$

Algorithm for numerical solution of (4.71). The equation (4.71), with its right-hand side term taking into account non-linear energy transfer and an additional term leading to spectrum oscillation is now solved numerically. It is necessary to ensure that the temporal step Δt of the numerical integration is much smaller than the oscillation period τ. In this case, the spectral density quasi-oscillation can be accurately taken into account in the numerical solution. The period τ should be less than the specified time of the non-linear spectrum evolution $\Delta t \ll \tau < \tau_4$. It is necessary to solve (4.71) numerically for the large time scale $(\tau_4-\tau_5)$ using a sufficiently small time step Δt. This is a rather difficult problem because the numerical computation of the non-linear energy transfer requires considerable CPU time. This means that the algorithm has to be efficient. That is why the algorithm, described earlier in Sect. 4.1, is applied to solve this problem. Using a small number of grid points the collision integral can be quite accurately calculated, taking little CPU time.

The equation (4.71) is solved in this case with the help of the two-step predictor–corrector method similar to that used by Lavrenov (1991b), allowing us to obtain an accurate numerical solution. The numerical integration time step is equal to 12.5 s, which makes it possible to conduct numerical integration up to the 10^5 s time interval.

Numerical modelling results. In numerical computation the value of the parameter μ is assumed to be equal to 0.9, which provides periodic oscillations of spectrum enhancement (4.70) ranging from 1.1 to 9.7 with mean value $\gamma = 3.3$. The oscillation periods are 10, 20 and 40 min respectively. The results of calculations of the frequency spectra for $\mu = 0.9$ at different moments t (for different values of spectrum peakness $\tilde{\gamma}$) are shown in Fig. 4.24a. The spectral density values are normalized by the maximum spectrum value at $\tilde{\gamma} = 3.3$.

The non-linear transfer function normalized by its maximum value at $\tilde{\gamma} = 3.3$ is shown for different $\tilde{\gamma}$ in Fig. 4.24b. Relative values of the non-linear transfer function differ significantly compared to the corresponding spectral

4.4 Influence of Mesoscale Effects on Wind Wave Evolution

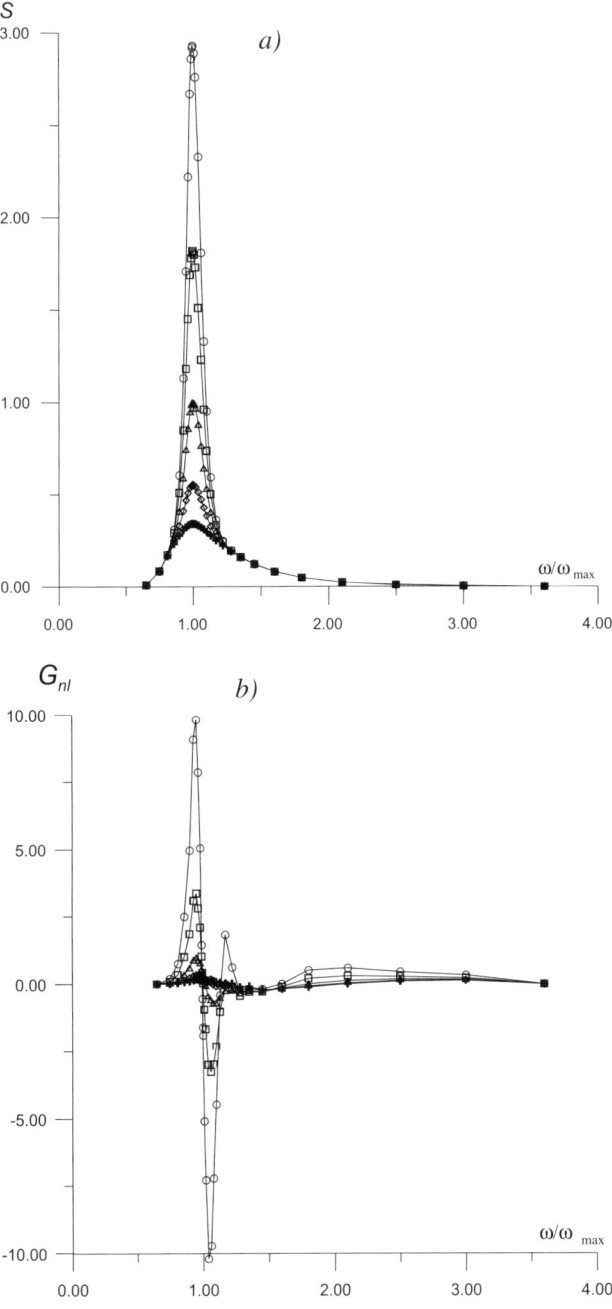

Fig. 4.24. Numerical results of different quasi-oscillation stages for $\tilde{\gamma} = 1.1(+)$; $\tilde{\gamma} = 1.8$ ($\Diamond$); $\tilde{\gamma} = 3.3$ (Δ); $\tilde{\gamma} = 5.9$ ($\square$); $\tilde{\gamma} = 9.7$ ($\circ$): (**a**) frequency spectra normalized by maximum spectrum value at $\tilde{\gamma} = 3.3$; (**b**) non-linear energy transfer functions normalized by its maximum value at $\tilde{\gamma} = 3.3$

148   4 Physical Mechanisms Forming the Wave Spectrum in Deep Water

density values. It is evident that the dependence between the spectral density and the non-linear energy transfer function is significantly non-linear. The spectrum variation not only significantly influences the non-linear energy transfer maximum, but also results in changes of the non-linear transfer function in general as well. The magnitudes of the non-linear energy transfer and also the specific frequencies where the non-linear energy transfer reveals itself in maximum, minimum or zero change depend on $\tilde{\gamma}$. The non-linear transfer maximum is shifted to the low-frequency range for greater spectrum peakness value.

Based on results of the numerical solution of the spectrum $S^n(\omega,\varphi)$ at each temporal step t_n, the spectrum maximum S^n_{max} and its frequency ω_{max} are determined. The changes of the frequency spectral maximum ω_{max} for four calculation versions are shown in Fig. 4.25. Results are shown for the case of the absence of oscillation (i.e. $\mu = 0.0$) and for three oscillation cases with different periods equal to 10, 20 and 40 min respectively.

A data group with the same symbols having similar values of ω_{max} for different time t ("stairs") indicates no changes of the frequency ω_{max} within the considered time interval. The value of the frequency ω_{max} is discontin-

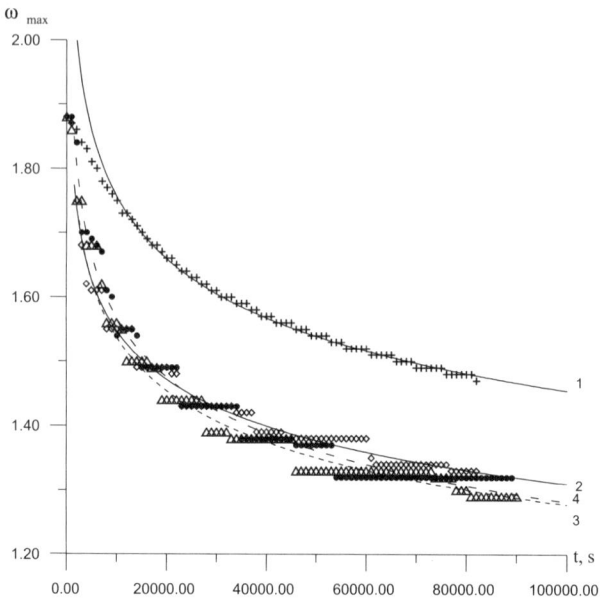

Fig. 4.25. Frequency evolution of the spectrum maximum ω_{max} created by non-linear energy transfer with different quasi-oscillation periods of: 1 (+) – without oscillations (approximation $\omega_{max} \sim 3.73\, t^{-0.082}$); 2 ($\Diamond$) – 10 min oscillation period (approximation $\omega_{max} \sim 3.01\, t^{-0.0724}$); 3 ($\Delta$) – 20 min oscillation period (approximation $\omega_{max} \sim 3.39\, t^{-0.085}$); 4 ($\bullet$) – 40 min oscillation period (approximation $\omega_{max} \sim 3.51\, t^{-0.0875}$)

uously changed at the next time interval. This cannot be explained by the rough frequency discretization used in the numerical scheme. The frequency variation of ω_{max} is much larger than the frequency discretization step. All in all the evolution of the frequency ω_{max} is not smooth. The discontinuous character of the non-linear evolution of the spectrum maximum frequency reveals that quasi-oscillations can serve as a "starting mechanism" for the low-frequency spectrum evolution. It occurs only at a specific moment, but not at every quasi-oscillation period. It takes place in cases when the non-linear spectrum evolution is accumulated and it changes the spectrum form in such a way that it is sufficient to make a small push (increasing the spectral density with quasi-oscillations). Thus, the spectral maximum frequency is transferred to another level.

It should be noted that the value ω_{max} is not changed with $\gamma = 1$ (i.e. the P–M spectrum) in the case of "oscillation absence".

Non-linear energy transfer causes an average spectrum shift to the low-frequency range. However, the mean evolution rate of the frequency maximum depends significantly on the given oscillation spectrum peakness. Thus, in the "oscillation absence" case the spectrum maximum frequency is monotonically decreased from the initial value of $\omega_{max}^0 = 1.88\,\mathrm{rad\,s^{-1}}$ to $\omega_{max} = 1.75\,\mathrm{rad\,s^{-1}}$ at $t = 10^4$ s, to $\omega_{max} = 1.60\,\mathrm{rad\,s^{-1}}$ at $t = 3 \times 10^4$ s, and to $\omega_{max} = 1.47\,\mathrm{rad\,s^{-1}}$ at $t = 10^5$ s. Oscillations result in a faster decrease of the spectrum maximum frequency. The oscillation period (for the used values) does not significantly influence the general tendency. In the presence of oscillations the spectrum maximum frequency is decreased from its initial value to $\omega_{max} = 1.60\,\mathrm{rad\,s^{-1}}$ at $t = 10^4$ s, to $\omega_{max} = 1.43\,\mathrm{rad\,s^{-1}}$ at $t = 3 \times 10^4$ s and to $\omega_{max} = 1.30\,\mathrm{rad\,s^{-1}}$ at $t = 10^5$ s.

In view of the fact that the spectrum maximum frequency is the most conservative parameter, the comparison of these results indicates a considerable influence of the oscillations on the non-linear spectrum evolution rate. In this case the average rate of the spectral maximum displacement is increased more than three-fold. As shown, the average rate of the spectral maximum displacement depends directly on the quasi-oscillation amplitude of the spectral maximum.

Parameterization of the influence of quasi-oscillation on non-linear energy transfer. The approximated frequency-angular spectrum averaged by a period of the order τ_4 is produced with the help of wind wave mathematical models. In this case the spectrum variations occurring at smaller time periods τ_3 are not taken into consideration. There appears the problem of taking into account adequately the quasi-oscillation effect and parameterization of its corresponding contribution to source functions in the mathematical models describing the formation of the wind wave spectrum.

It is important to point out that wind wave field fluctuations, observed in the quasi-stationary state interval, result in significant changes of the spectrum form, its peakness and wave steepness. That is why local variations of the non-linear energy transfer appear in the wave spectrum. There is a cubic proportion between the non-linear transfer and the spectral density. Only a double spectral density enlargement within its frequency maximum range is able to increase the non-linear transfer intensity almost by an order of magnitude. The periodic changes of the non-linear energy transfer occur within the quasi-cyclic alterations of its peakness. Should the peakness within the periods become larger compared with its average value, then the intensity of the non-linear energy transfer is increased. The non-linear transfer is decreased in the case of small peakness values.

Now an average analytical estimate of the non-linear energy transfer in the cyclically changed peakness will be obtained. According to (4.1), the first estimation of the non-linear transfer value is evaluated as $G_{nl}(S) \sim \omega_{max}^{11} S_{max}^3 / g^4$. It should be noted that more accurate numerical simulation results (see Fig. 4.24) carried out for the spectrum (4.56) reveal that G_{nl} is proportional to S_{max}^2 rather than to S_{max}^3. Assuming that the maximum frequency ω_{max} is not changed within one cycle and the spectral peak value is changed as $S_{max}(t) \sim \gamma^{\mu \sin(2\pi t/\tau)}$, integration of the non-linear transfer value within one quasi-oscillation period produces the average as follows:

$$\langle G_{nl} \rangle = \frac{1}{\tau} \int_0^\tau G_{nl}(S(t)) \, dt = \tilde{G}_{nl} \, I_0(p) \,, \tag{4.75}$$

where $\tilde{G}_{nl}$ is a non-linear transfer without quasi-oscillations ($\mu = 0$), I_0 is the modified first-order Bessel function, and the parameter p is determined as $p = 2\mu \ln(\gamma)$. Assuming that $\mu = 1.0$ and $\gamma = 3.3$, it is possible to derive $p \approx 2.39$ (i.e. $p > 1$). The Bessel function (4.75) is estimated by the following asymptotic formula:

$$I_0(p) \approx \frac{e^p}{\sqrt{2\pi p}} \left(1 + \frac{1^2}{1! \, 8p} + \frac{1^2 \cdot 3^2}{2! \, (8p)^2} + ... \right). \tag{4.76}$$

For the specific parameter values it can be found that $I_0 \approx 3.05$.

It can be pointed out that the spectrum mean over the period of quasi-oscillations determined as (4.70) differs from its corresponding value computed for the case of "no oscillation" ($\mu = 0$). The mean spectrum for the same period is equal to:

$$\langle S \rangle = \tilde{S} \cdot I_0(p/2) \,, \tag{4.77}$$

where $\tilde{S}$ is the spectrum value without quasi-oscillations ($\mu = 0$).

The factor is equal to $I_0(p/2) \approx 1.38$ in (4.77). If the non-linear energy transfer for the mean spectrum is recalculated, the value is 1.92 times greater, i.e. 1.59 times less than the non-linear energy transfer with quasi-oscillations.

Thus, it can be concluded that the averaged non-linear energy transfer within the quasi-oscillation period is essentially greater than the average spectrum value for the same period of time. An intensive energy flux towards the

4.4 Influence of Mesoscale Effects on Wind Wave Evolution

low-frequency spectrum range produces a larger displacement of the spectral maximum frequency to the low-frequency range.

A simplified approximation of the relative increase of non-linear energy transfer due to quasi-oscillations can be derived from the ratios (4.76) and (4.77):

$$\langle G_{\mathrm{nl}} \rangle \approx F(p) G_{\mathrm{nl}}(\langle S \rangle) , \qquad (4.78)$$

where $F(p) = \frac{I_0(p)}{I_0^2(p/2)}$, $p = 2\mu \ln(\gamma)$.

Thus, it is possible to assume that the relative increase of the non-linear energy transfer does not depend on the quasi-oscillation period, but is governed by the oscillation amplitude of the spectral maximum μ and the average spectrum peakness γ. The function $F = F(\mu)$ for various values of the parameter γ is shown in Fig. 4.26. It is monotonically increased with the growing value μ and the parameter γ. The value F is equal to 1.0 at $\gamma = 1.0$.

Since the peakness spectrum parameter $\gamma \geq 1$ decreases with developing waves (Babanin & Soloviev, 1998), the JONSWAP spectrum for the great fetch X transforms to the Pierson–Moskowitz spectrum with $\gamma = 1$. The effect of quasi-oscillation becomes smaller for a developed wind sea. For the wind

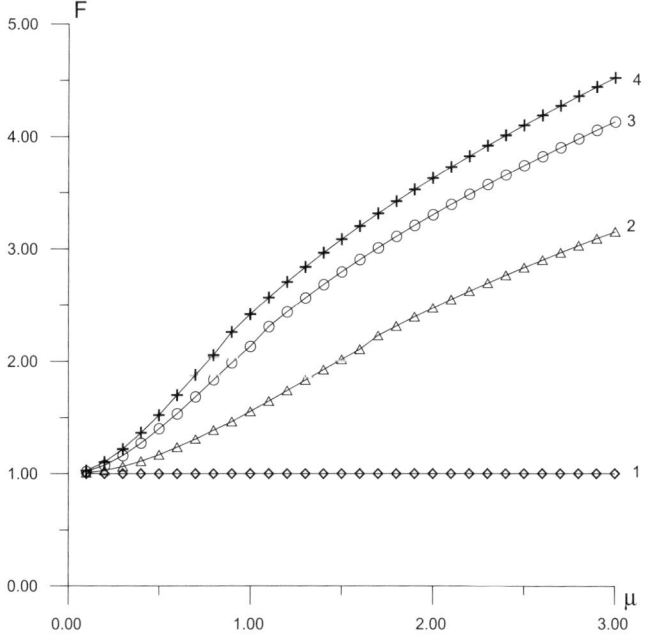

Fig. 4.26. Relation between function F and parameter μ for different average spectrum peakness values of γ: 1 ($\Diamond$) – $\gamma = 1.0$; 2 (Δ) – $\gamma = 3.3$; 3 ($\circ$) – $\gamma = 7.0$; 4 (+) – $\gamma = 10.0$

wave the initial development of the quasi-oscillation strongly influences the non-linear evolution.

Conclusions. Wind waves appear to be a random hydrodynamic process. To simplify wind wave investigation the quasi-stationary state interval (of the order of 20 min) was usually used (Davidan et al., 1978). The main statistical wave parameters were commonly considered as being approximately constant. However, investigations performed during the last few decades revealed that dispersion and the spectrum of the wave process change significantly within this interval. It is important to note that these variations are larger than the sample variability.

In the present book, numerical simulations of the energy balance equation are carried out to estimate the influence of the wind wave parameter fluctuation on the non-linear spectrum evolution. The non-linear energy transfer in the wave spectrum is calculated with the help of the original numerical integrating method of highest accuracy. The results of numerical simulations reveal that wind wave parameter fluctuation produces a significant increasing effect on the non-linear wave spectrum evolution. Its contribution at the initial stage of wind wave development is the most significant. The effect of quasi-oscillation becomes smaller for a developed wind sea.

The discontinuous character of the non-linear evolution of the spectrum maximum frequency with oscillations (see Fig. 4.25) reveals that quasi-oscillations can be a starting mechanism of low-frequency spectrum evolution. However, it does not operate within each quasi-oscillation period, but only at a specific moment. It takes place when the non-linear spectrum evolution is accumulated and changes the spectrum in such a way that it is sufficient to make a "small push" to transfer the spectral maximum frequency to another level.

The four-wave energy transfer is a non-linear mechanism depending on the spectral density in powers of three, contrary to the wind wave energy input or quasi-linear wave dissipation (Komen et al., 1994). Therefore, wind wave models usually underestimate the contribution of the mechanism to the wave spectral structure development. The non-linear energy transfer computed for the averaged spectrum is much less than the same total value for the spectrum with quasi-oscillations. The parameterization taking into account the effect of quasi-oscillations in the spectral models of wind waves is proposed in the present investigation.

5 Wave Evolution in Non-uniform Currents in Deep Water

5.1 Formulation of the Problem in the Local Coordinate System

The general formulation of the wind wave mathematical modelling problem in the ocean under the action of different factors forming the wind spectrum is presented in Chap. 1. The effects associated with wave spectrum transformation in a horizontal non-uniform current are investigated in this chapter.

Fundamental theoretical results for the problem were obtained by Longuet-Higgins and Stewart (1960, 1961, 1962, 1964). According to their investigations, waves and horizontal non-uniform currents can interact. As a result, wave energy can be received or transferred to the current. With the help of their theory a number of wave dynamic problems can be explained, although the field of application of the theory is limited. Thus, the wave energy is increased when the wave propagation is opposite to the flow, whose velocity is increased along its direction. The drawback of this theory is that the wave amplitude is estimated as having infinitely large values at the point where the wave group velocity is equal in value and opposite in direction to the current velocity. In fact, the results obtained by Longuet-Higgins and Stewart become invalid in the vicinity of this special point (caustic). The spectral approach used in this chapter eliminates such features near the caustics and provides an adequate description of wave behaviour in a non-uniform current.

As noted above, the influence of the current and bottom on waves is local. It is reasonable to consider the problem in the local coordinate system, using its formulation in the general form (1.84)–(1.90). In this case the balance equation of the wave action spectral density in the plane rectangular coordinate system can be written as:

$$\frac{\partial N}{\partial t} + \frac{\partial N}{\partial \boldsymbol{r}}\frac{\mathrm{d}\boldsymbol{r}}{\mathrm{d}t} + \frac{\partial N}{\partial \boldsymbol{k}}\frac{\mathrm{d}\boldsymbol{k}}{\mathrm{d}t} + \frac{\partial N}{\partial \omega}\frac{\mathrm{d}\omega}{\mathrm{d}t} = G \,, \tag{5.1}$$

where $\boldsymbol{r} = \{x, y\}$ is the horizontal spatial vector; $\boldsymbol{k} = \{k_x, k_y\}$ is the wave vector; G is the source function for different physical mechanisms forming the wind wave spectrum. The Hamiltonian equations are the characteristics of (5.1). Using the geometrical optics approximation, the wave packet propagation in the non-uniform medium is:

$$\frac{\mathrm{d}\boldsymbol{r}}{\mathrm{d}t} = \frac{\partial \omega}{\partial \boldsymbol{k}}, \quad \frac{\mathrm{d}\boldsymbol{k}}{\mathrm{d}t} = -\frac{\partial \omega}{\partial \boldsymbol{r}}, \quad \frac{\mathrm{d}\omega}{\mathrm{d}t} = \frac{\partial \omega}{\partial t}, \tag{5.2}$$

where ω is the wave frequency measured in an immovable coordinate system.

The frequency ω can be different from the wave frequency σ measured in a moving coordinate system connected with the current $\sigma^2 = (\omega - \boldsymbol{k}\boldsymbol{V})^2$, where $\boldsymbol{V} = \boldsymbol{V}(\boldsymbol{r},t)$ is the current velocity. For surface gravity waves, the dispersion relation can be written as $\sigma^2 = gk\,\mathrm{th}(kH)$, where $H = H(\boldsymbol{r})$ is the water depth. These equations are valid for the case of the current velocity, which does not vary vertically. The cases of vertical non-uniform current influence on waves are considered in Sect. 5.9.

In order to solve the problem of determining the wave action spectral density $N(\boldsymbol{k},\boldsymbol{r},t)$, it is necessary to define the current velocities $\boldsymbol{V}(\boldsymbol{r},t)$ and the water depth $H(\boldsymbol{r})$ and solve (5.2) and (5.1).

It should be noted that the solution is delivered in the phase space $\{\boldsymbol{k},\boldsymbol{r},t\}$.[1] Only one phase trajectory can pass through each space point, i.e. the phase trajectories do not intersect. In fact, this quality is a consequence of the uniqueness theorem for a set of ordinary differential equations with given initial conditions. There are some remarkable properties of propagating wave packet trajectories (Maslov & Fedoryuk, 1976; Vaynberg, 1982; Arnold, 1989).

It follows from (5.1) that when the source function is equal to zero (at $G = 0$): $\mathrm{d}N(x,y,k_x,k_y,t)/\mathrm{d}t = 0$. This means that the spectral density of the energy wave action is preserved along the ray:

$$N(x,y,k_x,k_y,t) = N_0(x_0,y_0,k_{x0},k_{y0}) \,. \tag{5.3}$$

The Jacobian of the transition from the initial to current values is absent in this case, unlike the wave description in physical space (see Sect. 1.3). This can be explained with the help of the wave description in phase space, dealing with canonical variables. The preservation of the wave action value in unit phase volume can be written as $N\Delta x\Delta y\Delta k_x\Delta k_y = N_0\Delta x_0\Delta y_0\Delta k_{x0}\Delta k_{y0}$. When the system motion in phase space is described by the Hamiltonian equations, the preservation of the phase space element volume follows from the Liouville theorem (Landau & Lifshits, 1973). The Jacobian of the transition from the initial phase volume element to the current one is $\partial(x_0,y_0,k_{x0},k_{y0})/\partial(x,y,k_x,k_y) = 1$. Thus, the relation (5.3) is proved. There are no caustic peculiarities associated with the vanishing Jacobian, as shown in the case of wave behaviour in physical space (see Sect. 1.3).

It follows from the condition (5.3) that the preservation of the wave action spectral density occurs along the trajectory of propagating wave packets without any sources or wave dissipation effects. It should be noted that this

---

[1] This space is also called the coordinate-impulse or simplex, and some sections of modern geometry (Arnold, 1989) are devoted to research of the behaviour of Hamiltonian systems in such space.

property was first described by Longuet-Higgins (1957) for the case of spatial wave spectrum transformation in shallow water.

5.2 Frequency-Angular Spectrum Evolution in a Current

Expression describing frequency-angular wave spectrum refraction.
The simplest case of the kinetic equation (5.1) is considered, neglecting the source function in the right-hand side of the equation. The spectral density N is preserved along the trajectory of wave packet propagation. The problem of determining N, using the initial conditions $N_0(\boldsymbol{k}_0, \boldsymbol{r}_0, t)$, is reduced to integrating the Hamiltonian equations (5.2) and defining the dependencies $\boldsymbol{k}_0 = \boldsymbol{k}_0(\boldsymbol{k}, \boldsymbol{r}, t)$, $\boldsymbol{r}_0 = \boldsymbol{r}_0(\boldsymbol{k}, \boldsymbol{r}, t)$. In most cases it is possible to solve the set of equations (5.1) only numerically.

The study proceeds from the wave action spectral density $N(\boldsymbol{k})$ to the energy spectral density $S = S(\omega, \beta)$, which is dependent on the frequency ω and the angle $\beta = \arctan(k_y/k_x)$. The transition from one spectral dependence to another can be easily made applying the dependence between components of the wave vector $\boldsymbol{k}$, frequency ω and angle β:

$$k_x = k(\omega, \beta) \cos(\beta) \, ; \quad k_y = k(\omega, \beta) \sin(\beta) \, .$$

The Jacobian of transition from k_x, k_y to ω, β is equal to:

$$\frac{\partial(k_x, k_y)}{\partial(\omega, \beta)} = k \frac{\partial k}{\partial \omega} \, .$$

The value of the spectrum S, depending on the initial conditions, can be written

$$S(\omega, \beta, \boldsymbol{r}, t) = \frac{\partial k^2}{\partial \omega} \sigma \left(\frac{\partial k_0^2}{\partial \omega_0} \sigma_0 \right)^{-1} S_0(\omega_0, \beta_0, \boldsymbol{r}, t) \, . \tag{5.4}$$

The expression (5.4) is obtained under sufficiently general assumptions. The wave refraction both in the presence of a non-uniform depth and in current can be described. Unlike shallow water, currents result not only in wave refraction, but also in additional effects connected with temporal and spatial non-uniformity of the current velocity. The temporal variability results in the Doppler frequency shift and the spatial non-uniformity leads to wave–current interactions. The first of these effects is considered in detail below.

Doppler frequency spectrum shift. As for waves propagating in a uniform non-stationary current, the wave vector $\boldsymbol{k}$ is constant according to (5.2), i.e. $\boldsymbol{k} = $ const, and the frequency ω varies in accordance with the condition: $\mathrm{d}\omega/\mathrm{d}t = \boldsymbol{k}\partial \boldsymbol{V}/\partial t$. The spectrum expression (5.4) can be written in this case as follows:

$$S(\omega, \beta, \boldsymbol{r}, t) = \frac{\partial \omega_0}{\partial \omega} S_0(\omega_0, \beta, \boldsymbol{r}, t), \qquad (5.5)$$

where S is the spectrum in the immovable coordinate system, and S_0 is the wave spectrum in a current. The frequencies ω and $\omega_0 = \sigma$ are connected with the dispersion relation for waves in the current: $\omega = \sigma + \boldsymbol{kV}$. The relation (5.5) can be rewritten for deep-water waves, when S_0 is independent of $\boldsymbol{r}$, in the following form:

$$S(\omega, \beta) = (1 + 2\sigma V \cos(\vartheta - \beta)/g)^{-1} S_0(\sigma, \beta), \qquad (5.6)$$

where ϑ is the angle between the direction of the Ox axis and the velocity $\boldsymbol{V}$.

The peculiarity of this relation is in the fact that the frequency σ does not have a single meaning for the function ω. The dependence of the value σ on ω can be given in the form:

$$\sigma_{\pm} = \frac{2\omega}{\tilde{V} \cos(\vartheta - \beta)} \left[-1 \pm \sqrt{1 + \tilde{V} \cos(\vartheta - \beta)} \right], \qquad (5.7)$$

where $\tilde{V} = 4V\omega/g$ is the non-dimensional current velocity. The $(+)$ sign in the relation (5.7) corresponds to straight waves; the $(-)$ sign to current reversed waves for $\tilde{V} \cos(\vartheta - \beta) < 0$ (Peregrine, 1976). The total spectrum $S(\omega, \beta)$ consists of the spectral sum corresponding to different branches of the relation (5.7). The relative contribution of reversed waves (with period $\tau < 4\pi V/g$) to the general range of the wind wave spectrum is comparatively small for real typical ocean current velocities. That is why only straight waves are usually considered. In this case the relation (5.5) can be rewritten in the form:

$$S(\omega, \beta) = S_0(\sigma_+, \beta)/\sqrt{1 + \tilde{V} \cos(\vartheta - \beta)}. \qquad (5.8)$$

Using the relations (5.7) and (5.8) it can be shown that the Doppler shift results in displacement of spectral components, especially in the large-frequency range. At the same time neither the mean wave height, nor the wave spatial spectrum are changed. The spectral maximum is displaced to higher frequencies in fair currents, whereas the high-frequency area of the spectral density becomes sloppier. The opposite change takes place in the countercurrents.

Frequency-angular wave spectrum transformation in a non-uniform current. The wave spectrum transformation in a horizontal non-uniform current occurs differently. The relation (5.4) is used for obtaining an expression for the wave spectrum evolution, with waves propagating in deep water ($\sigma^2 = gk$) under conditions of a horizontally non-uniform stationary current $\boldsymbol{V}(\boldsymbol{r})$. In this case the frequency ω remains constant along wave propagation rays and the final solution in explicit form can be derived.

5.2 Frequency-Angular Spectrum Evolution in a Current

Wave propagation from the area without the current ($V = 0$) to the area with the current velocity directed along the Ox axis $V = \{V(x,y); 0\}$ is considered. The initial wave spectrum (i.e. at $V = 0$) is assumed to be uniform and stationary $S_0 = S_0(\omega, \beta)$. The spectrum in the current can be written in accordance with the relation (5.4) as follows:

$$S^{\pm}(\omega, \beta, V) = \frac{16 S_0(\omega, \beta_0)}{\sqrt{1 + \tilde{V}\cos(\beta)}\left(1 \pm \sqrt{1 + \tilde{V}\cos(\beta)}\right)}, \quad (5.9)$$

where $\tilde{V} = 4V\omega/g$ is the non-dimensional current velocity. The ($\pm$) sign in this expression indicates the ambiguity of determining the wave spectrum in the current depending on the frequency ω, angle β and velocity V. An analogous ambiguity also occurs in determining the wave number $k = |\boldsymbol{k}|$ in a current:

$$k(\omega, \beta, V) = \frac{4\omega^2}{g\left(1 \pm \sqrt{1 + \tilde{V}\cos(\beta)}\right)^2}. \quad (5.10)$$

In order to determine the spectral value in (5.9), it is necessary to define β_0. This can easily be done when the velocity V depends only on one of two coordinates. The coordinate x is cyclic at $V = V(y)$. According to (5.2), the component k_x is constant with wave packet propagation and β_0 can be defined as:

$$\beta_0 = \arccos\left[\frac{4\cos(\beta)}{\left(1 \pm \sqrt{1 + \tilde{V}\cos(\beta)}\right)^2}\right]. \quad (5.11)$$

In the other case, for $V = V(x)$, $k_y = $ const, β_0 can be written as:

$$\beta_0 = \arcsin\left[\frac{4\sin(\beta)}{\left(1 \pm \sqrt{1 + \tilde{V}\cos(\beta)}\right)^2}\right]. \quad (5.12)$$

The first case (5.11) corresponds to the situation of wave propagation in shear horizontal non-uniform flow. It is considered below.

The second of the aforementioned cases is studied in detail. Variations of the angle β are considered as then dependend on the non-dimensional velocity $\tilde{V}$ for wave propagation in the increasing current $V(x)$. Proceeding from the preservation conditions of the frequency ω and the wave vector component k_y along the trajectory, the wave motion integral in the variables $\tilde{V}$ and β can be written in the form:

$$a\tilde{V}\cos(\beta) = 4\left(\sin(\beta) - \sqrt{\alpha\sin(\beta)}\right), \qquad (5.13)$$

where $\alpha = gk_y/\omega^2$ is a non-dimensional parameter, which is constant along the trajectory. The value of the parameter α is less than one and equal to $\sin(\beta)$ for wave packets going out of the area with no current ($V_0 = 0$). The value α can be greater than one for waves first generated in the current.

The relation (5.13) can be considered as the trajectory of wave packet propagation in the plane $\{\tilde{V}, \beta\}$. This trajectory with $\alpha < 1.0$ is shown as the curve II^a (or curve II^b for $\alpha > 1$) in Fig. 5.1a. The dependence $\beta = \beta(\tilde{V})$ does not have a single meaning, i.e. two values of the angle β correspond to the same value $\tilde{V}$, with $\alpha < 1$. When $\beta > 1.0$, there are three values β corresponding to the one value $\tilde{V}$. If the wave packet propagates from the area with no current ($V_0 = 0$) to the increasing countercurrent V, then the motion goes to the right upper part along the left curve II up to the point A (the curve II^a in Fig. 5.1a or in Fig. 5.1b). The projection of the group velocity on the axis Ox is positive, i.e.:

$$C_{gx} = \frac{1}{2}\sqrt{\frac{g}{k}}\cos(\beta) + V \geq 0 \qquad \text{at } V < 0.$$

The greatest current velocity is reached at the point A within the trajectory, whereas (5.13), as the function $\tilde{V}(\beta)$, is maximal. The point A is the turning point, after passing which the value C_{gx} becomes negative. Then the motion along the lower part of the curve II starts from the point A to the left (Fig. 5.1a, curve II^b), i.e. to the area of decreasing current velocities. The ambiguity of determining the angle β disappears at the turning point A. The angle can be found by solving the proper algebraic equation of the fourth degree (5.13). The real part of the solution is the following expression:

$$\beta_A = \arcsin\left\{\frac{1}{4}\sqrt[3]{\frac{2}{\alpha}}\left(\sqrt[3]{1-\nu} + \sqrt[3]{1+\nu}\right)\right.$$

$$\left. \times \left[1 \pm \sqrt{2\sqrt{1 - \frac{3\sqrt[3]{1-\nu^2}}{\sqrt[3]{1-\nu} + \sqrt[3]{1+\nu}}} - 1}\right]^2\right\}, \qquad (5.14)$$

where $\nu = \sqrt{1 - 16/27\alpha^2}$. It is necessary to use $(-)$ with $\alpha < 1$ in this expression.

There are two turning points: A' and A'' (see Fig. 5.1a) for waves with $\alpha > 1$. The wave blocking current velocity depends on the parameter α (for $\alpha < 1$, $V_A < -1.038g/4\omega$, $\beta_A < 0.096\pi$). The turning point disappears completely for $\alpha > 3\sqrt{3/4}$.

5.2 Frequency-Angular Spectrum Evolution in a Current

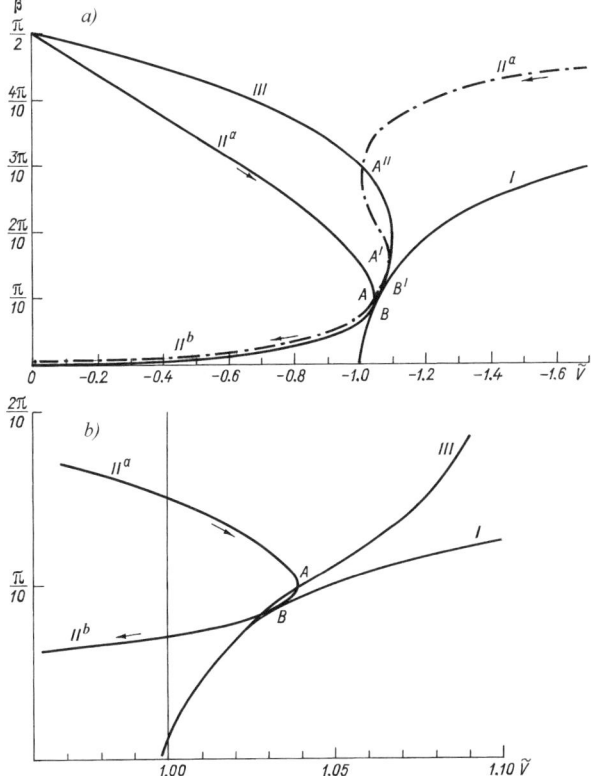

Fig. 5.1. Transformation of spectral parameters in increasing counter current (**a**) and around blocking point (**b**) at larger scale (in variables $\tilde{V}$, β) with arrow indicating propagation direction: where I – curve corresponding to relation $(\boldsymbol{k}\boldsymbol{C}_g) = 0$; II – wave packet propagation trajectory; III – curve with $C_{gx} = 0$

The determination of straight and reversed waves in a current is presented by Basovich & Talanov (1977). Wave packets with a positive scalar product of the wave vector $\boldsymbol{k}$ and the group velocity $\boldsymbol{C}_g$, $(\boldsymbol{k}\boldsymbol{C}_g) > 0$, are called straight waves. Correspondingly, the reverse waves are those with $\boldsymbol{C}_g$ negative. The transfer of straight waves to reversed ones takes place at the point with $(\boldsymbol{k}\boldsymbol{C}_g) = 0$. It is easy to show that this occurs at the trajectory point with $1 + \tilde{V}\cos(\beta) = 0$, rather than at the blocking point, as this was supposed earlier (Basovich & Talanov, 1977). In order to illustrate this fact, the relation is shown on the plane $\{\tilde{V}, \beta\}$ as the curve I (see Fig. 5.1a,b). The curve I touches the curve II at the point B, where $\tilde{V}_B = -1/\sqrt{1 - (\alpha/4)^2}$. Curve I presents an envelope of the trajectory varieties corresponding to different values of the parameter α. The coincidence of the blocking point A with the point of transfer from straight waves to reversed waves takes place only in the one-dimensional case, for $\alpha = \sin(\beta_0)$. The divergence between the points A and B becomes greater with increasing parameter α.

It should be noted that straight waves roll back and downstream ($C_{gx} < 0$) in the trajectory segment from the blocking point A to point B. They become reversed only later at the point B with the current velocity being equal to $V_B = -\omega \left(4g\sqrt{1-(\alpha/4)^2}\right)^{-1}$. The waves are taken downstream after the point B. The inequality $C_{gx} < 0$ is always satisfied. The angle β is decreased to zero with wave propagation to the area with small current velocities.

As has been noted, the group velocity component C_{gx} of straight waves can be both positive and negative. The condition $C_{gx} = 0$ can be written in the form $\tilde{V}\cos^3(\beta) - 2\cos(\beta) = 0$. The curve III, corresponding to this relation, is also plotted in Figs. 5.1a,b. It divides the plane $\{\tilde{V}, \beta\}$ into two domains with different values of the group velocity component C_{gx}, with $C_{gx} > 0$ being to the left of the curve III for straight waves. There are waves with $\alpha > 1$, having positive values C_{gx} in the segment of the trajectory II between two blocking points A' and A''. The maximum of the function $|\tilde{V}|$ is reached at $\beta = \arccos(\sqrt{2/3})$ being equal to $4/3\sqrt{2/3}$. There are no existing waves with $C_{gx} > 0$, for non-dimensional velocities $\tilde{V}$ greater than $4/3\sqrt{2/3}$, although straight waves can exist for $\tilde{V} > 1$.

At the point B the ambiguity is removed in the relations (5.9), (5.10) and (5.12). In these expressions the $(+)$ sign corresponds to the straight waves, and the $(-)$ sign to the reversed waves. The wave number k at the point B is equal to $4\omega^2/g$, i.e. depending neither on the current velocity, nor on the angle β.

In the wave spectrum expression (5.9) a singularity occurs at $1 + \tilde{V}\cos(\beta) = 0$. The value of the frequency-angular spectrum tends to infinity at the point B. This singularity appears as a result of variable substitution in the expression (5.4). The Jacobian (5.4) becomes infinite at $1 + \tilde{V}\cos(\beta) = 0$. In the case $(\mathbf{k}\mathbf{C}_g) = 0$, i.e the projection of the group packet velocity onto the chosen direction, determined by the angle β, is equal to zero. In this direction the measured packet time of the frequency $\omega = -g/4V\cos(\beta)$ is increased indefinitely. The wave packet is presented as a regular, monochromatic wave. Its spectrum is approximated by the delta function $S \sim \delta(\omega + g/4V\cos(\beta))$, having a singularity at $\omega = -g/4V\cos(\beta)$. Thus, the value of the time spectrum $S^{\pm}(\omega, \beta)$ is arbitrarily large for this component. Further it is shown that if the spatial spectrum were used instead of the temporal one this singularity would not occur. The singularity of the spectrum (5.9) is integrable at the point B.

The spectral value of the reversed waves $S^-(\omega, \beta)$ is increased with decreasing current velocity. A non-integrable singularity occurs at $V \to 0$, indicating the unlimited increase of gravity wave amplitudes. The lengths of these waves are decreased with decreasing current velocity according to (5.9).

5.2 Frequency-Angular Spectrum Evolution in a Current

The wave steepness carried away by the current is sharply increased. It could exceed the maximal permissible value that should lead to wave breaking.

An example with the initial spectrum angular distribution, in the absence of current, being quite narrow is considered. The spectrum can be written in the form:

$$S_0 = S_0(\omega)\delta(\beta_0),$$

where $\delta(\beta_0)$ is the Dirac delta function. Using (5.9), the one-dimensional spectrum is estimated as:

$$S^{\pm}(\omega) = \int_{-\pi}^{\pi} S^{\pm}\left(\omega, \beta_0(\beta, \tilde{V})\right) d\beta$$

$$= \frac{4S_0(\omega)}{\sqrt{1+\tilde{V}}\left(1 \pm \sqrt{1+\tilde{V}}\right)^2}. \quad (5.15)$$

The expression for the spectrum $S^{\pm}(\omega)$ coincides with the relation obtained by Huang et al. (1972). In the case of regular waves, the spectrum $S^{+}(\omega)$ is transformed into the known relation of wave amplitude evolution in a current obtained by Longuet-Higgins and Stewart (1961).

The total wave spectrum consists of straight and reverse wave spectra in a countercurrent. As mentioned above, the latter appears as a reflection of straight waves from a horizontal non-uniform current. They are carried downstream. In the case of the current velocity $V(x)$ increasing monotonically along the ordinate x up to some value and remaining constant within the interval, the wave spectrum is presented only by the straight wave.

The transformation of the wave frequency-angular spectrum is considered. The initial value of the wave spectrum is assumed to be described by the approximation:

$$S(\omega, \beta_0) = Q(\beta_0)(n+1)m_0 \frac{\omega_{max}^n}{\omega^{n+1}} \exp\left[-\frac{n+1}{n}\left(\frac{\omega_{max}}{\omega}\right)^n\right], \quad (5.16)$$

where m_0 is the spectrum zero moment and $Q(\beta_0)$ is the initial angular distribution approximated by the fourth power of a cosine. The spectral maximum frequency ω_{max} is assumed to be $0.86\,\mathrm{rad\,s}^{-1}$, $n = 4$.

In the countercurrent the wave angular distribution is determined by the relation:

$$Q^+(\beta,\tilde{V}) = \frac{8}{3\pi}\left[1 - \frac{16\sin^2(\beta)}{\left(1+\sqrt{1+\tilde{V}\cos(\beta)}\right)^4}\right]^2 \tag{5.17}$$

$$\times \Theta\left[1 - \frac{16\sin^2(\beta)}{\left(1+\sqrt{1+\tilde{V}\cos(\beta)}\right)^4}\right]\Theta\left(2\cos(\beta) - \cos^3(\beta) + \tilde{V}\right),$$

where $\Theta(\beta,\tilde{V})$ is the Heaviside function taking into account the absence of the wave component in the corresponding plane areas $\{\beta,\tilde{V}\}$. The spectral values transformed in the current calculated by (5.9), (5.16) respectively for the velocities $V = -1.0$ and $3.0\ \mathrm{m\,s^{-1}}$ are shown in Fig. 5.2a,b. The counter-current results in a sharp spectrum increase. The spectrum is decreased in

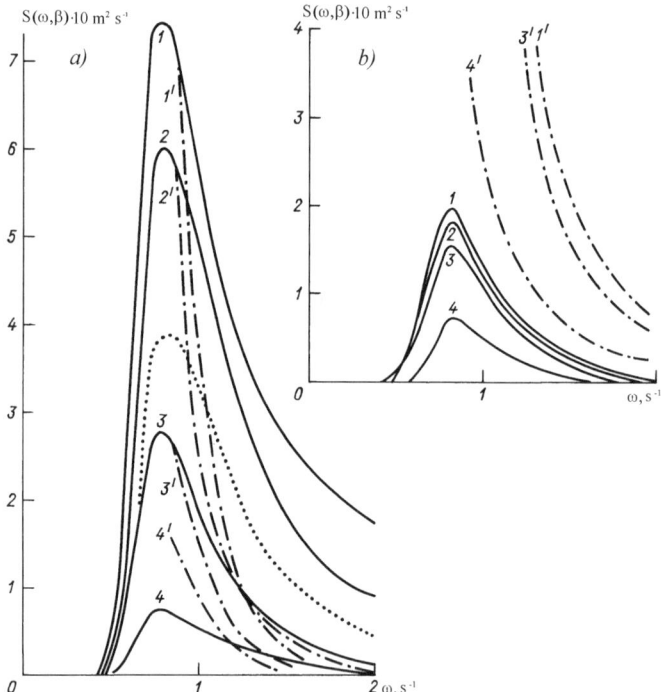

Fig. 5.2. Wave spectrum $S(\omega,\beta)$ evolution in countercurrent with $V = -1\ \mathrm{m\,s^{-1}}$ (**a**) and in fair current with $V = 3\ \mathrm{m\,s^{-1}}$ (**b**) for different angles β: $1 - 0°$; $2 - 15°$; $3 - 30°$; $4 - 60°$. The *dot-and-dash line* denotes corresponding equilibrium interval values; the *dotted line* denotes initial spectrum, at $V = 0\ \mathrm{m\,s^{-1}}$, $\beta = 0°$

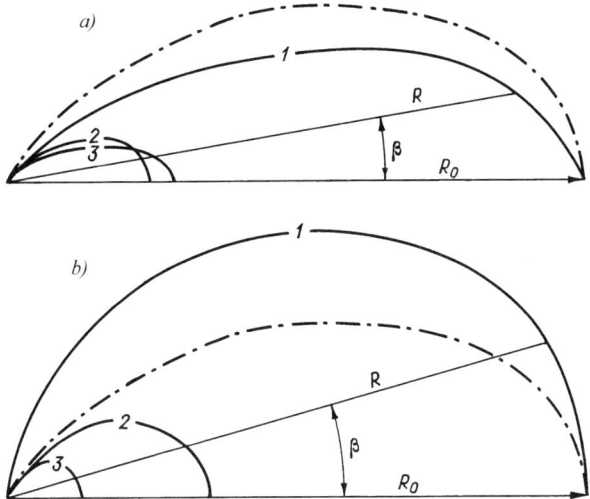

Fig. 5.3. Angular energy distribution function in countercurrent with $V = -1\,\text{m s}^{-1}$ (**a**) and in fair current with $V = 3\,\text{m s}^{-1}$ (**b**) for different frequencies ω: 1 – $0.85\,\text{rad s}^{-1}$; 2 – $0.65\,\text{rad s}^{-1}$; 3 – $1.5\,\text{rad s}^{-1}$. *Dot-and-dash lines* denote the initial angular energy distribution

a fair current. At the same time, the frequency of the spectrum maximum is almost unchanged.

The range of spectral directions for different frequencies is shown in Fig. 5.3a,b. It is normalized by the spectral maximum at a given current velocity. As can be seen, the direction range is narrowed by the countercurrent. It is narrowed to a greater extent at high frequencies. The process is vice versa in a fair current.

If waves propagate to the area with a current opposite to the wave direction the wave energy per unit area is increased due to interaction with the countercurrent. This can be limited by dissipation connected with wave breaking. There is an equilibrium interval appearing in the spectrum, i.e. the range of frequencies ω and the angles β, in which the energy influx from a current is balanced by breaking.

Generally speaking, there is no reason to assume that the equilibrium interval of the wave frequency spectrum in a current is approximated by the well-known Phillips formula $S_\infty(\omega) \sim \omega^{-5}$ (1980). The frequency-angular interval spectrum in a current can be found by recalculating the corresponding dispersion relation using the idea of the universally applied spatial spectrum within the equilibrium interval (Kitaigorodskii et al., 1975). Thus, the frequency-angular spectrum of the equilibrium interval can be written in the following form:

$$S_\infty(\omega,\beta) = \frac{\left(1 + \sqrt{1 + \tilde{V}\cos(\beta)}\right)^5}{32\sqrt{1 + \tilde{V}\cos(\beta)}} \tilde{\alpha} g^2 \omega^{-5} \Theta(\tilde{V},\beta) , \qquad (5.18)$$

where $\tilde{\alpha}$ is the Phillips constant and $\Theta(\tilde{V},\beta)$ is the corresponding angular distribution. The equilibrium interval (5.18) is denoted by a dot-and-dash line in Fig. 5.2a,b. The equilibrium interval curve intersecting the corresponding spectral dependence obtained earlier means that the spectral approximation (5.9) can be justified for the equilibrium interval area and to its right part. The descending spectrum part should not be higher than the equilibrium interval when the wave transformation in a countercurrent occurs slowly in comparison with the time of the wave spectrum stabilizing under the breaking effect. The spectral maximum is shifted to lower frequencies and the spectrum itself becomes narrower. The experimental results are confirmed by field observations (Peregrine, 1976).

The reverse picture is observed in a fair current area. In this case the current absorbs a part of the wave energy. The amplitude of the wave components becomes smaller than the stability limit. Wave breaking ceases, and the spectral dependence becomes lower than the equilibrium interval.

5.3 Spectral Model of Rips

Phenomenological description and experimental data. Rips are usually defined as irregular sea waves generated in water areas with the current flowing around unevenness of the bottom relief in shallow water or in the case of waves running countercurrent or due to some other similar reasons. The phenomenon of rips was observed in detail by investigators at the Research Institute of Oceanology of the Russian Academy of Sciences (Barenblatt et al., 1985; Leykin & Monin, 1985). The waves and currents were measured in the strait connecting Onega Bay and the White Sea (rip generation is schematically shown in Fig. 5.4). Rips were always observed during tide currents with their maximum speed and very often in cases of wind and current being opposite in direction. The rips were eliminated by strong wind waves. The rip waves were shorter and steeper in comparison with usual wind waves and swell. Furthermore, they were more asymmetric with sharpened crests and gentle troughs.

The spectra of rips were estimated by wave records measured in a mobile readout system moving with the current. The spectra were characterized by great variability. In most cases they were "two-humped", i.e. there were two maxima near the frequencies of $f_1 = \sigma/2\pi \approx 0.25$, and $f_2 \approx 0.5$ Hz. The spectral density $S(\sigma)$ decreased quickly in the high-frequency area behind the second peak, and it could be approximated by the Phillips equilibrium spectrum $S = \beta g^2 \sigma^{-5}$ in some frequency interval (see Fig. 5.5).

5.3 Spectral Model of Rips   165

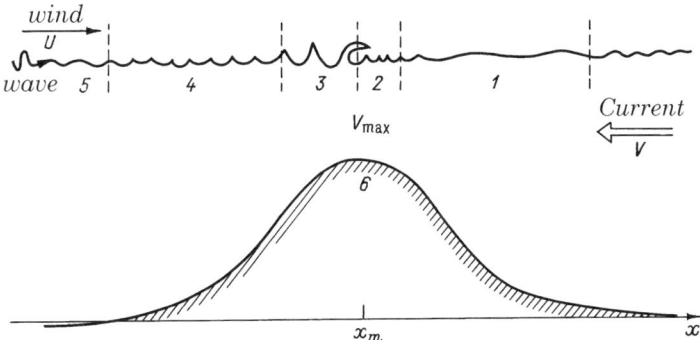

Fig. 5.4. Scheme of frontal type rip generation: 1 – calm water; 2 – transition zone; 3 – breaking zone; 4 – whirlpool ; 5 – background waves; 6 – submarine bank; x_m – maximum current speed point

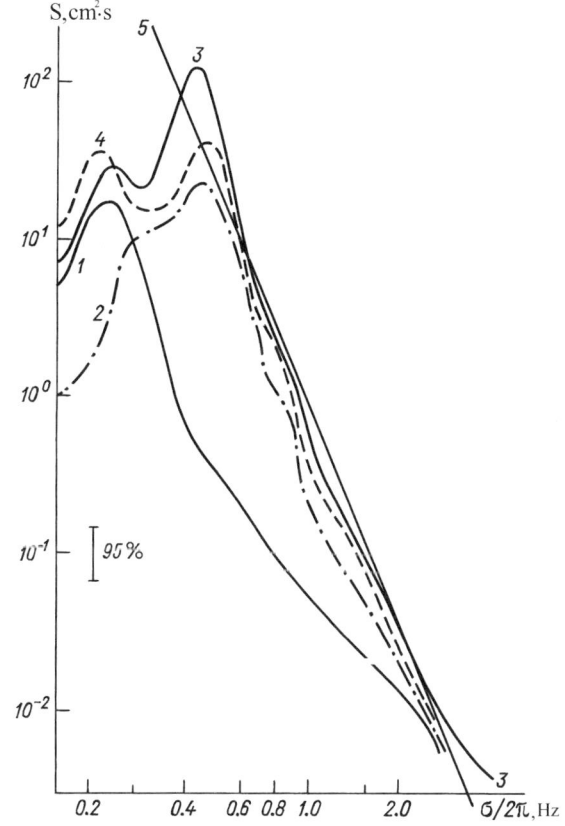

Fig. 5.5. Surface wave spectra in frontal type rips (Barenblatt et al., 1985): 1 – calm water; 2 – transition zone; 3 – breaking zone; 4 – whirlpool; 5 – Phillips' equilibrium spectrum

From the theoretical point of view, the rips are suggested to be due to either wind wave transformation in non-uniform currents or wave formation in the water surface with a current flowing around a submarine obstacle (Barenblatt et al., 1985). It is also believed (Leykin & Monin, 1985) that a suitable wave field model is a "soliton gas", i.e. a system of solitons with different direction propagation and amplitudes. The waves do not interact, but simply superimpose each other.

The first of these hypotheses is discussed in this section. The rip model can be considered as a result of wave transformation in a large-scale horizontal non-uniform current. The indirect role of different underwater obstacles is not excluded. They have a specific influence on surface waves, creating a strong non-uniform current.

Problem formulation. The problem is considered in the geometrical optics approximation. It is based on the wave action density balance equation in the spectral form (5.1). The main difficulty is insufficient knowledge of the source function G. There are theoretical ratios for the wind input generating mechanism G_{in} and the non-linear energy transfer G_{nl}, but unfortunately there is still no generally accepted expression for the wave dissipation G_{ds}. The limitation of the wave spectrum increase is often achieved in wind wave models as follows: the source function (without dissipation) is multiplied by the term $1 - \mu = 1 - f(S/S_\infty)$. The correcting function $1 - \mu$ establishes the ultimate possible value of the spectral density S_∞. Thus, it takes the wave energy dissipation into account indirectly. The Pierson–Moskowitz spectrum is used as an ultimate spectrum S_∞, formed by the constant wind, to calculate the wave in deep water without currents in wind wave models of the first and second generations. The approximation of such an ultimate spectrum is unknown in the case of a horizontal non-uniform current. Thus, Basovich et al. (1982) take the source function in the action density balance equation in the form $G = \beta_U N(1 - N/N_0)$, where N_0 is the action density in the absence of a current and β_U is the wave increase increment. However, the wave spectrum increase is insufficiently limited by the source function. For example, in the case of the wind action being neglected ($\beta_U = 0$), the spectral energy density $S = N\sigma$ is changed significantly. It can greatly exceed the equilibrium interval spectrum (Kitaigorodskii et al., 1975; Phillips, 1958).

An attempt should be made to determine the term limiting the increase of the wave action density spectrum in the kinetic equation (5.1) with waves in a horizontal non-uniform current and wind input. For this purpose the hypothesis of the equilibrium interval can be used as some ultimate state of the spatial energy spectrum (Kitaigorodskii et al., 1975). The equilibrium interval is considered to be invariant and its approximation can be written as $S_\infty(k) = (1/2)\alpha_F k^{-4} Q(\beta)$. Proceeding from the wave energy density balance equation (which takes into account the wind and current influence on waves) to the wave action density equation with the correcting function in the form $1 - \mu = 1 - (S/S_\infty)^q$, the following equation is obtained:

$$\frac{dN}{dt} = \beta_u N \left[1 - \left(\frac{N}{N_\infty}\right)^q\right] - \frac{N}{\sigma}\left(\frac{N}{N_\infty}\right)^q \frac{d\sigma}{dt}, \quad (5.19)$$

where $N_\infty = S_\infty/\sigma$ and q is a non-linear parameter characterizing the efficiency of the introduced restriction to the wave spectrum increase.

In (5.19) it is convenient to pass from the variables $\boldsymbol{k} = \{k_x, k_y\}$ to $\{k, \beta\}$, where $\beta = \arctan(k_y/k_x)$.

The characteristics of (5.19) are written in the form:

$$\frac{dx}{dt} = C_{gx} = C_g \cos(\beta) + V_x ; \quad \frac{dy}{dt} = C_{gy} = C_g \sin(\beta) + V_y ; \quad (5.20)$$

$$\frac{dk}{dt} = -\cos(\beta)\frac{\partial\omega}{\partial x} - \sin(\beta)\frac{\partial\omega}{\partial y} ;$$

$$\frac{d\beta}{dt} = \frac{1}{k}\left(\sin(\beta)\frac{\partial\omega}{\partial x} - \cos(\beta)\frac{\partial\omega}{\partial y}\right) ; \quad (5.21)$$

$$\frac{d\omega}{dt} = \frac{\partial\omega}{\partial t}, \quad (5.22)$$

where $\omega = \sigma + k\cos(\beta)V_x + k\sin(\beta)V_y$; V_x and V_y are the vector components of the current speed $\boldsymbol{V}$; and C_g is the absolute value of the wave group speed.

Solution of the one-dimensional problem. The ratios (5.19)–(5.22) are used to describe the rip spectrum. According to the scheme discussed by Barenblatt et al. (1985), it can be assumed (see Fig. 5.4) that the waves propagate from the area where the current can be neglected ($V_0 \approx 0$), to the area with speed $\boldsymbol{V} = \{-V(x), 0\}$, directed towards the waves. The current speed $V(x)$ is monotonically increased to some maximum value $V_{\max}$ (at $x = x_m$), then decreasing to zero. The wind role in the formation of the rips is assumed to be insignificant. Thus, the first term on the right side of (5.19) can be neglected.

The equation (5.19) is easily integrated in the one-dimensional case for $S(k,\beta) = S(k)\delta(\beta)$. Its solution can be presented in the analytical form:

$$N = N_0 \left[1 - \frac{1}{9}N_0^q \left(N_{\infty 0}^{-q} - N_\infty^{-q}\right)\right]^{-1/q} \quad (5.23)$$

where N_0 and $N_{\infty 0}$ are initial values of the wave action density and the equilibrium interval, prescribed at $t = t_0$. The arguments of the functions $N_0(k_0, \boldsymbol{r}_0, t_0)$ and $N_{\infty 0}(k_0, \boldsymbol{r}_0, t_0)$ are the values of k, considered at the moment t at a specific point $\boldsymbol{r}$: $k_0 = k_0(k, \boldsymbol{r}, t), \boldsymbol{r}_0 = \boldsymbol{r}_0(k, \boldsymbol{r}, t)$. The latter dependencies are solutions of the equation system (5.20)–(5.22).

As can be seen from the obtained solution (5.23) with $N_0 \ll N_\infty$ and $N_{\infty 0} \sim N_\infty$, the action density is preserved along the trajectories of wave packet propagation, i.e. $N(k, \boldsymbol{r}, t) \approx N_0(k_0, \boldsymbol{r}_0, t_0)$. This solution is considered above. When $N \sim N_\infty$, the equilibrium interval influences the solution N and in the limit $N_0 \gg N_\infty$, $N \approx \sqrt[9]{9}N_\infty$. As shown below, this situation can be fulfilled with waves propagating at increasing countercurrent. The spectrum increase is limited by wave breaking. There is the opposite case – the wave energy density is decreased due to interaction with a fair current. The solution is not influenced by the equilibrium interval (5.23).

One can pass from the action density spectrum $N(k)$ to the energy frequency spectrum $S(\sigma)$ measured in the readout system connected with the current $S(\sigma) = \sigma k \partial k/\partial \sigma N(\sigma)|_{k=k(\sigma)}$. It should be noted that the wave spectrum measurements at the current (Barenblatt et al., 1985) were made precisely in that coordinate system. As shown below, it is significantly different from the spectrum $S(\omega)$ measured in a fixed coordinate system, described in Sect. 5.2. Based on the ratio (5.23), the solution of the spectrum $S(\sigma)$ can be written in the form:

$$S(\sigma) = S_0(\sigma_0) \frac{\sigma k}{\sigma_0 k_0} \frac{\partial k}{\partial \sigma} \left(\frac{\partial k_0}{\partial \sigma_0}\right)^{-1}$$

$$\times \left\{ 1 + \frac{1}{9} \left[\frac{S_0(\sigma_0)}{\alpha_F g^2 \sigma_0^{-5}}\right]^q \left[\left(\frac{\sigma}{\sigma_0}\right)^{9q} - 1\right]\right\}^{1/q}, \quad (5.24)$$

where $S_0(\sigma_0)$ is the initial wave spectrum in the absence of current.

The expression for initial wave frequency spectrum in the absence of current (i.e. $\omega = \sigma$) is taken in the form (5.16), where n is the parameter determining the spectrum form. It is assumed to be equal to 5.5.

It is necessary to solve the equation system (5.20)–(5.22) for the given current speed profile $V(x)$ in order to determine the value σ_0 as a function of the variables σ, V, r in (5.24). In its exact form, this system is not integrated even for the one-dimensional case. That is why the following approaches are taken into consideration.

It follows from (5.22) that the frequency ω is preserved for the stationary current speed V along the wave packet propagation trajectory. It can be written as $\sigma - Vk = \sigma_0$ in the one-dimensional case. This ratio determines the wave number k_0 depending on k and the current speed V. However, the same speed value V can correspond to two different wave numbers. Both straight ($C_{gx} > 0$) and reverse waves ($C_{gx} < 0$) can exist in the countercurrent at the same point (with $x < x_m$). It is a result of straight waves reflected from the horizontal non-uniform current. The straight waves can only exist after having passed the maximum current speed (in the segment $x > x_m$). Due to this reason, the motion integral is insufficient for an unambiguous solution. It is necessary to apply kinematic schemes in this case.

Thus, a wave packet is reflected from the current and rolled down propagating in the increasing countercurrent until it reaches the point $V = V^*$, where the blocking condition $C_{gx} = C_g - V^* = 0$ is fulfilled. It is necessary that the conditions $V^* < V_{\max}$ should be formally met. Otherwise, the wave packet passes through the "barrier", i.e. the area with maximum current speed. Both straight and reverse waves can exist in the area $(x < x_m)$ for $V^* < V_{\max}$. There are no reverse waves for $V^* > V_{\max}$. Only straight waves, passing through the "barrier", can exist in the other area $(x > x_m)$. Thus, in the area $(x < x_m)$, the frequency spectrum $S(\sigma)$ can be written in the form:

$$S_1(\sigma) = S_0\left(\sigma(1-\tilde{y})\right) \left\{ 1 + \frac{1}{9} \left[\frac{S_0\left(\sigma(1-\tilde{y})\right)}{\alpha_F g^2 \sigma^{-5}(1-\tilde{y})^{-5}} \right]^q \left[(1-\tilde{y})^{-7q} - 1\right] \right\}^{-1/q}$$

$$\times \frac{1}{(1-\tilde{y})^2} \left\{ \Theta\left[\tilde{y} - \frac{\gamma}{4(1-\tilde{y})}\right] \Theta\left(\tilde{y} - \frac{1}{2}\right) + \Theta\left(\frac{1}{2} - \tilde{y}\right) \right\}, \tag{5.25}$$

where $\gamma = V/V_{\max}$; $\tilde{y} = V\sigma/g$ is a non-dimensional frequency; and $\Theta(\tilde{y})$ is the Heaviside function.

A similar expression can be written for waves passing through the point with maximum current speed $(x > x_m)$:

$$S_2(\sigma) = S_0\left(\sigma(1-\tilde{y})\right) \left\{ 1 + \frac{1}{9} \left[\frac{S_0\left(\sigma(1-\tilde{y})\right)}{\alpha_F g^2 \sigma^{-5}(1-\tilde{y})^{-5}} \right]^q \left[(1-\tilde{y})^{-7q} - 1\right] \right\}^{-1/q}$$

$$\times \frac{1}{(1-\tilde{y})^2} \left\{ \Theta\left[\frac{\gamma}{4(1-\tilde{y})} - \tilde{y}\right] \Theta\left(\frac{1}{2} - \tilde{y}\right) \right\}. \tag{5.26}$$

As can be seen, the expression (5.25) is transformed into (5.26) at the point $x = x_m$.

Now the wave spectrum described by the obtained solution will be investigated. In order to do this, it is necessary to prescribe the values of the parameters in (5.25) and (5.26). The parameter q is assumed to be equal to 2. It corresponds to the value which is frequently assumed by different authors in wind wave calculations (Ocean Wave Modeling, 1985).

The spectra (5.25) and (5.26) are written in a more general form by normalizing the spectra S_1 and S_2 by the maximum value of the initial spectrum $S_0(\sigma_{\max}) = (n+1)m_0 \exp\left(-\frac{n+1}{n}\right)/\sigma_{\max}$. The normalized spectra $\tilde{S}_1$ and $\tilde{S}_2$ are written as a function of the non-dimensional frequency $\tilde{y}$. The following non-dimensional parameters appear is these spectra: $\nu = V\sigma_{\max}/g$ is a non-dimensional current speed and $\nu_m = V_{\max}\sigma_{\max}/g$ is the effectiveness of the influence of the selected current profile on the major energy spectrum components. It can be shown that the current influence on the wave spectrum is negligible for $\nu_m < 0.1$. On the other hand, the wave spectrum is blocked completely by the current ($\tilde{S}_2 \approx 0$) for $\nu_m > 0.5$. That is why there is no sense in considering the case with ν_m being greater than the indicated value.

Firstly, the case of wave blocking at $\nu_m = 0.5$ is considered. It is necessary to define the dispersion m_0 to determine the initial spectrum value. The solution (5.25) and (5.26) includes the dispersion m_0 as a non-dimensional product of $m_0(n+1)\sigma_{\max}^4/\alpha_F g^2 = \delta_0$, which can be connected with the mean initial wave steepness $h_0/\lambda_0 = 2.722 \times 10^{-2}\sqrt{\delta_0}$ (where h_0 and λ_0 are the initial mean wave height and length, respectively).

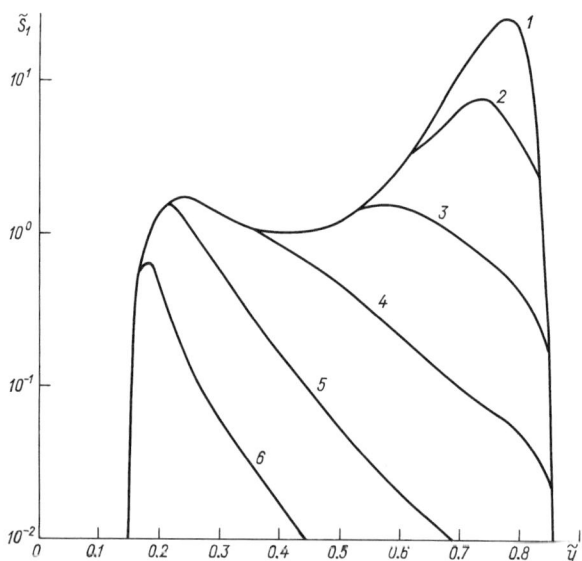

Fig. 5.6. Solution $\tilde{S}_1(\tilde{y})$ on a logarithmic scale at $\nu_m = 0.5$, $\nu = 0.175$ and for different parameter δ_0 values: 1 – 0.0; 2 – 0.001; 3 – 0.01; 4 – 0.1; 5 – 1.0; 6 – 10.0

The spectrum $\tilde{S}_1(\tilde{y})$ is shown in Fig. 5.6, on a logarithmic scale with different values of the parameter δ_0 at the point with current speed $\nu = 0.175$. Curve 1 denotes the spectrum $\tilde{S}_1$ at $\delta_0 = 0$, corresponding to infinitely small amplitude waves. The main spectrum feature, distinguishing it from the initial spectrum (5.16), is the existence of two maxima at the frequencies $\tilde{y}_1 = 0.240$ and $\tilde{y}_2 = 0.787$. Although the second spectral maximum is much greater that the first one, the spectrum becomes more symmetric. Such spectrum bifurcation is connected with the formal transfer from σ to $\tilde{y}(1 - \tilde{y})$ in the spectrum arguments. With decreasing speed ν the spectral maxima move apart, whereas the first spectral peak value tends to one, and the second spectral peak is increased, tending to infinity.

The spectra for different values of the parameter δ_0 are shown by the curves 2–6 (see Fig. 5.6). The second spectral maximum is decreased and completely disappears at $\delta_0 > 0.05$ during this parameter increase. Thus, the state of the high-frequency spectrum area is controlled by the parameter δ_0. The second spectral maximum value is also dependent on the parameter n in the approximation (5.16). In this case the parameter n is assumed to be equal to 5.5. The division of two maxima in the spectrum becomes more significant with increasing n.

Now the spectrum solution is described for the case of waves propagating in the horizontally non-uniform countercurrent. In order to estimate solutions (5.25) and (5.26), the value $\delta_0 = 0.01$ is used, corresponding to the initial

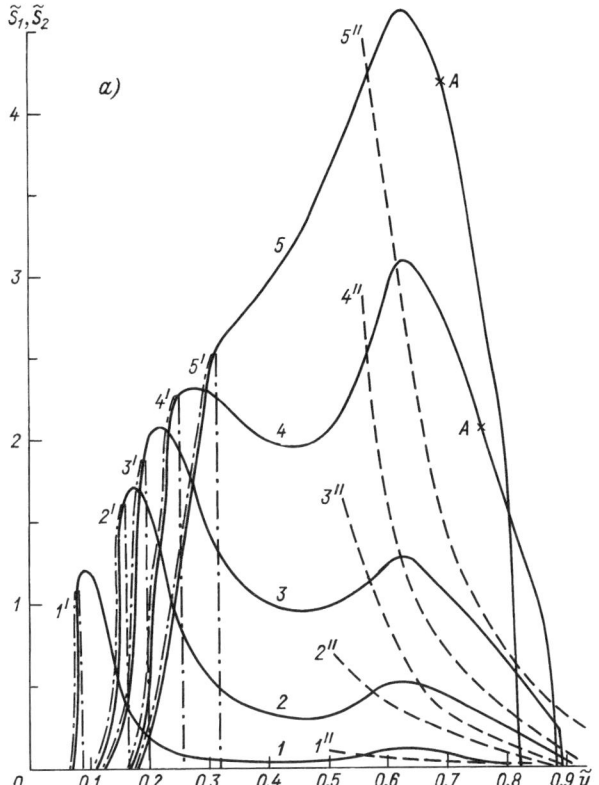

Fig. 5.7. (a) Wave spectra evolution in countercurrent with $\nu_m = 0.25$, $q = 2$ and $\delta_0 = 0.01$ at different speed values ν: 1 – 0.1; 2 – 0.15; 3 – 0.175; 4 – 0.20; 5 – 0.225 (where the *solid line* is the spectrum $\tilde{S}_1$, the *dot-and-dash line* is the spectrum $\tilde{S}_2$, and the *dotted line* is the equilibrium interval) (Fig. 5.7(b) see next page)

waves in the form of a mildly sloping swell. The wave spectra $\tilde{S}_1$ and $\tilde{S}_2$ in the current (Fig. 5.4) are shown in Fig. 5.7a. The undimensional maximum of the current speed is equal to $\nu_m = 0.25$. There is no blocking of all energy spectral components at the given current speed, whereas the spectrum $\tilde{S}_2$ is comparable to $\tilde{S}_1$ by value.

A definite increase of the spectrum initial maximum $\tilde{S}_1$ at the frequency $\tilde{y} = 0.12$ and the second maximum at the frequency $\tilde{y} = 0.64$ is observed at the initial wave propagation stage (at $\nu = 0.1$). The second spectral maximum appears at frequencies greater than 0.5. These frequencies correspond to reverse waves as a reflection of initial waves caused by the horizontally non-uniform current.

The spectrum evolution $\tilde{S}_1$, occurring at points with current speed ν equal to 0.15, 0.175 and 0.2, respectively, are also shown in Fig. 5.7a. A fur-

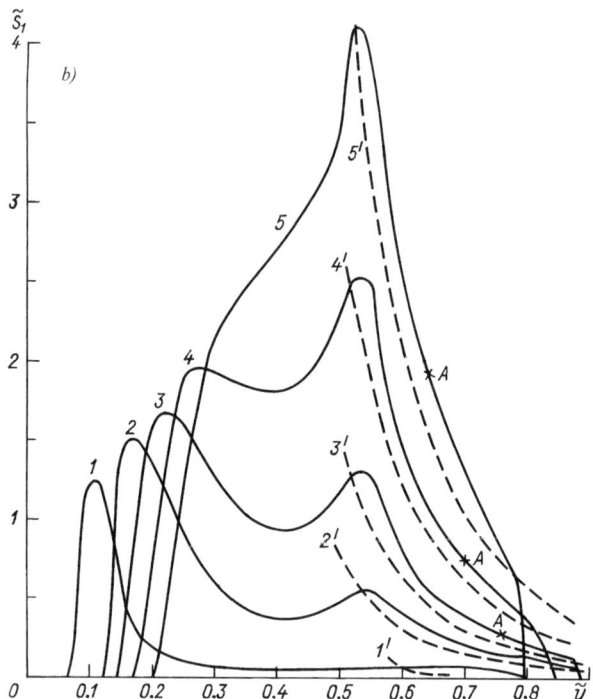

Fig. 5.7. (b) Wave spectra evolution in countercurrent with $\nu_m = 0.25$, $q = 10$ and $\delta_0 = 0.01$ at different speed values: 1 – 0.1; 2 – 0.15; 3 – 0.175; 4 – 0.20; 5 – 0.225 (where the *solid line* is the spectrum $\tilde{S}_1$, and the *dashed line* is the equilibrium interval)

ther increase of the spectrum $\tilde{S}_1$, especially near the second peak, as well as the approach of spectral maxima, are observed in this case. The second spectral maximum becomes much greater in comparison with the first one. The consecutive spectral evolution, shown for the indicated values of ν, is in sufficiently good agreement with the observation data (see Fig. 5.5). The dimensional frequency maximum $\sigma_1 = (1.20 - 1.45)\sigma_m$ of the first spectral peak remains practically unchangeable whereas the spectral density value is essentially changed. A similar conclusion is made by Barenblatt et al. (1985) based on experimental data analysis. However, it should be noted that the maximum speed ν_m does not exceed 0.2 in the experiment.

Both spectral peaks continue to draw together, merging into one with further wave propagation into an area with the larger current speed ν (see Fig. 5.7a, curve 5) increasing up to its maximum value. The second peak becomes much greater than the first one, "absorbing" the first spectral maximum. At the point $\nu = \nu_m$ the maximum of the spectrum $\tilde{S}_1$ is equal to 3.9 at the non-dimensional frequency $\tilde{y}_1 = 0.5$, corresponding to the frequency $\sigma_1 \approx 2\sigma_m$. It should be noted that the second maximum frequency σ_2 of the spectrum $\tilde{S}_1$ varies during the evolution within a considerably greater range from $6.5\sigma_m$ to $2\sigma_m$.

The dot-and-dash line (lines $1'$–$5'$) denotes the spectrum $\tilde{S}_2$ of waves passing from the area of maximum current speed ("barrier") to the area with current speed being the same as in case of the spectrum $\tilde{S}_1$. Unlike $\tilde{S}_1$, there is only one peak in the spectrum $\tilde{S}_2$. This coincides completely with the spectrum $\tilde{S}_1$ being then cut off to zero in the small-frequency area. This cut-off effect follows from the kinematic conditions for waves being blocked by the current speed and not able to overcome the "barrier". But a sharply expressed cut-off effect is hardly possible in field conditions. It appears due to neglecting the effect of non-linear energy transfer in the wave spectrum and the wind input. The wind input would have necessarily resulted in additional generation of high-frequency wave components and, accordingly, in a smoother spectrum decrease at these frequencies. However, rapid spectrum decrease is really observed within the given frequency range (curve 1 in Fig. 5.5) in the experiment (Barenblatt et al., 1985). It confirms the obtained solution qualitatively.

It should be noted that Curves 1–5 for the spectra $\tilde{S}_1$ correspond to the consecutive stages of spectra evolution with a wave propagating in the direction of the countercurrent speed increase, while Curves $1'$–$5'$ for the spectrum $\tilde{S}_2$ reflect the reverse situation, i.e. the spectrum $5'$ precedes the spectrum $1'$. In this case the waves move in the area with decreasing countercurrent speed. As a result, the wave energy is absorbed by the radiation tensions, and the spectrum $\tilde{S}_2$ is decreased. The wave intensity behind the barrier (at $x > x_m$) is sharply decreased attaining a more regular character. The waves become longer and gentler.

The spectrum cut-off location due to the blocking of the high-frequency component is marked by the point A in the spectrum $\tilde{S}_1$, following from the solution (5.25). The spectrum solution follows from the initial assumptions, as described for the aforesaid spectrum $\tilde{S}_2$. The point A moves to the high-frequency area with increasing maximum current speed ν_m. It results in "filling up" the spectrum with components, without being blocked at smaller values ν_m.

The Phillips equilibrium interval normalized by the maximum of the initial spectrum $S_R = (\nu/\tilde{y})^{-5} \exp\left(\frac{n+1}{n}\right)/\delta_0$ is marked with a dashed line in Fig. 5.7a (Curves $1''$–$5''$). As can be seen, the second spectral maximum and the spectrum of high frequencies are 2–3 times higher than the equilibrium interval value. It should be noted that the notion of the equilibrium interval was introduced by Phillips (1980) as some ultimate spectrum state, controlled by wave breaking in the presence of wind input. If the wave generation mechanism were connected not with the wind, but with the wave interaction with a horizontally non-uniform current, the ultimate spectrum state would not be supposed to coincide precisely with the Phillips equilibrium interval. The level of the spectral high-frequency components can be diminished significantly in the framework of this model. In order to do this, it is necessary to take a much greater value q instead of $q = 2$, which is typically connected with the quadratic energy dissipation in wind wave models. Thus, the dissipation effect becomes greater for $q = 10$ in the model (see Fig. 5.7b). The spectrum excess over the equilibrium interval is no more than 25 per cent. The second spectrum maximum is decreased compared with the previous case, and its frequency is slightly displaced to the low-frequency area.

Physical explanation of rips. Taking into consideration the aforementioned results, the following explanation of the origin of rips could be suggested. Waves propagating in the horizontally non-uniform current are increased in height and decreased in length. The blocking process begins for the shorter and, later on, for longer waves. At the same time reverse waves appear, and they are carried back and down stream. The free surface presents a superposition of two systems: straight and reverse waves. The height of the reverse waves is sharply increased while their length is decreased. The wave action density is preserved and the energy and the σ frequency are increased before breaking occurs. The crest wave breaks after the wave steepness has exceeded a definite ultimate value. The wave amplitude is sharply decreased, and the wave itself is carried away by the current with its height increasing and the length decreasing until the next breaking.

The balance can take place in the spectrum at a larger spectral density level compared to ordinary wind waves due to the efficient current influence on waves. It should be recalled that the mean steepness of wind waves is $1/36$, while the ultimate wave steepness can exceed $1/7$. The number of waves with a steepness near to the critical one is much larger in the spectrum of rips, resulting in strong water surface instability. In this case, the water surface presents itself as a whirlpool, i.e. a large quantity of relatively short steep waves subjected to intensive breaking.

In conclusion, it should be noted that the main feature of the evolution of the wave spectrum $\tilde{S}_1$, propagating in the increasing countercurrent, is the appearance of a second spectral maximum, its rapid growth and shift of the maximum frequency to the low-frequency area. This second maximum is caused by reverse waves, appearing as a result of the straight wave blocking in a current. The wind intensity increase would result in smoothing over the described effects. It would be due to two reasons: firstly, due to the decrease of the spectral maximum frequency $\sigma_{\max}$, i.e. decreasing the current non-dimensional speed ν_m and, secondly, increasing the energy level in the spectral maximum and in the range of high frequencies. This leads to an increase in the parameter δ_0 (see Fig. 5.6).

There is not only qualitative, but also quantitative agreement between the obtained theoretical solution with full-scale rip spectral observations (Barenblatt et al., 1985; Leykin & Monin, 1985). At the same time it is possible to describe with the help of the model a wave evolution not observed in the experiment due to a relatively small maximum current speed. This refers, pri-

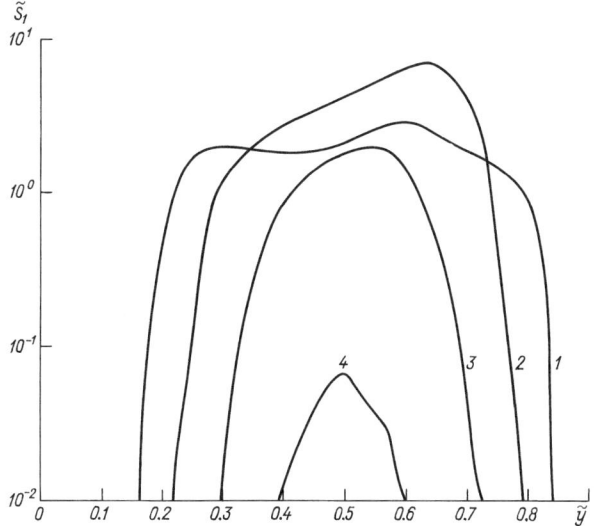

Fig. 5.8. Wave spectra evolution (on logarithmic scale) with $\nu_m = 0.5$ at different values of current speed ν: 1 – 0.225; 2 – 0.25; 3 – 0.30; 4 – 0.35

marily, to the phenomenon of the convergence of two spectral maxima when $\nu_m > 0.2$, which is theoretically described above.

This spectral convergence continues with further increase of the current speed ν_m and wave propagation to a larger speed area, as shown for the maximum current speed $\nu_m = 0.5$ in Fig. 5.8. The wave spectrum is seen to attain a symmetrical form relative to $\tilde{y} = 0.5$. It is decreased in width and, especially, in its value, while the latter is changed by two orders of magnitude. The low-frequency spectral components penetrate the area of higher current values compared to the components of the spectral maximum vicinity. There is an analogous phenomenon of "under-barrier leakage", described in quantum mechanics (Landau & Lifshits, 1974).

5.4 Estimation of Non-Linear Wave Interaction in the Rip Spectrum

Statement of the problem. Although the predominant influence of a horizontal non-uniform current on rip formation is shown in the previous section, the problem of the non-linear wave interaction in the rip spectrum remains open. There is a hypothesis (Leykin & Monin, 1985) that a "soliton gas" is suitable as a model of such wave field formation, i.e. a system of solitons with different propagation directions and amplitudes, which do not interact, but simply superimpose on each other. The authors are of the opinion that it is possible to approximate the spectral experimental data with the help of the Kolmogorov spectrum $S \approx g^{4/3}\sigma^{-11/3}$. This allows them to put forward the suggestion that small-scale rip excitations and their further enlargement are a result of non-linear wave action transfer to low frequencies up to the limits of the Phillips breaking. However, this explanation of rip formation seems to be doubtful, when the Kolmogorov spectrum is obtained for the transparency interval in the wind wave spectrum (Zakharov & Zaslavskii, 1982, 1983a,b). The influence of the non-uniform current on waves occurs over the entire frequency range; that is why the suggestion about the existence of the transparency interval in the rip spectrum can hardly be fulfilled in reality.

In addition, the asymmetrical form of rip waves ("sharpened crests and gentle troughs" as described by observers) and their intensive breaking ("boiling water") indicate the role of strong non-linear effects and dissipation over a wide range of spectral frequencies. These facts are confirmed by the derived solution (5.25), with the parameter q, determining the spectrum in the high-frequency range. It should be remembered that the non-linear degree of the initial equation (5.19) is described by the parameter q. As shown, the spectral solution $\tilde{S}_1$ corresponds more to field data with increasing parameter q. This indicates the significance of strongly non-linear effects on the formation of the high-frequency spectrum range. But the role of non-linear wave interaction in forming the rip spectrum remains open.

Formation of two-dimensional rip spectrum. It seems that the non-linear wave energy transfer could be estimated using the frequency spectrum approximation derived in the previous section. But in order to do this, it would be necessary to get not only a one-dimensional frequency spectrum, but also the spatial or frequency-angular spectrum $S(\sigma, \beta)$, which is not obtained in the experiment (Barenblatt et al., 1985). It is possible to obtain its analytical solution using (5.19)–(5.22), with the parameter q tending to infinity. Practically this means that the spectral density value is assumed to be equal to the Phillips equilibrium interval in the high-frequency area.

Thus, using (5.19) of the wave action spectral density $N(\mathbf{k})$ when $N \ll N_\infty$, the spectral density of the energy S, depending on the frequency σ and the angle $\beta = \arctan(k_y/k_x)$ can be easily found:

$$S(\sigma, \beta) = S_0(\sigma_0, \beta_0) \frac{\sigma k}{\sigma_0 k_0} \frac{\partial k}{\partial \sigma} \left(\frac{\partial k_0}{\partial \sigma_0} \right)^{-1}, \quad (5.27)$$

where $\sigma_0 = \sigma\left(1 - \tilde{y}\cos(\beta)\right)$; $\beta_0 = \arcsin[(\sigma/\sigma_0)^2 \sin(\beta)]$; and $S_0(\sigma_0, \beta_0)$ is the initial wave energy spectral density in the area with no current. Its angular energy distribution function is assumed to be proportional to cosine squared.

The frequency ω and the wave vector component k_y are preserved along the ray in the case of wave propagation from the area without any current towards a non-uniform stationary current, whose velocity is changed along its direction. In this case, the solution (5.27) can be written in the form:

$$S(\sigma, \beta) = S_0 \left[\sigma\left(1 - \tilde{y}\cos(\beta)\right) \right] \left(1 - \tilde{y}\cos(\beta)\right)^{-2}. \quad (5.28)$$

5.4 Estimation of Non-Linear Wave Interaction in the Rip Spectrum

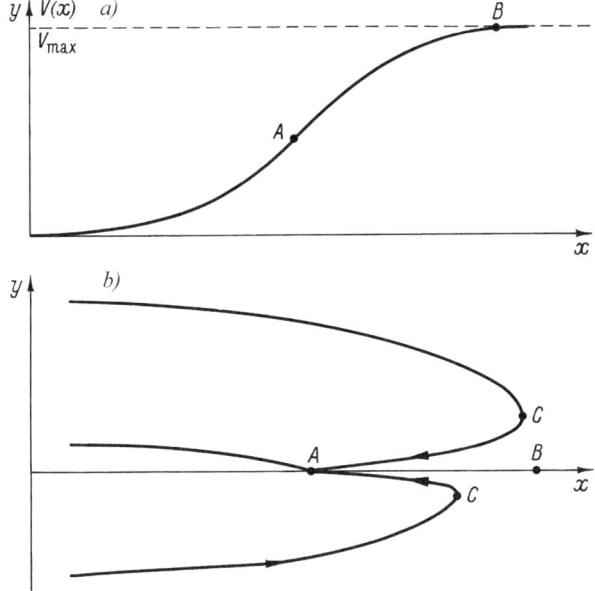

Fig. 5.9. Wave packet trajectories in countercurrent: (a) current velocity profile, B is the limiting point with $V = V_{\max}$; (b) wave packet trajectories propagating to point A; C is the turning point

The angular wave distribution is narrowed in the countercurrent due to the wave kinematic parameter transformation. The following inequality is fulfilled:

$$\cos^2(\beta_0) = 1 - \frac{\sin^2(\beta)}{(1 - \tilde{y}\cos(\beta))^4} \geq 0 . \qquad (5.29)$$

In order to create a full spectrum, it is necessary to "collect" all the rays incoming at the estimated point. But not all initial wave packets are able to reach the considered point. The trajectories of wave packets on the background of the countercurrent velocity (see Fig. 5.9a) are shown in Fig. 5.9b. Two types of rays can be distinguished. The first of them leaving the initial boundary reaches the calculated point with a positive value of the group velocity component:

$$C_{gx} = 0.5\sqrt{g/k}\cos(\beta) + V \geq 0 .$$

These are straight waves propagating upstream. In addition, the same point can be reached by packets of reversed waves ($C_{gx} < 0$). The two-dimensional area of changes in the variable $\{\tilde{y}, \beta\}$ for the case of a countercurrent ($V < 0$) is shown in Fig. 5.10. The curve II corresponding to the condition $C_{gx} = 0$ divides the integrating area into two parts, with $C_{gx} > 0$ or $C_{gx} < 0$. When the maximum velocity of the chosen current profile (see

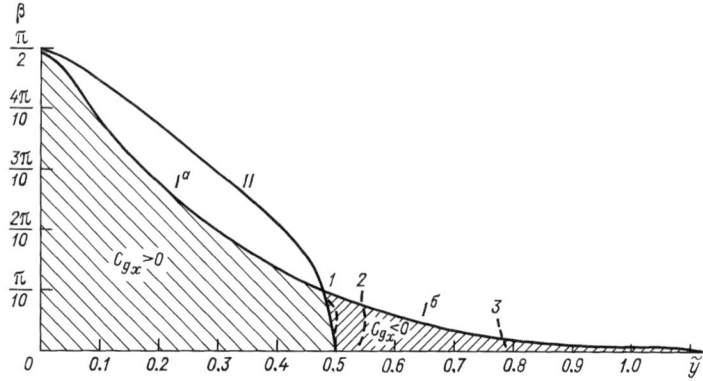

Fig. 5.10. Integration area limited by curve I, according to the transformation of angular distribution function (5.29); II – line with a group velocity turning to zero $C_{gx} = 0$. Dotted lines denote right-hand boundaries of the integrating area corresponding to $\gamma = 0.995$ (1); $\gamma = 0.95$ (2); $\gamma = 0.67$ (3)

Fig. 5.9a) is limited $|V_{max}| < \infty$, there are some straight waves, passing the given point, which are not reflected by the countercurrent. They do not create any reverse waves. The condition for the absence of these waves at the given point can be written with the help of the results (see Sect. 5.2) as following:

$$0.5\cos(\beta) - \tilde{y} \leq 0 ; \tag{5.30}$$

$$0.5\gamma\sqrt{(1 - \sin^2\beta_A)\sin(\beta_A)/\sin(\beta)} - \tilde{y} \geq 0 , \tag{5.31}$$

where $\gamma = V/V_{max}$, $(0 \leq \gamma \leq 1)$ and the angle β_A is determined in accordance with the ratio (5.14).

The conditions (5.30) and (5.31) cut off an area of the permissible function argument variations (see Fig. 5.10), depending on the ratio of the velocities γ. Taking into account the kinematic conditions, the wave spectrum can be presented in a current in the following form:

$$S_1(\sigma, \beta, V) = \frac{2V}{\pi g}\nu^n m_0(n+1)(1-\mu^2) \tag{5.32}$$

$$\times \exp\left\{-\frac{n+1}{n}\left[\frac{\nu}{\tilde{y}(1-\tilde{y}\cos\beta)}\right]^n\right\}(1-\tilde{y}\cos\beta)^{-n-5}$$

$$\times \tilde{y}^{-n-1}\Theta(1-\mu^2)\left\{\Theta\left(\frac{1}{2}\cos\beta - \tilde{y}\right)\right.$$

$$\left. + \Theta\left(-\frac{1}{2}\cos\beta + \tilde{y}\right)\Theta\left(\tilde{y} - \frac{1}{2}\gamma\sqrt{\frac{1-\sin^2\beta_A}{\sin\beta/\sin\beta_A}}\right)\right\},$$

where $\mu = \sin\beta\,(1 - \tilde{y}\cos\beta)^{-2}$.

5.4 Estimation of Non-Linear Wave Interaction in the Rip Spectrum

This expression presents the angular spectrum distribution narrowing in a countercurrent and the presence of its two-peak structure connected with the existence of straight and reverse waves reflected by blocking.

The estimation of wave breaking in the countercurrent can be introduced using the Phillips equilibrium interval. If the spectral density is assumed to be greater than the equilibrium interval as an ultimate value, the spectrum is taken to be equal to the equilibrium interval value. However, the equilibrium interval is determined by the angular distribution accuracy. It is quite natural to assume it to be equal to the similar value of ordinary wind waves. The angular distribution is assumed to expand in the equilibrium area due to the tendency for spectrum isotropy connected with wave breaking. According to Davidan et al. (1985), the equilibrium interval is approximated by the quadratic cosine formula ($\approx \cos^2(\beta)$). Thus, a two-dimensional wave spectrum in a current is described by the ratio:

$$S(\sigma, \beta, V) = \min\{S(\sigma, \beta, V) \, ; \, S_R(\sigma, \beta)\} \, , \tag{5.33}$$

where

$$S_R(\sigma, \beta) = \begin{cases} \alpha_F g^2 \sigma^{-5}(2/\pi)\cos^2(\beta) & \text{with } |\beta| \leq \pi/2 \\ 0 & \text{with } |\beta| > \pi/2 \end{cases} \tag{5.34}$$

The evolution of the frequency-angular wave spectrum $S(\sigma, \beta)$, integrated over the directions and normalized by the maximum of its initial value in the absence of current is shown in Fig. 5.11:

$$\tilde{S}(\sigma/\sigma_{\max}) = S_0^{-1}(\sigma_{\max}) \int_{-\pi}^{\pi} S(\sigma, \beta, V) \, \mathrm{d}\beta \, . \tag{5.35}$$

The values of the determining parameters in (5.35) are the same as in the previous section. The graphs of the spectra along the stream at points of different current velocities ν (see Fig. 5.4 for the area $x < x_m$) are given in Fig. 5.11. The relative value $\sigma/\sigma_{\max} = \tilde{\sigma}$ is plotted along the horizontal axis, often used in the presentation of the spectrum rather than in the non-dimensional frequency $\tilde{y}$. There is no symmetrical form in the spectrum relative to the central frequency $\tilde{y} = 0.5$ (see Fig. 5.7a), as soon as the point $\tilde{y} = 1$ tends to infinity in these variables. In this case the spectrum $\tilde{S}(\tilde{\sigma})$ can be compared to the experimental value presented in Fig. 5.5. There is a two-peak structure in the spectrum $\tilde{S}(\tilde{\sigma})$, as shown in Fig. 5.11. An increase of the wave energy spectral density takes place in waves propagating to the larger countercurrent velocity area, whereas the spectral density in the vicinity of the high-frequency spectral maximum is increased to a greater extent. It is shifted to the low-frequency area. At the same time the low-frequency maximum is shifted to higher frequencies to a lesser extent. Both maxima converge into one at $\nu > 0.2$.

180   5 Wave Evolution in Non-uniform Currents in Deep Water

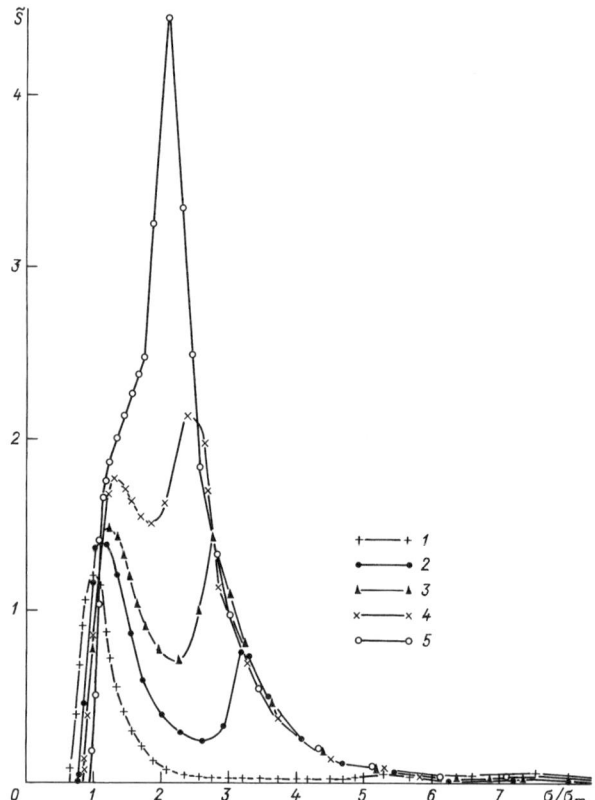

Fig. 5.11. Spectra evolution in countercurrent with $\nu_m = 0.25$ at points with different velocities ν: 1 – 0.10; 2 – 0.15; 3 – 0.175; 4 – 0.20; 5 – 0.225

Numerical estimation results of non-linear energy transfer in the rip spectrum. In order to estimate numerically the non-linear energy transfer in the rip spectrum in the form (5.33) the algorithm described in Sect. 4.1 can be used. Numerical estimations of the non-linear energy transfer (4.1) are integrated over the directions for different stages of spectrum development. The results for the following current points: $\nu = 0.10$; $\nu = 0.15$; $\nu = 0.175$; $\nu = 0.20$; $\nu = 0.225$ are shown in Fig. 5.12.

The function value G_{nl} (see Fig. 5.12) is normalized in the following way:

$$\tilde{G}_{\mathrm{nl}}(\tilde{\sigma}) = G_{\mathrm{nl}}(\sigma) \left(S^3(\sigma_{\max})\sigma_{\max}^{11}/g\right)^{-1}, \qquad (5.36)$$

where $\tilde{\sigma} = \sigma/\sigma_{\max}$.

The form of the function $\tilde{G}_{\mathrm{nl}}$ is similar to the non-linear transfer functions in the wind wave spectrum. A positive energy transfer is mainly observed at frequencies smaller than the second spectrum maximum. There is a negative function value, becoming positive again for larger frequencies. The function

5.4 Estimation of Non-Linear Wave Interaction in the Rip Spectrum

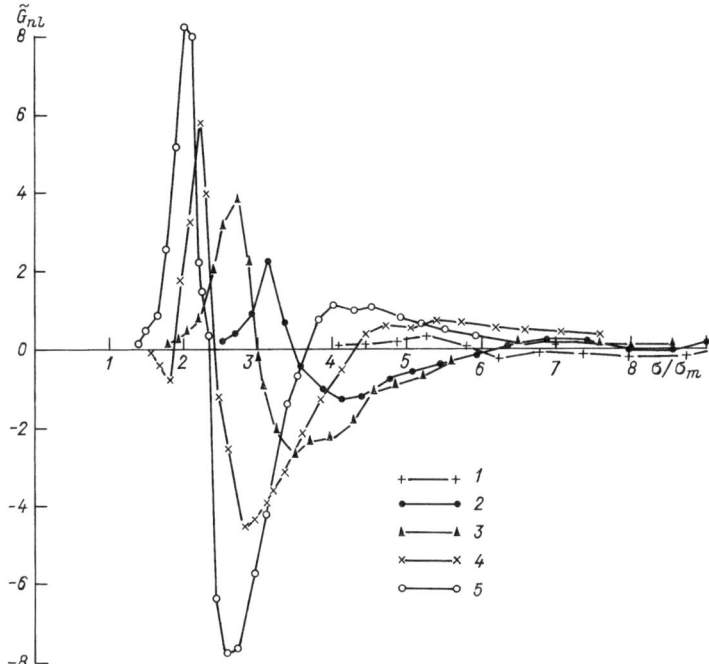

Fig. 5.12. One-dimensional function of non-linear transfer in wave spectrum at different velocity points ν: 1 – 0.10; 2 – 0.15; 3 – 0.175; 4 – 0.20; 5 – 0.225

shows that non-linearity creates a tendency for a shift of the second spectrum maximum to the low-frequency area due to energy outflow from the spectrum at high frequencies. The value $\tilde{G}_{nl}$ is increased with developing waves. Its maximum and minimum values are shifted to low frequencies simultaneously with the second spectrum maximum. The renormalized function G_{nl} by the value $S^3(\sigma'_{max})\sigma'^{11}_{max}/g$ becomes more stable, with σ'_{max} being the frequency of the second high-frequency spectrum maximum. It is rather weakly dependent on the wave development stage.

It should be noted that the function value in the vicinity of the first low-frequency maximum is smaller by 3–4 orders of magnitude in comparison with the second maximum for the initial stage of $\nu < 0.175$. This proves that the non-linear energy transfer does not practically influence the formation of the first maximum spectrum.

The convergence of the spectral maxima continues with further wave development $\nu = 0.20$–0.225. The area ($\tilde{\sigma} < 1.75$) with the non-linear transfer becoming negative appears at frequencies smaller than the second maximum frequency at $\nu = 0.20$. This non-typical phenomenon of the function $\tilde{G}_{nl}$ shows energy pumping from this area to the second spectrum maximum, i.e. in "the opposite direction". This is connected with the influence of the first frequency maximum on the non-linear energy transfer. In this case there is a tendency to make the spectral form smoother. The negative area disappears with the closing of further spectral maxima ($\nu = 0.225$) and their merging into one. The two-dimensional function $\tilde{G}_{nl}(\tilde{\sigma}, \beta)$ for different directions ($\beta = 0;\ 15;\ 30;\ 45;\ 60°$) is shown in Fig. 5.13 in order to analyse values of the low-frequency function $\tilde{G}_{nl}$. The negative area of the non-linear transfer at low frequencies is typical only for directions close to the general one $\beta = 0°$. There are only positive values of the function $\tilde{G}_{nl}(\tilde{\sigma}, \beta)$ in other directions of the area. At the same time the function $\tilde{G}_{nl}(\tilde{\sigma}, 0)$ changes its sign before the negative area in the general direction. There is a positive maximum at frequency $\tilde{\sigma} < 1.3$ somewhat to the left of the spectral frequency maximum.

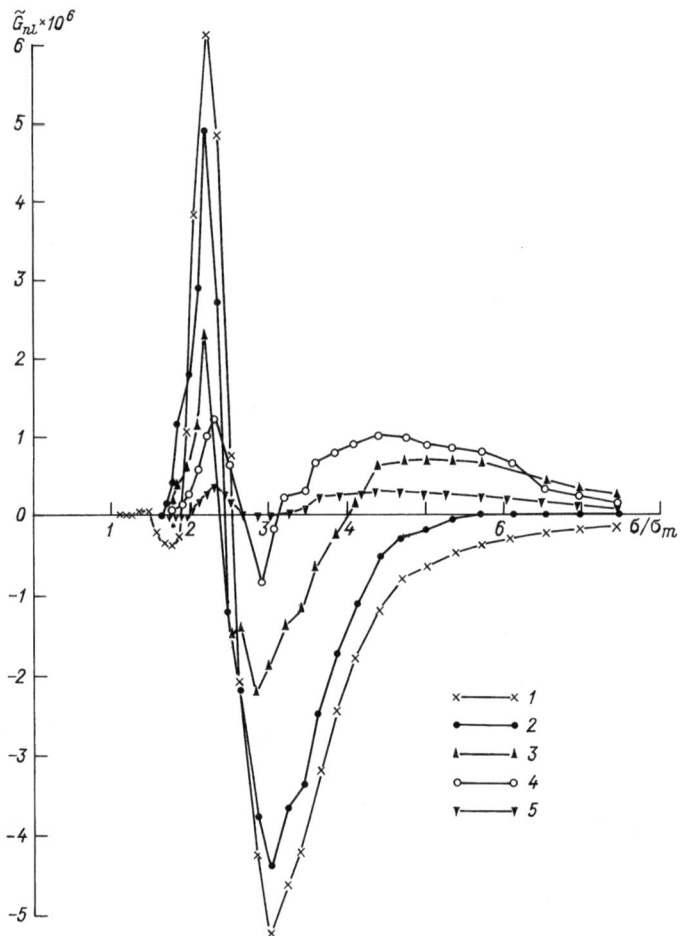

Fig. 5.13. Two-dimensional non-linear energy transfer function at the point $\nu = 0.20$ for different angle directions β: $1 - 0°$; $2 - 15°$; $3 - 30°$; $4 - 45°$; $5 - 60°$

Similar calculations performed for a narrower angular distribution $\left[\approx \cos^4(\beta)\right]$ in the equilibrium spectrum area reveal the negative energy outflow at low frequencies. This becomes more intensive and exists not only in the general direction $\beta = 0°$, but also in the direction $\beta = 15°$. Thus, the statement of Leykin & Monin (1985) that the rip wave spectrum is formed as a result of its small-scale excitation and subsequent enlargement due to non-linear wave action transfer to low frequencies is not confirmed even for the area of the second frequency maximum. As can be seen, non-linear transfer is of a more complicated character. It can also transfer wave action in the opposite direction at some frequency ranges.

In conclusion it should be noted that the rip spectrum can be conventionally divided into two areas, taking into account the peculiarities of the non-linear energy transfer, namely, the areas of the first and second maxima. The intensity of the non-linear energy transfer in the area of the energy-carrying frequency range of the second maximum is greater than its value in the vicinity of the low-frequency one. For this reason non-linear energy transfer can be considered as one of the most efficient spectrum formation mechanisms in the vicinity of the second maximum. Practically, it does not influence low-frequency spectrum formation. The spectrum evolution can be considered to be a simple superposition of independent harmonics subjected to the influence of a horizontally non-uniform current. The rip formation in this frequency range can be described with the help of the "soliton gas" notion. As soon as two frequency spectrum peaks are merged, a change of character of the non-linear energy transfer takes place. Being still negligible in the low-frequency area, a negative energy flux can be observed in the frequency range between the spectral peaks, testifying the energy being pumped in the opposite direction, i.e. from the lower frequencies to the higher ones.

5.5 Wave Element Transformations in Current, Varying Along its Direction

At present there are in fact no well-grounded practical recommendations for estimating wave parameters in a non uniform current. The results of Longuet-Higgins and Stewart, obtained in the early 1960s (1960, 1961, 1962, 1964) are often used to estimate the current effect on waves. It should be remembered that the wave energy balance equation, taking into account the radiation stresses due to the influence of current non-uniformity, is deduced in these papers. A wave height change in the one-dimensional case of regular wave propagation in deep water from the area without current ($V_0 = 0$) to the area with the countercurrent ($V < 0$) or the fair current ($V > 0$) is described by the formula (Longuet-Higgins, Stewart, 1964):

$$\frac{h}{h_0} = \frac{c_0}{\sqrt{c(c+2V)}}, \qquad (5.37)$$

where h_0 is the initial wave height; h is the wave height in the current V; and c and c_0 are the corresponding wave phase velocities. Using the frequency preservation conditions, valid for the stationary current velocity, the phase velocity c can be written in the following form:

$$\frac{c}{c_0} = \frac{1}{2}\left[1 \pm \sqrt{1 + 4V/c_0}\right]. \tag{5.38}$$

The ($\pm$) sign in this relation indicates the ambiguity of determining the wave parameters in the current V. The variations of the height ratio h/h_0 and the length ratio λ/λ_0 are shown in Fig. 5.14. The change of wave parameters corresponding to the (+) sign is designated as J_1, and to the (−) sign as J_2. The evolution of the so-called straight wave with height $h^{(+)}$, increasing monotonically with countercurrent velocity, is described by the

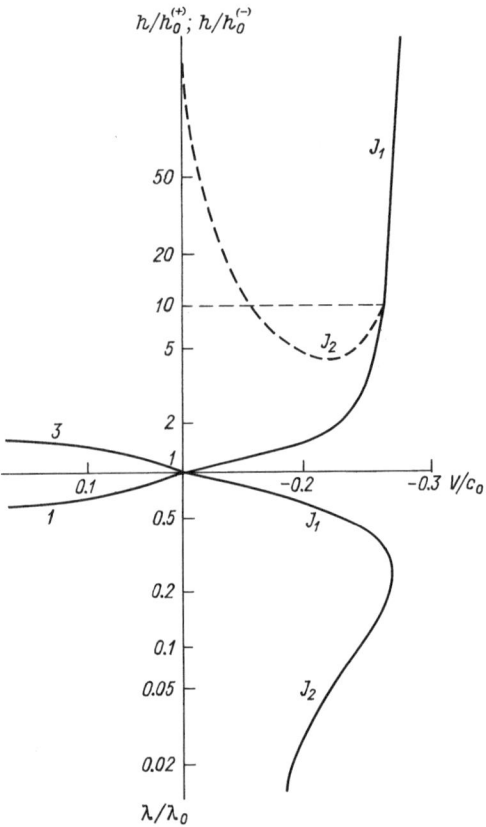

Fig. 5.14. Variation of height h/h_0 and length λ/λ_0 of a monochromatic wave depending on the counterflow velocity: J_1 is the transformation of straight wave components; J_2 is the transformation of reversed wave components; V/c_0 is the current velocity normalized by the initial phase speed

5.5 Wave Element Transformations in Current, Varying Along its Direction

curve J_1. It takes an infinitely large value (singularity) at the special point $(V/c_0 = -1/4)$. The straight wave length is decreased in a countercurrent, being four times less at the special point than the initial value λ_0. There is also a change of the wave parameters, corresponding to the $(-)$ sign in the formula (5.38), i.e. for a reverse wave. It is described by the curve J_2, which is also of infinitely large value at the point $V/c_0 = -1/4$.

The second singularity appears in the ratio h/h_0 for a reverse wave with decreasing countercurrent velocity. Its wavelength tends to zero (although, in reality, the capillarity does not allow the wave number to increase up to infinity as $V \to 0$). The wave height h/h_0 is decreased in a fair current, while the wavelength λ/λ_0 is increased.

It is necessary to note that for a long time different researchers used the relation (5.38) only with the $(+)$ sign. In this case the reverse waves were not taken into consideration in the countercurrent. The ambiguity of this expression was considered to be purely formal, whereas the $(-)$ sign was supposed to have no physical sense. However, reverse waves really do exist, as shown by the experimental (Pokazeyev & Rozenberg, 1983) and theoretical (Basovich & Bakhanov, 1979; Lavrenov, 1986, 1998; Peregrine, 1976) studies. These are waves reflected from a non-uniform current and carried away downstream. The relation (5.37), applied near a special point (caustic), admittedly leads to a wrong result, as soon as there is a violation of the Longuet-Higgins and Stewart theory.

In order to solve the problem the idea of using a spectral approach was proposed by Lavrenov (1986, 1988b). The approach eliminates the aforementioned singularity and produces an adequate wave field description.

A regular monochromatic wave discussed above is an idealization that does not occur in natural conditions. Usual wind waves consist of different spectral components, corresponding to different wave vectors $\boldsymbol{k} = \{k_x, k_y\}$. That is why there can exit a separate caustic for every component $\boldsymbol{k} = \{k_x, k_y\}$. Thus, the entire horizontal plane $\boldsymbol{r} = \{x, y\}$ could be covered with caustics. The statistical averaging of different waves results in a smoothing caustic. Thus, there is no need to introduce any corrections in the geometrical optics approximation for accurate wave field estimation (Krasitskii, 1974).

The kinetic equation (5.1) in the form (5.19) will be used to describe the wave evolution in a current. Unlike in Sect. 5.2, the wave number k (or the frequency σ) and the angle β will be used instead of the variables ω and β, because the dependence of the wave number k (or frequency σ) on them is ambiguous. Implementation of these variables avoids the aforementioned shortcoming. There is no singularity in the spectral energy density $S(k, \beta)$, occurring in (5.9). Furthermore, the variables k and β make it possible to use easily in the estimations the idea of Phillips about the invariance of the equilibrium interval (Kitaigorodskii et al., 1975).

The propagation of waves in deep water from an area where the current is absent, to a non-uniform stationary current will be considered with the help of (5.19). The current velocity is considered to be directed along the Ox axis and increases monotonically in the same direction up to some maximum value $V_{\max}$. It follows from (5.2) that the frequency ω and the wave vector component k_x are preserved along the ray. According to Sect. 5.4, the solution of the spectral energy density S, depending on the frequency σ and the angle $\beta = \arctan(k_y/k_x)$, can be found in the form (5.32).

The mean statistical elements of waves, depending on the current velocity V, can be easily obtained with the help of the solution (5.32). In particular, it is sufficient to integrate the spectral value $S(k,\beta)$ within the corresponding range of the argument variations to estimate the statistical moments m_{pq}:

$$m_{pq} = \int_0^\infty \int_{-\pi}^{\pi} S(k,\beta) k^{p+q} \cos^p(\beta)\, dk\, d\beta \,. \qquad (5.39)$$

In order to integrate (5.39), the spectrum $F(\tilde{y},\beta) = S(\sigma,\beta,V)(d\sigma/d\tilde{y})$ will be used. This is a function of the non-dimensional variable $\tilde{y}$ and it depends on the parameters ν, γ, n.

The change of mean wave components in a current will be estimated in the absence of reverse waves, using the spectrum (5.32). Such a situation is observed, firstly, in a fair current and, secondly, in a countercurrent, within the area, where the current velocity is no longer changed ($\gamma = 1$) (see Fig. 5.9a). The evolution of the wave parameter in the countercurrent transitional segment ($0 < \gamma < 1$) is to be considered after that.

The relative mean wave height in a current can be defined using the zero moment (5.39). It can be presented in the following form:

$$\frac{h}{h_0} = \left[\frac{2\pi}{m_0} \int_0^\infty \int_{-\pi}^{\pi} F(\tilde{y},\beta)\, d\beta\, d\tilde{y} \right]^{\frac{1}{2}} . \qquad (5.40)$$

Similarly, the mean wavelength can be written as:

$$\frac{\lambda}{\lambda_0} = \nu^2 \left(\frac{n+1}{n} \right)^{\frac{2}{n}} \sqrt{\Gamma\left(1 - \frac{4}{n}\right)} \qquad (5.41)$$

$$\times \left[\int_0^\infty \int_{-\pi}^{\pi} F(\tilde{y},\beta)\, d\beta\, dy \middle/ \int_0^\infty \int_{-\pi}^{\pi} y^4 F(\tilde{y},\beta)\, d\beta\, dy \right]^{\frac{1}{2}} ,$$

where $\Gamma(n)$ is the gamma function.

The ratio of the mean wave periods can be presented analogously:

$$\frac{\tau}{\tau_0} = \nu \left(\frac{n+1}{n} \right)^{\frac{1}{n}} \sqrt{\Gamma\left(1 - \frac{2}{n}\right)} \qquad (5.42)$$

$$\times \left[\int_0^\infty \int_{-\pi}^{\pi} F(\tilde{y},\beta)\, d\beta\, dy \middle/ \int_0^\infty \int_{-\pi}^{\pi} y^4 (1 \mp \tilde{y}\cos(\beta))^2 F(\tilde{y},\beta)\, d\beta\, dy \right]^{\frac{1}{2}} .$$

5.5 Wave Element Transformations in Current, Varying Along its Direction

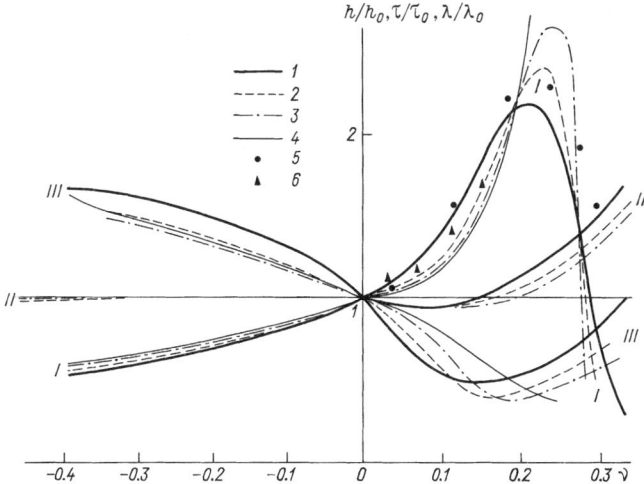

Fig. 5.15. Relative change of wave height h/h_0 (I), period τ/τ_0 (II) and length λ/λ_0 (III) for different values of parameter n: 1 – $n = 5$; 2 – $n = 8$; 3 – $n = 10$; 4 – $n = 15$; 5 – full-scale observations of Zhevnovatiy (1971) for h/h_0, 6 – Scripps' institute data

It can be seen from the obtained integral ratios (5.40)–(5.42) that the values: h/h_0, λ/λ_0 and τ/τ_0 are functions of the non-dimensional parameters ν and n, giving them a more universal character. The functions h/h_0, λ/λ_0 and τ/τ_0 are not expressed in analytical form. Their estimations are made numerically.

The numerical results for counter- ($\nu > 0$) and fair ($\nu < 0$) currents are shown in Fig. 5.15. The non-dimensional current velocity, i.e. the parameter ν, is plotted along the horizontal axis. The relative mean parameter values (5.40)–(5.42) are shown along the vertical axis. The calculations are made for different values of the parameter n, equal to 5, 8, 10 and 15, respectively, determining the spectrum width. As shown in the nomogram, the wave heights in a current are increased monotonically up to some maximum value defined by the parameter n. Thus, with $n = 5$ the maximum wave height is increased up to 2.13 in a countercurrent, achieving it's maximum at the point $\nu = 0.21$. After that the height is decreased. The wave energy is practically equal to zero in the range $\nu > 0.37$ due to the large value of the countercurrent velocity.

The wave height increases more slowly at small ν for the parameter values $n = 8$, 10 in comparison with that for $n = 5$. However, with further increase of the parameter ν the wave height is increased to a greater extent than for $n = 5$. The height maximum is achieved at larger values ν, with the wave height being sharply decreased after that. The transformation of the wave heights with $n = 15$ practically coincides with the solution of (5.37) and (5.38), obtained for a monochromatic wave.

At first the wavelength is decreased to some value with increasing current velocity. Thus, the minimum value of the relative wavelength λ/λ_0 is equal to 0.42 for $n = 5$. This is achieved at the point $\nu = 0.12$, and it is increased monotonically after that. It should be remembered that for a monochromatic wave (see Fig. 5.14), only the wavelength decrease takes place in a countercurrent. The monotonic increase obtained in our calculations is due to the spectral wave character. A specific filtration takes place in a countercurrent, resulting in elimination of short waves and penetration of only long waves to the larger current velocity areas.

As for the wave period, its value is slightly decreased at small ν and then begins to increase. Thus, the period is estimated as $\tau/\tau_0 = 1.8$ for $n = 5$, at $\nu = 0.35$. At the same time it should be remembered that the period is not changed for monochromatic waves.

Hence, the transformation of mean wave elements in a current is principally changed by the spectral wave structure. With the spectrum being narrowed, i.e. the parameter n being increased, the change of mean wave elements is more similar to the classic solution for monochromatic waves (5.37) and (5.38). For natural waves, characterized by values of the parameter $n = 4$–8, the change of mean wave elements is principally different from monochromatic wave elements.

The wave height is smoothly decreased in the fair current ($\nu < 0$), while the period remains constant and the wavelength is increased. There is a smaller influence of the fair current on the change of mean wave elements, compared to the countercurrent. In this case the spectral wave structure is not so important as in a countercurrent.

The full-scale observational data obtained by the Scripps Institute of Oceanography (USA), as well as by Zhevnovatiy (1971) in 1968–1969 in the White Sea onboard ships of the USSR Hydrometeorological Service are shown in Fig. 5.15. While comparing the observations and calculation results, the value of the parameter ν is expressed through the initial mean wave elements and the current velocity (5.41):

$$\nu = \frac{V}{\sqrt{g\lambda_0}} \frac{\sqrt{2\pi}}{\sqrt[4]{\Gamma(1-4/n)}} \frac{1}{[(n+1)/n]^{1/n}}. \tag{5.43}$$

As shown, there is good correspondence between the observations and calculation results (see Fig. 5.15).

Now the wave evolution in the transition area ($0 < \gamma < 1$) will be considered with the current velocity gradient being different from zero, i.e. $\partial V/\partial x \neq 0$. In this case not only the straight waves ($C_{gx} > 0$), but also the waves with a negative group velocity projection ($C_{gx} < 0$) propagate to the given point. As noted, the latter appear to be as a result of straight waves reflected from a non-uniform current. The height of the reverse waves carried away downstream is increased, while the length is decreased.

5.5 Wave Element Transformations in Current, Varying Along its Direction

There is a second singularity (as $V \to 0$) in the solution with the reverse wave heights, attaining an infinitely great value, with their lengths tending to zero. The spectral approach implementation to the solution does not eliminate this singularity as soon as it occurs for all reverse waves at one point, when at $V = 0$. The wave height cannot be infinitely increased in a current, because the waves are to be broken down. The breaking can be taken into account indirectly, assuming that the wave spectrum does not exceed the wave spectrum equilibrium interval. This condition is fulfilled with the wave transformation in a current taking place sufficiently slowly compared with the time of the establishment of the wave spectrum under breaking. As mentioned in Sects. 5.3–5.4, the wave breaking solution (5.33) can be described, assuming the existence of the equilibrium interval. Its expression can be written as:

$$F_\infty(\tilde{y}) = \frac{\beta \nu^4}{g^2 \tilde{y}^{-5} \sigma_{\max}^4} Q(\beta) , \qquad (5.44)$$

where

$$\frac{\alpha_F \nu^4 g^2}{m_0 \sigma_{\max}^4} = \frac{\alpha_F \nu^4}{\delta^2 2\pi} \left(\frac{n+1}{n}\right)^{\frac{4}{n}} \Gamma\left(1 - \frac{4}{n}\right),$$

and $\delta = \frac{h_0}{\lambda_0}$ is the mean initial wave steepness. Now the wave spectrum solution is also dependent on the parameter δ, i.e. on the mean steepness of the initial waves.

Taking the reverse waves into account the area of integration over the variables $\{\tilde{y}, \beta\}$ is changed, depending on the parameter γ ($\gamma < 1$). The integration area is increased with decreasing γ from one to smaller values (see Fig. 5.10) and the mean wave parameters are changed accordingly. As shown by numerical calculations, changes in the wave element practically stop at $\gamma < 0.67$, if the current velocity is $V < 0.67 V_{\max}$ at the given point; the mean wave elements are not effected by the point $V_{\max}$ in the current velocity profile. The relative contribution of reverse waves is decreased with increasing the parameter γ from 0.667 to 1.00. At the same time the change of mean parameters h/h_0, λ/λ_0 and τ/τ_0 approaches the earlier described ones at $\gamma = 1$.

The results of calculations of the mean wave height and mean wavelength transformation with $n = 5$ at $\gamma = 0.67$ in increasing countercurrent are shown in Fig. 5.16. The nomograms of the element changes are given for various mean initial wave steepness. Thus, the wave height increases to the maximum value and then decreases similarly to the previous case. The maximum wave heights increase with decreasing steepness δ, i.e. the smaller the initial steepness, the smaller is the probability of wave breaking, and a larger relative increase of the wave heights is observed.

It should be noted that due to the presence of reverse waves, the mean heights in the transition current area ($0 < \gamma < 1$) can significantly exceed the wave heights in the maximum current velocity area ($\gamma = 1$). At first

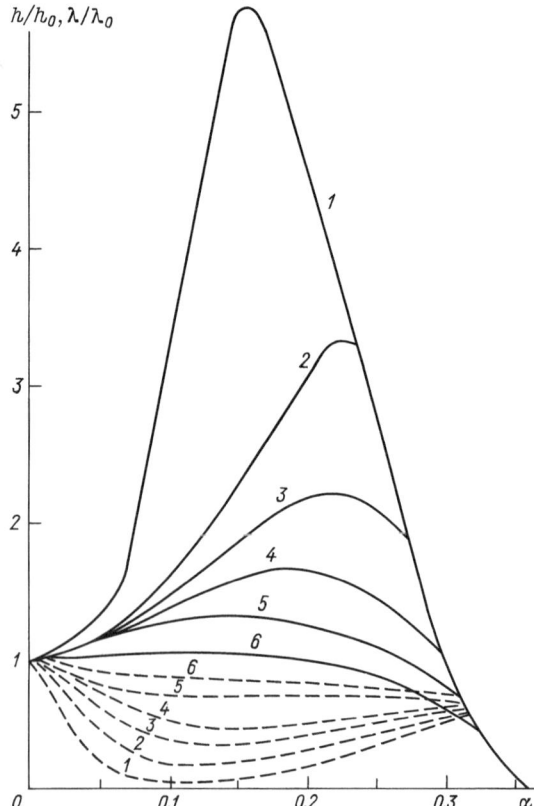

Fig. 5.16. Change of wave heights and lengths in countercurrent taking into account reverse waves with $n = 5$, $\gamma = 0.67$ and different values of initial wave steepness: $1 - 0.75 \times 10^{-3}$; $2 - 7.52 \times 10^{-3}$; $3 - 20.8 \times 10^{-3}$; $4 - 40.6 \times 10^{-3}$; $5 - 62.7 \times 10^{-3}$; $6 - 200.3 \times 10^{-3}$. *Solid lines* denote mean wave heights h/h_0 and *dotted lines* wavelengths γ/γ_0

the mean wavelength decreases with increasing parameter value and then it starts increasing. The least wavelength is observed for waves with small initial steepness. This occurs due to the presence of reverse waves, whose influence is greatest at small values of the initial wave steepness δ.

5.6 Wave Diffraction Around a Caustic

Wave height evaluation in the vicinity of the caustic. As shown in the previous section, as soon as the spectrum becomes narrowed, the maximum wave height tends to a larger value in the countercurrent. It becomes infinitely large at the point $\nu \cong 0.25$ in the limit $n \to \infty$ (see Fig. 5.15). This is in agreement with the solution for the monochromatic wave (5.37) at the blocking point.

There exists a viewpoint (Phillips, 1980) that a monochromatic wave is broken earlier before reaching the blocking point. However, this does not seem to be quite correct. It depends on the steepness whether the wave is broken or not. The steepness may be not so large for breaking before or during the blocking. As is shown, reverse waves can also be broken, in the case of their rolling downstream to the point with a sufficiently small current velocity. On the one hand, large wave amplitudes and steepness require implementation of a non-linear theory. On the other hand, it is necessary to apply more accurate estimation methods even for a small wave steepness around the turning point (or the wave blocking point), since the unreasonably large values for such waves are also given with the help of the monochromatic wave solution (5.37).

These methods, describing the wave field, are based on the assumption of locally plane waves. However, this assumption is not always true. There are some situations in which changes to the small wave field, compared with the wavelength, are accumulated. This can result in a significant difference of the wave field on some segment from the local flat field. Such wave field changes take place near the caustic. The caustic is the boundary between the area with a complex wave pattern as a result of the interference of two wave groups and the neighbouring area without any waves. The caustics are certain local peculiarities of many different wave configurations in a fluid.

Definite attention is also paid in the scientific literature to wave fields of a different nature in the caustic areas. The classification of different types of caustics is given, for example, in the monograph by Kravtsov & Orlov (1980) with a large reference list or in the well-known monograph by Arnold *The Theory of Catastrophes* (1989). As for papers dealing with gravity waves in water, the publications of Dobrokhotov and Zhevandrov (1988a,b) should be pointed out. They are devoted to the use of asymptotic expansions and the Maslov canonical operator. However, no proper attention is paid to the problem of wave fields in a non-uniform current.

Following the historical succession of studying the problem of waves in the caustic area, the Helmholtz equation describing light propagation should be noted. It can be written in one-dimensional form as follows:

$$\frac{\partial^2}{\partial z^2}a + k_0^2 n^2(z)a = 0 , \qquad (5.45)$$

where $k = \sigma/C$ is the wave number, σ is the frequency, C is the speed of light in vacuum, $n(z)$ is the refraction coefficient characterizing the properties of the wave propagation medium, and a is the wave amplitude. The propagation of monochromatic electromagnetic waves with the wave function $ae^{-i\sigma t}$ is described with the help of the Helmholtz equation (5.45). Wave propagation of a different nature can also be described by this equation, correspondingly changing the designations of the parameters k_0 and n.

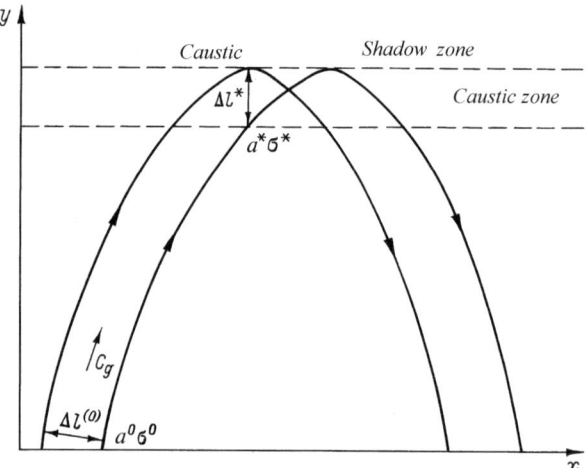

Fig. 5.17. Ray turning in caustic area

Taking a quantum-mechanical analogy (Kravtsov & Orlov, 1980; Landau & Lifshits, 1974a), it can be said that (5.45) is also the linear Schrödinger equation describing the stationary wave function for a particle with mass m and energy E in a force field with potential $U(z)$, where k_0 is the value $\sqrt{2mE/\hbar}$; $n(z)$ is the value $1 - U(z)/E$, so that $k_0^2 n^2 = 2m(E-U)/\hbar^2$ ($\hbar$ is the Planck constant).

The turning points $z = z^*$ are analogues to the caustic, with the coefficient $n(z)$ becoming zero in problems where a wave field is described by the ordinary differential equation (5.45). The term "turning point" corresponds to the case $n = 0$, when the particle kinetic energy is compared to the potential energy and the particle changes its direction of motion. At the same time there is also a turning point in the ray trajectory in coordinate space (see Fig. 5.17).

Smooth caustic surfaces without any special points are called ordinary. In the area of an ordinary caustic, two rays intersect, touching this caustic (see Fig. 5.17). The peculiarities in the caustic (points of sharpening, loops, etc.) are accompanied by a great number of rays intersecting at one point. Generally, the classification of different caustic types is based on the theory of differentiated reflection peculiarities, also called the "theory of catastrophes" (Arnold, 1989). An ordinary caustic is called "a fold" according to the terminology of this theory.

The important role of caustics in wave problems is determined by characterizing the ray family, permitting it to form the entire space field pattern. The family of the rays can be restored with the help of ray directions at the caustic surface. The areas where rays cannot penetrate are called the "caustic shadow zone". A zero field corresponds to the caustic shadow zone in the geometrical optics approximation. Actually, the field is different from zero there and arises from diffraction. The field exponentially decreases in the shadow zone with distance from the caustics in most ordinary cases. A complex generalization of geometrical optics (Voganov & Kantsenbaum, 1982; Kravtsov & Orlov, 1980) or other methods (Kravtsov, 1968; Ludwig, 1966) can be used for calculation of the field.

As mentioned above, the wave amplitude in the geometrical optics approximation becomes infinite due to ray convergence in the vicinity of the caustic. However, the Helmholts equation (5.45) and some other basic equations of wave theory do not permit infinite field values. That is why the amplitude singularity in the caustics indicates an inapplicability of the geometrical solution in the caustics themselves and in the area around them. Nevertheless, reasonable estimates of the field in caustics can also be obtained by using geometrical optics, when the area of the inapplicability of geometrical optics is identified in advance and the notions of conservation of wave action density flow are applied. This approach, proposed by Kravtsov & Orlov (1980) and in Non-linear Waves (1981), is used for the case with energy flow being preserved. The action density rather than the energy is preserved for waves in a non-uniform current. This approach also permits generalizations for the wave action conservation case.

The width of the ray tube formed by two rays at some distance from the caustic is assumed to be equal to $\Delta l^{(0)}$ (see Fig. 5.17). This value is equal to Δl^* near the caustic. It is the length of the perpendicular reconstructed from the point of the first caustic ray contacting the second ray. The value Δl^* is equal to the caustic zone width. Let the values $a^{(0)}, C_g^{(0)}, \sigma^{(0)}$ be the amplitude, group velocity and frequency at a distance from the caustic respectively, and let a^*, C_g^*, σ^* be the values near it. Assuming that the amplitude a is changed insignificantly in the caustic zone, the conservation of the action density flow can be written as:

$$\frac{(a^{(0)})^2 C_g^{(0)}}{\sigma^{(0)}} \Delta l^{(0)} \cong \frac{(a^*)^2 C_g^*}{\sigma^*} \Delta l^* . \qquad (5.46)$$

In the case of the ordinary caustic $\Delta l^{(0)} \sim (\Delta l^*)^{1/2}$, the wave amplitude a^* in the caustic is estimated as:

$$a^* \cong a^{(0)} \left(\frac{\sigma^*}{\sigma^{(0)}}\right)^{1/2} \varepsilon^{-1/6} , \qquad (5.47)$$

where $\varepsilon = (k_0 M)^{-1}$ is a small geometrical optics parameter.

For swell waves with length about $\lambda \sim 50$ m and typical transverse horizontal current size about 100 km, the wave amplitude increase is approximately $a^*/a^{(0)} \sim 10^{2/3}$ at $\sigma \sim \sigma^{(0)}$. The obtained estimation gives the notion of possible energy concentration of the swell field in the ocean current.

Wave field asymptotic around a caustic. In order to estimate the wave field around a caustic precisely it is necessary to do more accurate calculations. For this purpose a modified traditional scheme of the WKB approximation can be applied. In order to do so the Maslov method (Kravtsov, 1968; Maslov & Fedoryuk, 1976), describing the uniform wave field asymptotics in the entire space can be used. The Maslov method is often used in obtaining a solution of the Helmholtz equation and it can be briefly described as follows.

As mentioned above, the application of the geometrical optics method leads to an amplitude value singularity in the caustic. However, when the wave field is described by the integral representation the wave amplitude in the caustic is finite in value. The Fourier integral can be used as the representation. Integration is with respect to the wave vector component $\mathbf{k}$, in which the direction the wave medium properties are changed (for example, along the Ox axis). In other words, one proceeds from the spatial coordinates $\{x_1 = x, x_2 = y\}$ to the mixed spatial–momentum representation $\{k_1 = k_x, x_2 = y\}$. If the medium properties are changed in two directions simultaneously, it is necessary to proceed from the spatial to the momentum coordinates, i.e., to find a solution as the Fourier double integral.

However, the formal application of this method for waves in non-uniform currents would be wrong, if the waves were described by a more complicated equation than the Helmholtz equation (5.45), for which the aforementioned asymptotic methods were initially developed. In this case the wave action density is preserved rather than the energy. Voronovich and Goncharov (1982) proposed the idea of generalizing the Maslov method for propagating hydrodynamic waves in non-uniform currents. The problem of the influence of large-scale oceanic motions on short internal wave propagation is considered in their paper. This approach can also be generalized for describing waves in water near the caustic both in the presence of non-uniform currents and with uneven bottom relief.

Thus, the properties of the wave medium will be assumed to change along the Ox axis and they are not dependent on time t. It follows from general kinematic theory (see Chap. 1) that the wave vector component, directed along the Oy axis, is constant, i.e. $k_y = k_{y0}$. The frequency ω is persevered along the ray as well:

$$\omega = F(k_x, k_{y0}, x) = \text{const} . \qquad (5.48)$$

According to (1.24), the trajectory of the wave packet propagation can be parametrically presented in the following form:

$$\int_{x_0}^{x} C_{gx}^{-1} \, \mathrm{d}x = t - t_0 ; \quad y = y_0 + \int_{0}^{t} C_{gy} \, \mathrm{d}t , \qquad (5.49)$$

where the initial data are prescribed as $x|_{t=t_0} = x_0$; $y|_{t=t_0} = y_0$; $k_x|_{t=t_0} = k_{x0}$; $k_y|_{t=t_0} = k_{y0}$. The wave amplitude at the initial moment $a(x, y, t)|_{t=t_0} = a_0$ is also considered to be known.

5.6 Wave Diffraction Around a Caustic

According to the Maslov method, the wave integral presentation can be written as:

$$\eta(x,y,t) = \frac{1}{\sqrt{2\pi}} \int a_0 \sqrt{\frac{\sigma}{\sigma_0} \left(\frac{\partial(k_x, y)}{\partial(x_0, y_0)} \right)^{-1}} \exp\{i[k_x(x - \tilde{x}) + \psi(\tilde{x})]\} \, dk_x \,, \tag{5.50}$$

where the wave number k_x is the integration variable, $\tilde{x}$ is a function of k_x, obtained from (5.48) x, and $\psi(x,y,t)$ is the usual wave phase. σ, σ_0 are wave frequencies relative to immovable water at the given and initial time moments $t = t_0$.

The Jacobian function in (5.50) can be written as follows:

$$\frac{\partial(k_x, y)}{\partial(x_0, y_0)} = \frac{\partial k_x}{\partial x} \frac{\partial(x, y)}{\partial(x_0, y_0)} = \frac{\partial k_x}{\partial x} \frac{\partial x}{\partial x_0}. \tag{5.51}$$

Using (5.48) and (5.49) it can be found that $\partial k_x / \partial x = -\partial F / \partial x / C_{gx}$ and $\partial x / \partial x_0 = C_{gx}/C_{gx0}$. The integral value (5.50) can be estimated with the help of the saddle-point method. The saddle point k_x^{**} is found using the equation $\partial \Theta / \partial k_x = 0$, where $\Theta(x, k_x) = k_x(x - \tilde{x}) + \psi(x)$.

When $\frac{\partial \Theta}{\partial k_x} = x - \tilde{x} + \left(\frac{\partial \psi}{\partial \tilde{x}} - k_x \right) \frac{\partial k_x}{\partial x} = x - \tilde{x}$, then it turns out that $\tilde{x}(k_x^{**}) = x$ and $\partial^2 \Theta / \partial k_x^2 = -\partial \tilde{x} / \partial k_x = C_{gx}/(dF/\partial x)$. Application of the usual saddle-point method to the integral (5.50) leads to the following formula:

$$\eta(x,y,t) = a_0 \sqrt{\frac{\sigma}{\sigma_0} \frac{C_{gx0}}{C_{gx}}} \exp\{i\psi\} \,. \tag{5.52}$$

The insignificant phase factor is omitted in this ratio. The expression (5.52) makes up the condition of the preservation of the wave action density. It can be proved comparing the expressions (5.52) and (1.46).

Now the "turning" point of $x = x^*$ will be considered. The sign of the velocity C_{gx} is changed, and wave packet propagation begins moving in the opposite direction at this point. In this case $\partial^2 \Theta / \partial k_x^2 \big|_{k=k^*} = 0$, and the phase expansion $\Theta(k_x)$ starts with a term of order $(k_x - k_x^*)^3$ at the saddle point $k_x^{**} = k_x^*$:

$$\frac{\partial^3 \Theta}{\partial k_x^3}\bigg|_{x=x^*} = \frac{\partial}{\partial k} \left(\frac{C_{gx}}{\partial F / \partial x} \right)\bigg|_{x=x^*} = \frac{1}{\partial F / \partial x} \frac{\partial C_{gx}}{\partial k_x}\bigg|_{x=x^*} \tag{5.53}$$

Considering some area around the turning point $x = x^*$, with the saddle-point k_x^{**} being close to k_x^*, the expression for $\Theta(k_x)$ can be restricted to terms of the order $(k_x - k_x^*)^3$. The asymptotic expression can obtained using an expansion of the integral (5.50) and omitting the insignificant phase factor:

$$\eta(x,y,t) = a_0 \sqrt{2\pi \frac{\sigma}{\sigma_0} \frac{C_{gx}}{\partial F/\partial x}} \kappa^* \mathrm{Ai}\left[\kappa^*\left(x-x^*\right)\right]$$
$$\times \exp\left\{\mathrm{i}\left[k_x^*\left(x-x^*\right) + \psi\left(x^*\right)\right]\right\}, \qquad (5.54)$$

where $\mathrm{Ai}(X)$ is the Airy function, and

$$\kappa^* = \left[\frac{1}{2}\frac{1}{\partial F/\partial x}\frac{\partial^2 F}{\partial k_x^2}\right]^{-\frac{1}{3}}\bigg|_{x=x^*}.$$

The Airy function $\mathrm{Ai}(X)$ is estimated for large positive values $(X>0)$ as:

$$\mathrm{Ai} \approx \frac{1}{\sqrt{2\pi}} X^{-1/4} \exp\left(-\frac{2}{3}X^{3/2}\right), \qquad (5.55a)$$

and for large negative values $(X<0)$ as:

$$\mathrm{Ai} \approx \frac{1}{\sqrt{2\pi}} X^{-1/4} \cos\left(\frac{2}{3}|X|^{3/2} - \frac{\pi}{4}\right). \qquad (5.55b)$$

The Airy function can be written at $X<0$ as a superposition of two exponential dependencies of the type $\exp(\pm\mathrm{i}X)$, one of them describing the straight and the second the reflected waves. These two waves with close wave numbers superimposing each other create the interference pattern, with the total amplitude subjected to modulation.

The wave vector component is defined as $k_x = k_x^* \pm \sqrt{\kappa^{*3}(x^*-x)}$ in the vicinity of the turning point. The $(\pm)$ signs are referred to the straight and reflected waves, respectively. There is a shift by $-\pi/2$ in the reflected wave phase due to the ray touching the caustic. Taking into account the latter ratio in the formula (5.52), as well as applying the asymptotic of the Airy function (at $X<0$) in the formula (5.54), the results coincide, indicating the uniformity of the asymptotic solution. The solution around the caustic with asymptotics is shown in Fig. 5.18.

The maximum amplitude $a = |\eta|$ is reached with $\kappa^*(x^*-x) = 1.02$; at the same time $\mathrm{Ai} = 0.536$, and

$$a_{\max} \cong 1.69 a_0 \sqrt{\frac{\sigma}{\sigma_0} C_{gx}} \left|\frac{\partial F}{\partial x}\right|^{-\frac{1}{6}} \left(\frac{\partial^2 F}{\partial k_x^2}\right)^{-\frac{1}{3}}\bigg|_{x=x^*}. \qquad (5.56)$$

The general case of propagating surface gravity waves in water is described with the help of the ratios (5.52)–(5.56). As a specific case estimates of monochromatic waves in a countercurrent can be obtained, i.e. a solution of the problem mentioned in the previous section will be considered using the diffraction approximation. Thus, the following kinematic relations at the turning point are obtained with the help of (5.38):

5.6 Wave Diffraction Around a Caustic

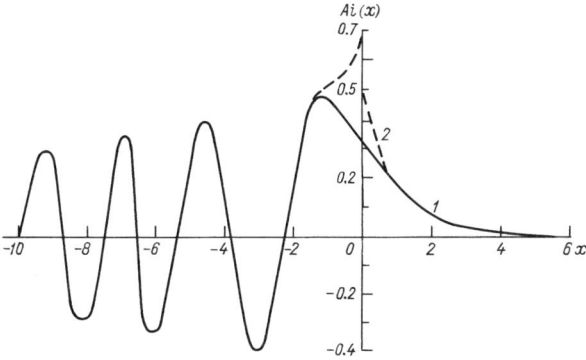

Fig. 5.18. Airy function (1) and its asymptotic approximations (2) (5.55a) for $X > 0$ and (5.55b) for $X < 0$

$$V = V^* = -c_0/2 \, ; \qquad k^* = 4k_0 \, ;$$

$$\sigma^* = 2\sigma_0 \, ; \qquad \partial^2 F/\partial k_x^2 \big|_{k_x=k_x^*} = -2\sqrt[-5]{gk_{x0}^3} \, .$$

The maximum wave amplitude around the caustic is found to be equal to

$$a_{\max} = 4.27 a_0 \left(\frac{\sigma_0}{\partial V/\partial x} \right)^{1/6}, \tag{5.57}$$

where a_0 is the initial wave amplitude and σ_0 is the frequency at $V = 0$, and the derivative $\partial V/\partial x$ is taken at the turning point. It should be noted that the maximum wave increase (5.57) is almost twice as small as the value determined by Peregrine (1976) using additional simplifying assumptions for an asymptotic expression. The wave steepness can be easily estimated at the turning point with the help of the ratio (5.57):

$$k^* a_{\max} = 17.1 k_0 a_0 \left(\frac{\sigma_0}{\partial V/\partial x} \right)^{1/6}. \tag{5.58}$$

For a wave with length $\lambda = 100$ m and current velocity gradient $\Delta V/\Delta x = 10^{-4}$ s^{-1}, the wave intensification is estimated as $a_{\max}/a_0 \cong 18.2$. This great increase in the steepness is possible only with the initial value being so small that the steepness does not exceed the ultimate value at the turning point, at which breaking takes place: $(a_{\max} k^*)^2 = 0.2$, i.e.

$$a_0 k_0 < 3.2 \times 10^{-2} \left(\frac{\sigma_0}{\partial V/\partial x} \right)^{1/6} < 0.615 \times 10^{-2} \, .$$

The waves are destroyed before reaching the turning point if the initial steepness is greater than this value.

Large wave steepness values, occurring in a countercurrent, indicate the necessity to take into account non-linear effects. Non-linearity can change the current speed blocking waves.

The correction for the current blocking speed can be easily obtained using the Stock wave expansion. Thus, the energy transfer speed can be estimated as:

$$C_g \cong V + \frac{1}{2}\sqrt{g/k}\left(1 + \frac{3}{2}(ak)^2\right). \qquad (5.59)$$

The value C_g is equal to zero at the blocking point. Then, the current speed is $V^* = -0.32\, c_0$ for the ultimate wave steepness $(ak)^2 = 0.2$, where c_0 is the phase wave velocity of an infinitely small amplitude in the absence of the current $V = 0$. Thus, it should be noted that the finite steepness wave value shows an increased blocking current speed for steeper waves $V^* = -0.3\, c_0$ as compared with sloping ones $V^* = -0.25\, c_0$.

The non-linear wave effects around the caustic can be described by the non-linear Schrödinger equation:

$$i\frac{\partial a}{\partial t} = \frac{\partial^2 a}{\partial r^2} - ra + b\left|a^2\right|a, \qquad (5.60)$$

where b is a non-linearity parameter. It is shown that the local wave amplitude a around the caustic for the incident wave is stable (Smith, 1976). The wave field increase is not practically influenced by non-linearity around the caustic. The wave profile is asymmetric due to the amplitude modulation. To some extent it is similar to waves in a water surface described by the Airy function. Depending on the sign of the derivative amplitudes, the front wave slope can be of greater steepness compared to the rear one, although the reversed phenomenon can also be observed.

5.7 Wind Wave Transformation in Cross-Velocity Shear Current

Solution of the spectral equation. The problem of wave spectrum transformation will be considered in a shear horizontally non-uniform current. Its solution can be obtained in analytical form. The current velocity V is assumed to be directed along the Oy axis with its changes along the x coordinate: $V = \{0; V_y(x)\}$. The depth $H = H(x)$ is also changed along this direction (see Fig. 5.19). This situation can appear, for example, in a coastal zone with an along shore current.

The equation (5.2) of wave packet propagation can be written in the following form:

5.7 Wind Wave Transformation in Cross-Velocity Shear Current

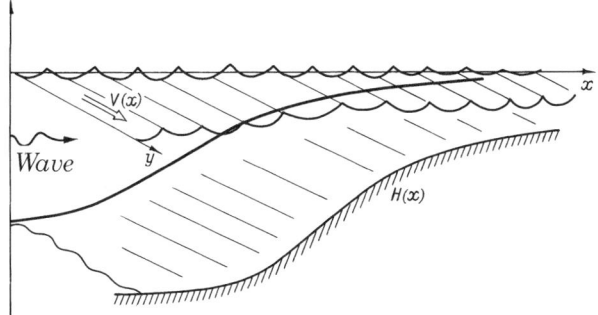

Fig. 5.19. Wave transformation in shear horizontally non-uniform current

$$\frac{dx}{dt} = C_{gx} = C_g \cos(\beta) \; ; \quad \frac{dy}{dt} = C_{gy} = C_g \sin(\beta) + V_y \; ; \tag{5.61}$$

$$\frac{dk_x}{dt} = \frac{1}{2}\sqrt{\frac{gk}{\text{th}(kH)}}\frac{1}{\text{ch}^2(kH)}\frac{dH}{dx} + k_y \frac{\partial V}{\partial x} \; ; \quad \frac{dk_y}{dt} = 0 \; . \tag{5.62}$$

As can be seen from the problem formulation, the coordinate y is cyclic. The component k_y of the wave vector $\mathbf{k}$ remains constant along the wave packet propagation trajectory. Thus, it be can written as:

$$k_y = k \sin(\beta) = k_0 \sin(\beta_0) \; . \tag{5.63}$$

The second motion integral is a constant of the frequency ω. It can be written for the given type of current and depth change as follows:

$$\sqrt{gk \, \text{th}\,(kH\,(x))} + kV(x)\sin(\beta) = \omega \; . \tag{5.64}$$

The relations (5.63) and (5.64) are sufficient to define the wave number k and angle β along the trajectory, depending on the depth $H(x)$ and current velocity $V(x)$.

This formulation of the problem should be used for further interpretation of the obtained experimental field data, discussed in the following section. It is assumed that $G = 0$ for the wave action balance equation (5.1). This corresponds to the case of wave propagation in a non-uniform medium without taking into account the non-linear wave interaction and wind input. This assumption follows from comparing the typical time of action of mechanisms forming the wind wave spectrum in the frontal zone, described below.[2]

---

[2] The time scale of the non-linear wave interaction in the main frequency range of the spectrum $S(\sigma)$ is estimated as $1/\tau_{nl} \approx G_{nl}/S \approx \sigma_{max}^{11} S^2(\sigma_{max})/g^4 \approx 4\times 10^{-4}\,\text{s}^{-1}$, i.e. $\tau_{nl} = 2.5\times 10^3$ s. The typical time scale of the wind input is τ_{in}: $1/\tau_{in} \approx G_{in}/S \approx 0.25\rho_a/\rho_w(Uk-\sigma) \approx 5\times 10^{-4}\,\text{s}^{-1}$, i.e. $\tau_{in} \approx 2\times 10^3$ s, while the typical time of the radiation tension action is τ_{cur}: $1/\tau_{cur} \approx G_{cur}/S \approx$

In this case the spectral density of the wave action N is preserved along the trajectory of wave packet propagation: $N(\mathbf{k},\mathbf{r},t) = N(\mathbf{k}_0,\mathbf{r}_0,t_0)$, where the initial wave vector $\mathbf{k}_0$ and coordinates $\mathbf{r}_0$ are functions of $\mathbf{k},\mathbf{r}$, and t at a given point. This is a transformation from the spectral density of the wave action N to the spectral density of the wave energy $F(k,\beta)$, which are dependent on the wave number k and angle β. The solution (5.27) can be written in the following form:

$$F(k,\beta) = F_0(k_0,\beta_0) \frac{\sigma k}{\sigma_0 k_0}. \tag{5.65}$$

The energy spectrum of waves propagating into the given area from the initial boundary can be used as the expression (5.16).

Thus, the spectral energy density $F(k,\beta)$ along the ray is determined by substituting the expression (5.16) into (5.65). The connection between the spectra $F_0(k_0) = S_0(\sigma_0) \cdot \partial\sigma_0/\partial k_0$ can be used. The values k_0 and β_0 are found with the help of the relations (5.63) and (5.64):

$$\sin(\beta_0) = \sin(\beta)\frac{k}{k_0}; \tag{5.66a}$$

$$\sqrt{gk_0 \operatorname{th}(k_0 H_0)} = \sqrt{gk \operatorname{th}(kH(x))} + k[V_0 - V(x)]\sin(\beta), \tag{5.66b}$$

where V_0 and H_0 are the velocity and depth, respectively, at the initial ray point. Using the results of Jolm (1979), the solution for the transcendental equation (5.66b) can be written with an accuracy of up to 10^{-6} in the following form:

$$q^2 = p^2 + \frac{p}{1 + \sum_{n=1}^{\infty} d_n p^n},$$

where the values of d are: $d_1 = 0.666666(6)$; $d_2 = 0.3555555(5)$; $d_3 = 0.16084656$; $d_4 = 0.0632098765$; $d_5 = 0.0217540484$; $d_6 = 0.0065407983$; $g^2 = k_0 H_0$; $p = \left(\sqrt{gk \operatorname{th}(kH(x))} - k\Delta V(x)\sin(\beta)\right)^2 H/g$ (here $\Delta V = V_0 - V$).

For the deep-water case it can be written that

$$\sqrt{gk_0} = \sqrt{gk} - k\Delta V \sin(\beta) \tag{5.67}$$

$2\Delta V/\Delta y \approx 6 \times 10^{-3}\,\mathrm{s}^{-1}$, $\tau_{cur} \approx 1.5 \times 10^2$ s. In this case the frequency of the spectrum maximum $\sigma_{\max}$ is equal to 2.2 rad s^{-1} with the wind speed $U = 6$ m s^{-1}. Thus, the typical evolution time of the spectrum due to wind input and non-linear wave interaction is one order of magnitude larger than the time of the current impact on the waves.

5.7 Wind Wave Transformation in Cross-Velocity Shear Current

The wave spectrum solution can be presented in the form:

$$F(k,\beta) = Q m_0 (n+1) \left(\frac{k_m}{k}\right)^{\frac{n}{2}}$$

$$\times \frac{\exp\left[-\frac{n+1}{n}\left(\frac{k_m}{k}\right)^{\frac{n}{2}}\left(1 - \Delta V\sqrt{kg}\sin(\beta)\right)^{-n}\right]}{k\left(1 - \Delta V\sqrt{k/g}\sin(\beta)\right)^{n+5}}, \quad (5.68)$$

where $Q = Q(\beta_0^0 - \beta_0)$ is the angular energy distribution, with β_0^0 being the initial general directions of wave propagation and β_0 being determined by the relation:

$$\beta_0 = \arcsin\left[\frac{\sin(\beta)}{\left(1 - \Delta V\sqrt{k/g}\sin(\beta)\right)}\right]. \quad (5.69)$$

It should be noted that (5.69) as well as (5.66a) are determined as:

$$|\sin(\beta)| = |\sin(\beta)k/k_0| \le 1 \quad (5.70)$$

Otherwise, as shown below, the wave packet parameters turn out to be indefinite in some areas, being in the caustic shadow zone. The boundary itself, where the equality $|\sin(\beta)k/k_0| = 1$ is fulfilled, makes up a caustic. In this case the monochromatic wave amplitude attains an infinitely large value in its classical approximation.

Investigation of wave packet kinematics. In order to construct the total spectrum (5.68) at the given point $\{x_0, y_0\}$, it is necessary to reduce all rays of different wave packets to this point. Not only should the waves propagating directly from the initial boundary to the given point be considered, but also the waves reflected from a horizontally non-uniform current and arriving at the point $\{x_0, y_0\}$.

In order to solve the problem the following kinematic considerations will be used. The wave packet trajectories are shown schematically in Fig. 5.20. There are trajectories of waves propagating from the initial boundary to the given point as well as reflected wave trajectories. If the initially given spectrum $F_0(k_0, \beta_0)$ is defined as non-zero within $|\beta_0| \le \pi/2$, the energy-supply components can appear in the entire angle range $0 \le |\beta_0| \le 2\pi$ due to their reflection by a current. Each of the four quadrants will be considered separately in order to analyse the possibility of spectral components appearing in the entire range of angles.

Thus, spectral components can exist at the point A $(x < x_m)$ in a current within quadrant I (see Fig. 5.20) within the entire range of angles $0 \le \beta < \pi/2$. The corresponding spectrum can be written in the following form:

$$F_A^I = F(k,\beta). \quad (5.71)$$

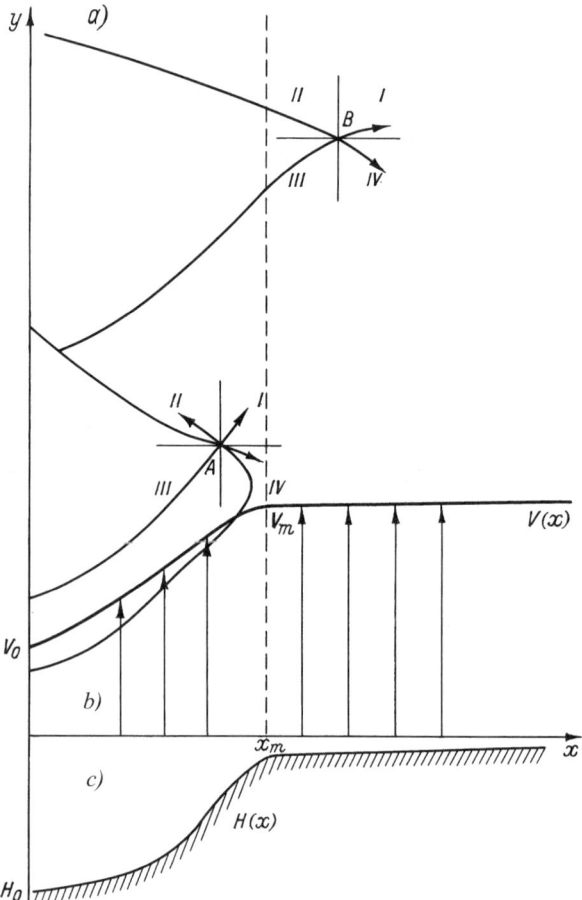

Fig. 5.20. Trajectories of wave packet propagating against current velocity profile (**a**) and (**b**); depth background (**c**) for different quadrants at points A and B; $V(x)$ – current velocity profile; $H(x)$ – depth profile

The wave components can exist at the same point A in quadrant II if they are reflected at the point x^*: $x < x^* < x_\mathrm{m}$. This means that there is a trajectory point x^* being a turning point of the wave packet propagation direction. The component of the group velocity C_{gx} is zero with $k_x = 0$ at this point. After having been reflected the wave packet can reach the given point x, in the case of the current velocity value formally satisfying the inequality

$$V < V^* < V_\mathrm{m} , \qquad (5.72)$$

where V is the current velocity at the given point x; V^* is the velocity at the turning point x; and V_m is the maximum current velocity.

5.7 Wind Wave Transformation in Cross-Velocity Shear Current

The current velocity V^* at the turning point can be written as:

$$V^* = \frac{\omega - \sqrt{gk_y \,\text{th}(k_y H)}}{k_y} . \tag{5.73}$$

In case of monotonic depth variations, it is necessary to prescribe the dependence $V = V(x)$ and $H = H(x)$ or $V = V(H)$, and then to solve the transcendental equation relative to the value V^* (or H^*): $\sqrt{gk_y \,\text{th}\,(k_y H(x^*))} + V(x^*) = \omega$. For the existence of the reflected wave solution, the equation x^* should be found within the range $x < x^* < x_m$.

It should be emphasized that a necessary condition for the existence of wave components in quadrant II is their presence in quadrant I. The former originate from the latter, as a result of their reflection from the current. In this case the angle turn takes place, i.e. an angle β is changed to the value $\pi - \beta$. The corresponding spectrum can be written as follows:

$$F_A^{II} = F(k, \pi - \beta)$$
$$\times \Theta \left[(V_m - V) k \sin(\beta) + \sqrt{gk_y \,\text{th}\,(k_y H)} - \sqrt{gk \,\text{th}\,(kH)} \right] . \tag{5.74}$$

Spectral components should always exist in the quadrant IV in case they are present at the initial boundary:

$$F_A^{IV} = F(k, \beta) . \tag{5.75}$$

The angular spectrum distribution of the these components becomes narrower due to the influence of current and shallow water (5.70).

In quadrant III, it is possible to observe spectral components if wave reflection takes place in the area $x > x_m$, when the corresponding difference of the current velocity values is present. The kinematic condition of the existence of waves in this quadrant is written as follows:

$$V_1 < V^* < V_0 , \tag{5.76}$$

where V_1 is the current velocity at the right boundary point of the current profile. The wave reflection can be increased in the presence of a local coastal countercurrent ($V_1 < 0$). It should be remembered that the spectral components in quadrant III arise from the components in quadrant IV due to their reflection. The wave spectrum in this quadrant can be presented as follows:

$$F_A^{III} = F(k, \pi - \beta) \,\Theta \left[\sqrt{gk \,\text{th}\,(kH)} - \sqrt{gk_y \,\text{th}\,(k_y H)} + (V - V_0) k \sin(\beta) \right]$$
$$\times \Theta \left[\sqrt{gk_y \,\text{th}\,(k_y H)} - \sqrt{gk \,\text{th}\,(kH)} - (V - V_1) k \sin(\beta) \right] . \tag{5.77}$$

It can be shown that the first of the Heavyside functions Θ is equal to one for the given spectrum in the current.

At the point B (behind the maximum current velocity, $x > x_m$) in quadrant I there are the same spectral components as at the initial boundary, except for the components reflected from the current (in the area $x < x_m$). In other words, such wave components exist when the inequality $V^* > V_m$ is fulfilled. The spectrum can be written as:

$$F_B^I = F(k, \beta)$$
$$\times \Theta \left[\sqrt{gk \, \text{th}\, (kH)} - \sqrt{gk_y \, \text{th}\, (k_y H)} + (V_m - V) k \sin(\beta) \right]. \quad (5.78)$$

In quadrant II the spectral components are absent in all cases, as they appear as a result of direct wave reflections (in quadrant I at $x < x_m$), whereas reflection does not take place at $x > x_m$. Thus,

$$F_B^{II} = 0. \quad (5.79)$$

In quadrant III, waves can exist if they are reflected in the area $V^* < V_0$ (which takes place at $V_1 < V_0$), thus:

$$F_B^{III} = F_A^{III}. \quad (5.80)$$

Some of the spectral components in quadrant IV are absent in the area with $V < V_0$ due to their reflection. The condition $V > V^*$ must be identically fulfilled for the existing components and there is no kinematic limitation in respect of the wave components, so:

$$F_B^{IV} = F(k, \beta). \quad (5.81)$$

Thus, the solution for wave spectrum evolution in a current with a finite water depth is described by the ratios (5.71), (5.74)–(5.81).

In conclusion it should be noted that the obtained spectral solution can also be applied for the more general case, with the velocity value $V(x)$ being a non-monotonic function of the argument x: $x_1 \leq x \leq x_2$ and having its maximum at the flow point x_m. The current velocity monotonically decreases on both sides with distance from the point x_m. It is necessary to keep in mind that there is no wave reflection occurring at the flow points $x > x_m$. Some wave spectral energy is reflected, whereas another part passes through at the flow points $x < x_m$.

The experimental results obtained in December 1975 near the Florida shores (Hayes, 1980) are a good proof of the aforesaid theoretical study. Observations and theoretical estimations reveal that swell waves propagating from the west to American shores can be reflected by the Gulf Stream or they pass through it depending on the initial wave direction and frequency.

Aerial photographic results of the sea surface. Analysing the obtained analytical solution, there arises a natural question, whether the solution corresponds to reality or not. The process of obtaining corresponding

5.7 Wind Wave Transformation in Cross-Velocity Shear Current

full-scale data is connected with the experimental complexity. It is necessary to conduct synchronous measurements of the non-uniform spatial field of current velocity and spatial wind wave spectra. This unique experiment was done by researchers from the St. Petersburg State University (Russia) by aerial photography.

The main purpose of aerial photography is to study the structure of water flow in the strait between two islands. The general strait area is of approximately rectangular form with a width in its narrow part of 22 km and a length of 25 km. There are two types of aerial photography: water surface survey and current survey, with floats used along the line between the nearest points of the islands' opposite shores. The route length is about 22 km. The aerial photography is made at the scale of 1 : 15,000. The straits' area is observed with the help of four parallel routes across the strait. Both ends of the islands are linked with edges of these routes.

The aim is to obtain a flare portion as full as possible on the aerial films of the strait surface survey. This allows one to obtain spectra of the sea surface optical representation using the Methods of Examining Marine Currents from an Airplane (1964). These spectra can be connected with spectra of sea surface slopes (Grushin et al., 1986).

The current surveys are made along single routes and only along one direction across the strait at its narrowest place. The routes are linked with both shore ends. The floats are dropped every 500 m. Then a direct survey of the entire float series is made twice with a 15–20 min interval. Diagrams of the float motion vectors are obtained for every route after the received data have been processed. The error of determining the current velocities is not more than 10 per cent under experimental conditions. A fragment of the current pattern diagrams is shown in Fig. 5.21. Its peculiarity is a small current velocity (0.2–0.3 m s^{-1}) on the right-hand side and a strong current velocity (1.3–1.6 m s^{-1}) on the left-hand side of the diagram.

The spatial current velocity change is revealed rather sharply within a relatively narrow area. It corresponds to the front location along the strait, seen

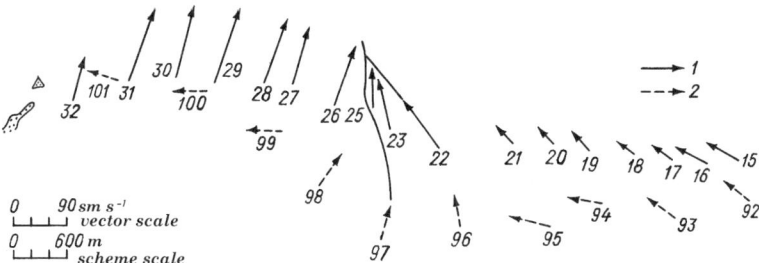

Fig. 5.21. Current pattern fragment: 1 – current velocity vector; 2 – visually estimated wave propagation directions (figures designate the numbers of aerial photographs)

206   5 Wave Evolution in Non-uniform Currents in Deep Water

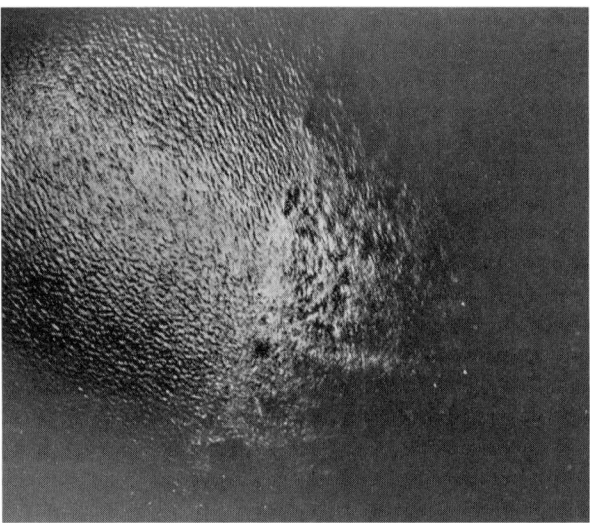

Fig. 5.22. Aerial photograph fragment depicting line between two flows

in the photo-diagram. In this diagram, the front depicted on the aerial film is shown by a solid line with concentrated "white caps" and intensive wave breaking.

While approaching the line dividing two flows, the wind waves are changed in the following way. To the right of the line, a system of two-dimensional wind waves is observed (see Fig. 5.21), where the direction of the prevailing wave system is shown with dotted arrows for the corresponding aerial photographs. While approaching the frontal line, the wave field is characterized by a more complicated pattern. It already consists of several wave systems and they cannot be separated in the direct vicinity of the flow frontal line. The corresponding fragment of sea surface aerial photograph is shown in Fig. 5.22. The flow frontal line is in the middle of the flash zone. To the right of the line there is no indication of any periodicity or prevailing wave direction at all. There are intensive rips noted in this case. To the left of the flow frontal line, a system of typical two-dimensional waves propagating leftward is observed.

The Fourier spectra obtained from the water surface survey are shown in Fig. 5.23. It should be noted that the relative values of this spectrum are comparable with the tensor component of the spectral density slopes (Grushin et al., 1986). Spectra are presented as isolines of the same intensity in the plane of the wave number k and the angle β, the latter being the angle between the wave vector and the chosen direction. It should be remembered that the spectra obtained with the optical method are 180° symmetric.

The wind wave features observed in the frontal zone are specified with the help of the spectra. A wind wave system, with energy bearing maximum

5.7 Wind Wave Transformation in Cross-Velocity Shear Current   207

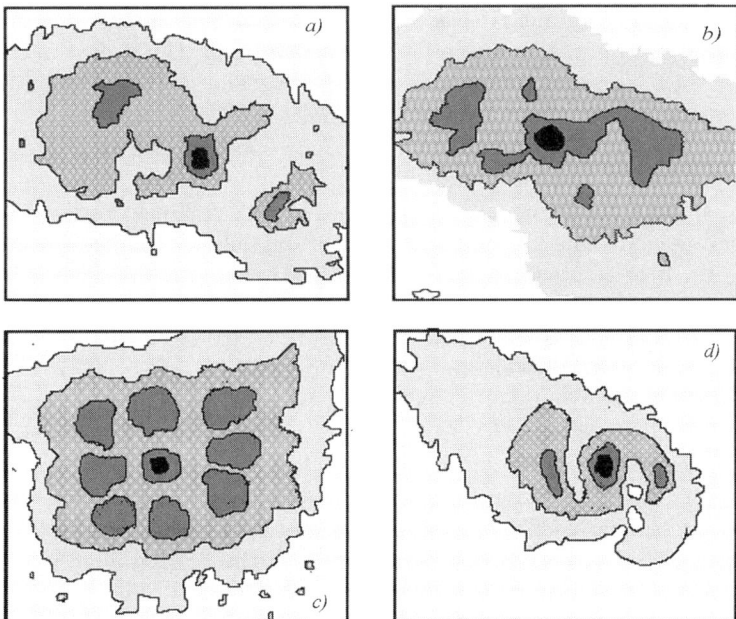

Fig. 5.23. Sea surface spectra obtained using photographs: (**a**) – 93 (see Fig. 5.21); (**b**) – 95; (**c**) – 96; (**d**) – 99

corresponding to a wave with 8.5 m length, is clearly seen in Fig. 5.23a, corresponding to the right hand part of the field diagram (see Fig. 5.21, Photo 93). The general direction of wave propagation is at an angle of 160° with the horizontal axis of the diagram (see Fig. 5.21). The spectrum closer to the front line (see Fig. 5.21, Photo 95) is shown in Fig. 5.23b. There are at least two clear wave systems with the propagation direction differing from each other by 120°. Whereas the first wave system can be connected with the wave system depicted in Fig. 5.23a, the presence of the second one is thought to be reflected from the first wave system from the line dividing the two flows. The wave spectrum in close proximity to the frontal zone (see Fig. 5.21, Photo 96) is shown in Fig. 5.23c. It is difficult to point out any wave systems with a clear propagation direction. The angular wave distribution is almost isotropic in the frontal zone. A single wave system can once more be clearly identified in the spectrum to the left of the dividing line (see Fig. 5.23d), with the wave propagation direction comprising an angle close to 180° with the horizontal axis of the diagram (see Fig. 5.21).

Interpretation of aerial photographic data. Comparison of the derived solution with full-scale data for deep water will be made for the case of constant current velocity in the area $x > x_m$ (i.e. $V_m = V_1$, see Fig. 5.20). The kinematic ratios obtained in the previous section indicate that the spectral components in quadrant III (see Fig. 5.20) are absent. In the area $x < x_\mathrm{m}$, the spectral components can be found in quadrants I, II, IV. But in the case $x > x_\mathrm{m}$ they are present in quadrants I and IV. The solution is described by the formula (5.68) with the kinematic conditions (5.71)–(5.81).

The tensor component of the wave slope spectral density $F(k,\beta)\,k^2$ is estimated by the formula (5.68). Its relative values can be compared to the optical spectra of the sea surface pattern (see Fig. 5.23), so long as the corresponding conditions for obtaining and processing are fulfilled (Grushin et al., 1986). It is assumed in the formula (5.68) that $n = 4$, which is usually used for a developed wind sea. It should be noted that the maximum slope spectrum $F(k,\beta)\,k^2$ is shifted by 1.4 times to a larger wave number area for $\Delta V = 0$, relative to the maximum spectrum $F(k,\beta)$, $k_\mathrm{max} \approx 0.47\,\mathrm{m}^{-1}$.

The slope spectrum can be calculated for different points of the current velocity profile at $\beta_0^0 = 160°$. It should be taken into account that the theoretical calculations are displaced by the angle $\pi - \beta$ relative to full-scale data. The results are presented in the $\{k, \beta\}$ plane as isolines of the slope spectrum $F(k,\beta)\,k^2$ standardized by its maximum. A two-dimensional spectral density (see Fig. 5.24a) is presented to the right of the line subdividing the two flows, where the current velocity is equal to $0.3\,\mathrm{m\,s}^{-1}$. The corresponding experimental spectrum is presented in Fig. 5.23b.

There are two pronounced maximuma seen in the two-dimensional spectrum. The first one deals with initial waves propagating at the angle $160°$ to the horizontal axis (see Fig. 5.21). There is an angle of $40°$ for the second maximum. The energy distribution of reflected waves becomes narrower. Some spectral components pass to the area of higher current velocities without being reflected. The isoline curves are explained with the help of the superposition of the kinematic condition for the existence of reflected waves (5.78) with the function of angular energy distribution in the initial wave spectrum.

The calculated spectrum on the left side of the subdividing line with the difference of current velocities $\Delta V = 1.4\,\mathrm{m\,s}^{-1}$ is shown in Fig. 5.24b. The corresponding experimental spectrum is in Fig. 5.23d. In this case the angular spectrum becomes narrower. One general wave direction is clearly evidently across the flow. The wavelength of the main system is increased two-fold.

Comparison of the experimental and theoretical spectra allows the identification of a number of common features. They indicate qualitative (in some cases quantitative) agreement of the obtained theoretical results with full-scale observations.

Thus, the frontal zone, as the boundary of two flows, appears to be a peculiar wave filter. There is a reflection of some spectral wave components in this zone. These components create rips superimposing with the initially propagated wave system. An intensive wave breaking appears, resulting in the isotropic angular spectrum distribution, seen in the sea surface spectra (see Fig. 5.23c). Some of the spectral components, without being reflected, propagate practically across the flow and manifest themselves as a quasi-regular wave system.

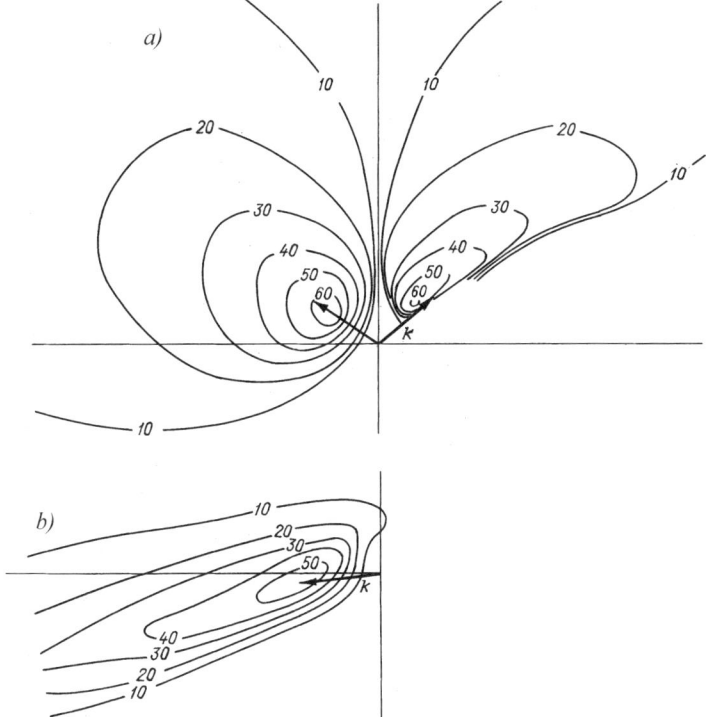

Fig. 5.24. Wave slope spectra obtained theoretically at flow points corresponding to photographs 95 (**a**) and 99 (**b**) (see Fig. 5.21)

Remote ocean sensing. Remote sensing of the spatial wind wave field of the ocean's surface is of special interest. Its non-uniformity can be an indicator of the ocean surface layer condition. Namely, it can show the presence of eddies, fronts, internal waves, concentrations of surface-active components, etc. The wind waves can be changed significantly under the influence of spatially non-uniform currents, revealing their presence.

It should be noted that alongside investigations of the influence of currents on waves, the non-uniform wind wave field is studied as a peculiar indicator of the current velocity distribution. It is shown that the root mean square value of the rough sea surface slope can be changed depending on the countercurrent (Huang et al., 1972). The non-uniformity of the wave field is investigated as a manifestation of the influence of large-scale internal waves in the form of non-uniform currents at the sea surface (Action of Large-Scale Internal Waves on the Sea Surface, 1982). Theoretical and experimental results reveal a relation between internal wave parameters and remote sensing measurement data of surface waves.

This direction is developed in the scientific work of the Marine Hydrophysical Institute of the Academy of Sciences, Ukraine (Bolshakov et al., 1988;

Grodskiy et al., 1989; Dulov & Kudryavtsev, 1989, 1992; Kudryavtsev et. al., 1995; Dulov et al., 1996, 1998). They propose a method of determining the wind wave spectrum, using photographic negatives (Bolshakov et al., 1988). The observed wave fields are compared with the solution of the refraction equations for waves in a current. In some cases of quasi-uniform wave developing conditions the current effect is identified in the fields of integral characteristics of the two-dimensional wind wave spectra. Although this problem is sufficiently difficult and not completely investigated, in general it seems to be quite promising.

In this section an attempt is undertaken to explain the spatial non-uniformity of a wind wave field in the frontal zone characterized by large current velocity gradients. The reverse problem, namely, the estimation of the current velocity with the help of wave spectra based on aerial photography data can be solved using the proposed mathematical model.

For example, using the direction of the spectrum maximum of reflected waves and the corresponding part of the direct wave spectrum, the current direction and velocity at which this reflection occurred can be easily estimated. The current direction can be determined with the help of the mirror reflection condition:

$$\beta_V = (\beta_1 + \beta_2)/2 , \qquad (5.82)$$

where β_2 is the propagation direction of the reflected waves, equal to $40°$, and β_1 is the direction of the corresponding direct waves, equal to $140°$.

Using the kinematic relations (5.63), (5.64), the current velocity variation at the two-flow boundary in the wave reflection area is estimated as:

$$\Delta V = \sqrt{\frac{g}{k}} \frac{1 - \sqrt{\sin(\beta_1)}}{\sin(\beta_1)} \approx 1.2\,\mathrm{m\,s}^{-1} . \qquad (5.83)$$

As can be seen, these estimations are in good agreement with full-scale observations. The error in determining the latter is within the range of 25 per cent, which is mainly connected with the discreteness of spectra presentation, their variability and error in determining the angle β_1.

So, if it were possible to identify the direct, reflected and transmitted spectral components not only in the spectral maximum, but also in a wider frequency range, the accuracy of determining the current velocity with the help of aerial photographs would be increased, by specifying the methods of initial data processing and analysis.

Wave element estimation in currents with transversal horizontal velocity shear. The numerical estimation of mean wave elements will be made in a shear horizontally non-uniform current with a maximum at the point x_m. In order to do so, the earlier obtained spectral solution is used. It should be taken into account that there is no wave reflection at the flow points $x > x_\mathrm{m}$. The wave elements are determined by numerically integrating the corresponding spectral moments (5.71), (5.74), (5.75), (5.77)–(5.81).

5.7 Wind Wave Transformation in Cross-Velocity Shear Current

The solution will be transformed into a more general form by introducing the following non-dimensional parameters: $\alpha = k_m H_0$ is the initial effective depth; $\nu_0 = V_0 \sqrt{k_m/g}$ is the initial effective current velocity; $\mu = V_{\max}/V_0$ is the maximum current velocity; $\gamma = H_0/H$ is the ratio of initial depth to the given value at the point considered; $\lambda = V/V_0$ is the relative current value; and $\lambda_1 = V_1/V_0$ (where V_1 is the velocity at the last point of the current profile).

The value $f = k_0/k$ is defined as the solution of (5.66). Proceeding to the new variable $\tilde{y}^2 = k/k_m$ and non-dimensional parameters, the spectrum $F(\tilde{y}, \beta)$ can be written in the form:

$$F(\tilde{y}, \beta) = Q(\beta_0) m_0 (n+1) \sqrt{\frac{\text{th}(\tilde{y}^2 \alpha_0/\gamma)}{\text{th}(f \tilde{y}^2 \alpha_0)}} f^{-\frac{n+5}{2}} \tilde{y}^{-(n+2)}$$

$$\times \exp\left[-\frac{n+1}{n}(f\tilde{y}^2)^{-\frac{n}{2}}\right], \qquad (5.84)$$

where $Q(\beta_0) = \frac{8}{3\pi} \cos^4(\beta_0 - \beta_0^0) \Theta(1 - \sin^2(\beta))$, and Θ is the Heaviside function.

The aforementioned kinematic conditions allow the solution for the described cases to be written in the following form.

The solution within the first quadrant ($0 < \beta < \pi/2$) can be written: at the point A (see Fig. 5.20) as:

$$F_A = F(\tilde{y}, \beta); \qquad (5.85)$$

at the point B:

$$F_B = F(\tilde{y}, \beta) \Theta\left[\sqrt{\text{th}(\tilde{y}^2 \alpha_0/\gamma)}\right.$$
$$\left. - \sqrt{\text{th}(\tilde{y}^2 \alpha_0/\gamma_0 \sin(\beta))} \sin(\beta) - \nu_0(\mu - \lambda)\sin(\beta)\right]. \qquad (5.86)$$

Within the second quadrant ($\pi/2 < \beta < \pi$) it is defined at the point A as:

$$F_A = F(\tilde{y}, \beta - \pi) \Theta\left[\sqrt{\text{th}(\tilde{y}^2 \alpha_0/\gamma_0 \sin(\beta))} \sin(\beta)\right.$$
$$\left. - \sqrt{\text{th}(\tilde{y}^2 \alpha_0/\gamma)} + \nu_0(\mu - \lambda)\sin(\beta)\right]; \qquad (5.87)$$

at the point B:

$$F_B = 0. \qquad (5.88)$$

Within the third quadrant ($\pi < \beta < 3\pi/2$) the solution is written at the point A as:

$$F_A = F(\tilde{y}, \beta - \pi)\,\Theta\left[\sqrt{\text{th}\,(\tilde{y}^2\alpha_0/\gamma_0\sin(\beta))}\sin(\beta)\right.$$
$$\left. - \sqrt{\text{th}\,(\tilde{y}^2\alpha_0/\gamma)} + \nu_0(\lambda - \lambda_1)\sin(\beta)\right]\;; \qquad (5.89)$$

at the point B:

$$F_B = F_A(\tilde{y}, \beta)\;. \qquad (5.90)$$

In the fourth quadrant $(3\pi/2 < \beta < 2\pi)$ it can be presented at the point A as:

$$F_A = F(\tilde{y}, \beta)\;;$$

at the point B as:

$$F_B = F(\tilde{y}, \beta)\;. \qquad (5.91)$$

With the help of the above solution, the moments of the spectra are calculated numerically. The mean relative values of the wave elements are determined as: height $\tilde{h} = h/h_0$, period $\tilde{\tau} = \tau/\tau_0$ and length $\tilde{\lambda} = \lambda/\lambda_0$.

The results of numerical wave element calculations in a shear horizontally non-uniform current in deep water are shown in Figs. 5.25a,b and 5.26. The relative value of the current velocity V/V_0 is marked on the horizontal axis. The left axis boundary corresponds to the initial point, whereas the right one coincides with a finite point of the current velocity diagram. It is seen (see Fig. 5.25a) that the wave height is greater than one at the left boundary, but it is less than one at the right boundary. The phenomenon is due to the wave reflection effect. The wave height is decreased two-fold at the angle $60°$.

The wavelength and the period are also changed in the current, but to a lesser extent. The general direction of wave propagation in a current is shown in Fig. 5.26. It is subjected to significant variations at different points of the current profile.

Diffraction estimations of monochromatic wave elements in a shear horizontally non-uniform current. The obtained solution (5.68) is a spectral generalization of the classic solution derived in the case of the initial spectrum being prescribed in the form of the Dirac delta function:

$$F_0(k_0, \beta_0) = 2\pi m_0 \delta(k_0 - k_0^0)\,\delta(\beta_0 - \beta_0^0)\;. \qquad (5.92)$$

The relative mean wave height $\tilde{h} = h/h_0$ can be calculated in a current, using the function (5.92) as the initial boundary spectrum:

5.7 Wind Wave Transformation in Cross-Velocity Shear Current   213

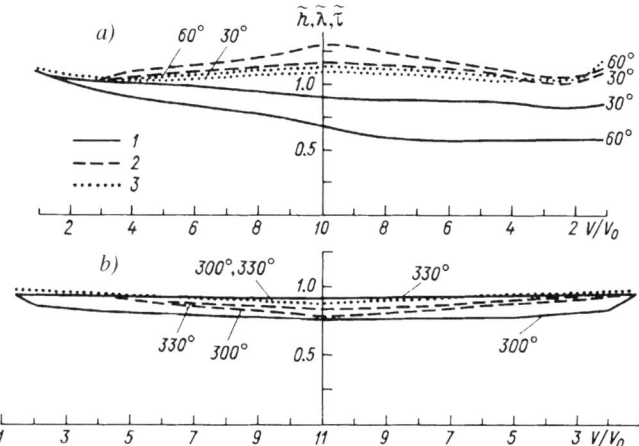

Fig. 5.25. Transformation of relative wave elements in shear horizontally non-uniform current, depending on current velocity for initial general wave directions of 30° and 60° (**a**) and for initial general wave directions of 300° and 330° (**b**): 1 – height $\tilde{h}$; 2 – period $\tilde{\tau}$; 3 – length $\tilde{\lambda}$

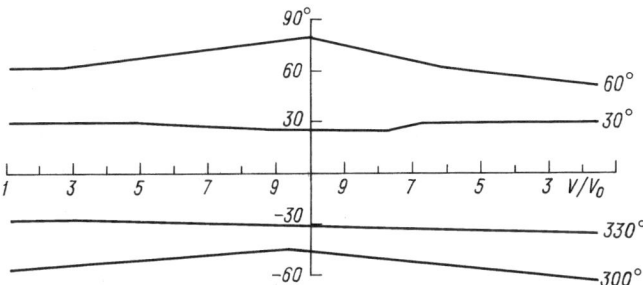

Fig. 5.26. Variation of general wave direction in shear horizontally non-uniform current for different initial angle values β_0: 30, 60, 300, 330°

$$\tilde{h}^2 = \frac{1}{2\pi m_0} \int_0^\infty \int_{-\pi}^\pi F(k,\beta)\, dk\, d\beta = \int_0^\infty \int_{-\pi}^\pi \frac{\delta\left(k_0 - k_0^0\right) \delta\left(\beta_0 - \beta_0^0\right)}{\left(1 - \Delta V \sqrt{\frac{k}{g}} \sin(\beta)\right)^3}\, d\beta\, dk$$

$$= \int_0^\infty \int_{-\pi}^\pi \delta\left[\arcsin\left(k/k_0 \sin(\beta)\right) - \beta_0^0\right]$$

$$\times \frac{\delta\left[k\left(1 - \Delta V \sqrt{\frac{k}{g}} \sin(\beta)\right)^2 - k_0^0\right]}{\left(1 - \Delta V \sqrt{\frac{k}{g}} \sin(\beta)\right)^3}\, d\beta\, dk. \quad (5.93)$$

The expression (5.93) can reduced to the following form, omitting intermediate calculations:

$$\tilde{h}^2 = \frac{\cos\left(\beta_0^0\right)\left(1 - \Delta V \sqrt{\frac{k_0}{g}}\sin\left(\beta_0^0\right)\right)^4}{\sqrt{\left(1 - \Delta V \sqrt{\frac{k_0}{g}}\sin\left(\beta_0^0\right)\right)^4 - \sin^2\left(\beta_0^0\right)}}. \quad (5.94)$$

The wave height evolution in a non-uniform current is described by the relation (5.94), depending on the velocity difference ΔV, the initial value of the wave number k_0 and the angle β_0. As can be seen, this expression coincides with the analogous one obtained using monochromatic classical wave theory (Peregrine, 1976). There is a singularity, arising in the ratio (5.94) at the point in which the denominator becomes zero. At this point the current velocity variation is equal to:

$$\Delta V = \sqrt{\frac{g}{k_0}} \frac{1 - \sqrt{\sin(\beta_0)}}{\sin(\beta_0)}. \quad (5.95)$$

This is a turning point of the wave packet propagation, as soon as the group velocity component $C_{gx} = \frac{1}{2}\sqrt{\frac{g}{k}}\frac{k_x}{k}$ becomes equal to zero, i.e. $k_x = 0$. As shown, this singularity is smoothed by the finite width spectrum, producing a finite value of the spectral average statistical wave height.

But in the case of a monochromatic wave it is necessary to obtain a more precise description of the solution. That is why a wave field should be considered in detail in a shear current around the turning point. It follows from the general kinematic relations that the wave packet trajectory can be written in the form (5.61) and (5.62). The component of the group velocity is equal to zero: $C_{gx} = 0$ at the turning point $\beta = \pi/2$ and wave packet propagation occurs in the opposite direction along the Ox axis. In order to obtain a wave pattern near the caustic the asymptotic method, described in Sect. 5.6, can be applied again.

To describe the waves in a shear current with the help of the general integral presentation (5.50), it is sufficient to use the following kinematic relations:

$$\omega = F\left(k_y^0, k_x, x\right) = \text{const};$$

$$\frac{\partial F}{\partial x} = k_y \frac{\partial V}{\partial x}; \quad \frac{\partial^2 F}{\partial k_x^2} = \frac{1}{2}\sqrt{g/k^3}\left(1 - \frac{3}{2}k_x^2 k^{-2}\right). \quad (5.96)$$

In the vicinity of the turning point the component of the wave vector is estimated as: $k_x = \pm\sqrt{\kappa_*^3 (x^* - x)}$, where the $(+)$ sign is referred to the direct wave and the $(-)$ sign to the reflected one. At the turning point $x = x^*$ the following values are defined: $\kappa_* = \frac{1}{2}\left(\frac{\partial F}{\partial x}\sqrt{\frac{\partial^2 F}{\partial^2 k_x}}\right)^{-1}\bigg|_{x=x^*}$, $C_{gx} = 0$, $k_x = k_x^* = 0$, $\sigma^*/\sigma_0 = \sqrt{k_x/k_0}$. Substitution of these relations into the asymptotic expression (5.50) results in the following free surface elevation:

5.7 Wind Wave Transformation in Cross-Velocity Shear Current

$$\eta = a_0 \sqrt[3]{4} \sqrt{\pi \cos(\beta_0)} (\sin(\beta_0))^{\frac{7}{12}} \left(\sigma_0 \frac{\partial V}{\partial x}\right)^{\frac{1}{6}} \bigg|_{x=x^*} \text{Ai}\left[\kappa_*(x-x^*)\right]$$

$$\times \exp\left[i k_x (x - x^*) + i\psi(x^*)\right]. \tag{5.97}$$

Using this formula it is easy to find the amplitude maximum, taking place at $\kappa_*(x - x^*) = 1.02$ for the initial angle $\beta_0 = 42.79°$:

$$a_{\max} = 1.04 a_0 \left(\frac{\sigma_0}{\partial V/\partial x}\right)^{\frac{1}{6}}. \tag{5.98}$$

It should be noted that the expression (5.98) coincides with the earlier derived ratio (Peregrine, 1976), with the exception of using the value $\sigma|_{x=x^*} = \sqrt{gk_y}$ instead of σ_0. As a result, the amplitude maximum is achieved at $\beta_0 = 45°$, and the numerical coefficient in the expression (5.98) makes up 1.065 (instead of $\beta_0 = \arccos\left(\sqrt{7/13}\right) = 42.79°$ and the coefficient of 1.04).

As an example the swell height increase in the field condition will be estimated. Thus, if the wavelength is equal to 50 m and the velocity gradient is $\Delta V/\Delta x = 10^{-4}$, then the swell height increase is equal to ≈ 4.64.

The wave steepness around the caustic can be easily determined using the expressions (5.63), (5.64) and (5.98) as:

$$a_{\max} k^* = 1.51 \, a_0 k_0 (\cos(\beta_0))^{\frac{19}{12}} (\sin(\beta_0))^{\frac{1}{2}} \left(\frac{\sigma_0}{\partial V/\partial x}\right)^{\frac{1}{6}} \bigg|_{x=x^*}. \tag{5.99}$$

The maximum of this value is achieved at $\beta_0 = 31.255°$ and is equal to:

$$(ak)_{\max} = 0.848 \, a_0 k_0 \left(\frac{\sigma_0}{\partial V/\partial x}\right)^{\frac{1}{6}} \bigg|_{x=x^*}.$$

Comparing the formulas (5.57) and (5.98), it is possible to conclude that the wave intensification is more effective in a countercurrent (see Sect. 5.6), although the wave parameters are significantly affected by the shear current as well.

Using the wave estimation method around a caustic, it is possible to take into account not only the current effect as it is, but also simultaneously the bottom relief change, occurring along the Ox axis. In order to do so, it is sufficient to use $F(k_{y0}, k_x, x)$ in the expression (5.96) in the form $\sqrt{gk \, \text{th}(kH(x))} + k_y V(x)$.

A peculiar wave pattern appears in a shear flow with the velocity value $V(x)$ not being a monotonic function of the argument x: $x_1 < x < x_2$. Consider the case when the velocity has its maximum at the centre of the flow $x_m = (x_1 + x_2)/2$ and decreases monotonically with distance from the centre

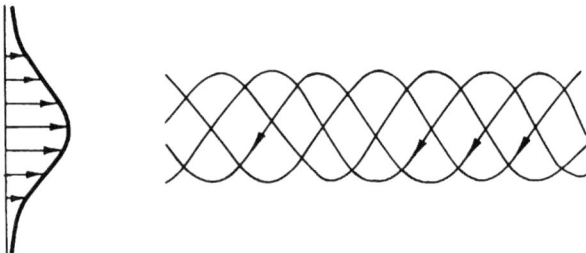

Fig. 5.27. Diagrams of captured wave rays in jet current. The profile of the current velocity distribution is shown on the left

to both sides. If the waves propagate along such a flow, they can be reflected from the caustic located either on one side or the other of the flow centre. The waves propagating in a flow can be entrained by the current in the case of their vector component k directed against the velocity V (see Fig. 5.27).

The waves propagate within the area limited on both sides by caustics. The results described above (5.97) can be applied for the caustics situated at quite a large distance from each other. However, the wave solutions for two caustics cannot be "joined". At the same time there appears the problem of the definition of the eigenvalue for the wave number k, which should satisfy the ratio:

$$\left(M + \frac{1}{2}\right)\pi = \int_{x_1^*}^{x_2^*} k \cos(\beta) \, dx , \qquad (5.100)$$

where x_1^* and x_2^* are the locations of the first and second caustics, and M is the number of zeros of the oscillation between them. In quantum mechanics, the condition (5.100) is known as the problem of captured waves described using the Schrödinger equation (Landau & Lifshits, 1974a). The phenomenon of counter-waves in a current, symmetrical relative to the central axis, is investigated by Peregrine and Smith (1975). The solution for small values $|\beta - \pi/2|$ and closely located caustics is expressed through the Ermith function. It turns out that the captured waves can only exist at current velocities close to the maximum when the range of the angle variations $|\beta - \pi/2|$ across the current is relatively small ($\pm 15°$).

Wave propagation counter to the river flow can be illustrated by the ray diagrams presented in Fig. 5.27. But in the opposite case shear currents with a central velocity maximum force the waves, propagating downstream, to leave the maximum velocity area. The dissipation of such waves on shoals is increased in rivers. The waves propagating against the current are concentrated near the flow centre due to refraction. As a result, they are subjected to smaller dispersion and dissipation, due to the fact that river waves, propagating against the current, are more evident than the waves propagating in the fair direction.

5.8 Freak Wave Problem

Phenomenological review. The size, power and technical equipment on modern ships would allow one to think that there is no danger for a ship sailing everywhere on the ocean. Actually this is not quite so. On June 13, 1968 the *Word Glory* tanker (built in the USA in 1954) under the Liberian flag, while travelling along the South African coast, encountered a freak wave, which broke the tanker into two pieces and led to the death of 22 crew members. A number of similar cases are well known (Mallory, 1974; Sanderson, 1974; Sturm, 1974; Davidan & Lopatukhin, 1982; Lavrenov, 1985a; 1998b).

The term "freak waves" (or abnormal waves, exceptional waves, killer waves, cape rollers and rogue waves) pertains to individual asymmetric waves with a crest of an extremely high slope, in front of which appears a longer and deeper trough than compared with ordinary wind waves. The height of such waves can reach 15–20 m and more, sometimes in a relatively calm sea. The wave often appears suddenly. That is why it is practically impossible for a ship's crew to take any precautionary measures.

Abnormal waves are often observed frequently in different regions of the World's oceans, with strong currents: for example, the Gulf Stream, the Kuroshio among others. Extremely large waves are observed near the southeastern shore of South Africa in the Agulhas Current between East London and Durban (see Fig. 5.28). That is why the region is considered to be very dangerous.

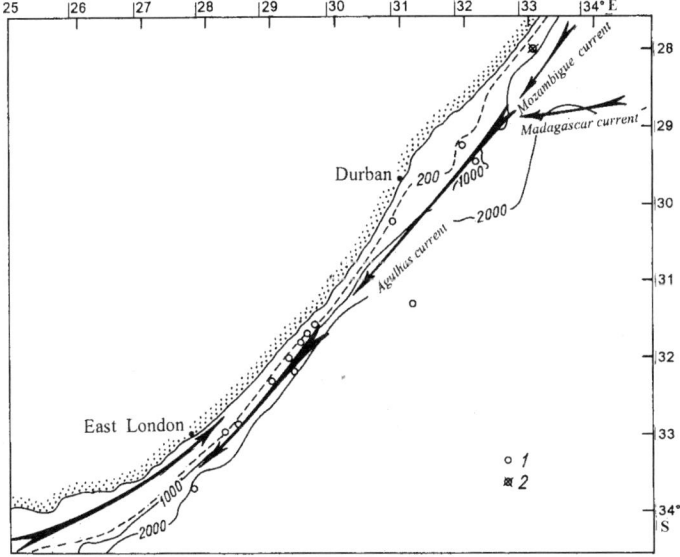

Fig. 5.28. Map of the southeast coast of South Africa. 1 – location of abnormal wave accident; 2 – Location of the "Taganrogsky Zaliv" tanker-refrigerator

A list of cases when abnormal waves were observed up to 1973 is given by Captain J. Mallory (Mallory, 1974), along with an analysis of the corresponding atmospheric and oceanographic conditions which accompanied the appearance of the abnormal waves. In his article, Captain Mallory mentioned 11 cases of catastrophic ship collisions with abnormal waves in the region (see Fig. 5.28). The conditions mainly came down to the conjunction of south-westerly waves with the passage of an atmospheric cold front. In fact, abnormal waves have appeared there more often and the consequences of their action were less dramatic. This could be explained by the weather conditions in the region, when ships had to reduce their speed thus diminishing the destructive effect of the wave action.

On April 27, 1985 the same thing happened to the *Taganrogsky Zaliv*, a ship of the former Soviet Union. Using the information offered by the captain of the ship the accident will be analysed and mathematical simulations will be produced with the help of the spectral theory of the interaction of wind waves and currents.

Description of the accident. On April 27, 1985 at 01.01 pm ship time (11.01 am GMT), the Soviet tanker-refrigerator *Taganrogsky Zaliv*[3] was subjected to an abnormal wave (the point of the accident is denoted by a circle with a cross in Fig. 5.28). As a result, a sailor working on the foredeck was mortally wounded and washed overboard.

The author of this monograph was invited as a wind waves expert to take part in the work of the Commission of the Crimean Transport Prosecuting Office, examining the incident (Lavrenov, 1985a). As a result of the Commission's investigation, a number of peculiar details concerning abnormal wave generation was revealed. They were used later in mathematical modeling of this phenomenon.

On the date under consideration, the *Taganrogsky Zaliv* tanker-refrigerator was sailing from the Indian Ocean to the south-east Atlantic. The possibility of encountering a weather storm is high enough near the Cape of Good Hope that the ship was prepared for sailing in stormy weather. The north-north-east wind was blowing at a speed of $7 \, \text{m s}^{-1}$. At 5 am, it changed direction to south-south-west with the same force. Compared to the previous day the atmospheric pressure was diminishing until the wind changed direction, after which it began increasing. At 8.00 am, the wind became stronger and at 11.00 am it reached $15 \, \text{m s}^{-1}$. At midday the wave impact on the ship was felt. A lifeboat was torn off, and two mooring-line reels were loosened and washed into the water.

After 12.00, the wind speed diminished to $12 \, \text{m s}^{-1}$, and the wind sea became calmer. The wind force did not change during the next three hours,

---

[3] *Taganrogsky Zaliv* is a ship of unrestricted sailing radius. The vessel's length is 164.5 m, the largest width is 22 m, the displacement during the accident was 15,000 tons, and the board height above water was 7 m.

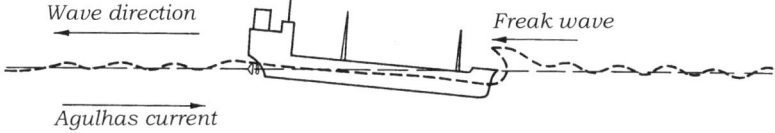

Fig. 5.29. Freak wave collision with the *Taganrorsky Zaliv* described by witnesses

and the wave height did not exceed 5 m and the length was 40–45 m. To overcome the results of wave impact, the boatswain and three seamen were sent to the foredeck. The speed of the ship was diminished to a minimum in order to give safer control of the ship's movement. The ship rode well on the waves. The fore and main deck were not flooded with water.

By one o'clock, the job was almost done on the foredeck. At that moment, the front part of the ship suddenly dipped, and the crest of a very large wave appeared close to the foredeck. It was 5–6 m higher over the foredeck. The wave crest fell down on the ship. One of the seamen was killed and washed overboard. All attempts to save him were in vain. The location of the ship is shown by a cross in Fig. 5.29 (Lavrenov, 1985a).

The weather conditions were suitable for fulfillment of deck repair work. Nobody was able to foresee the appearance of such a wave. When the ship went down, riding on the wave, and its frontal part was stuck into the water, nobody felt the wave's impact. The wave easily rolled over the foredeck, covering it with more than two metres of water (see Fig. 5.29). The length of the wave crest was not more than 20 m.

The weather conditions. It should be noted that the incident occurred in the region of the south-eastern coast of South Africa in the Agulhas Current. From time to time ships encounter abnormal waves there. The situation will be analysed using the synoptic information offered by the captain of the *Taganrogsky Zaliv* tanker.

The synoptic maps of surface atmospheric pressure received onboard the ship on April 26 and 27 are shown in Fig. 5.30. It follows from these charts that the meteorological situation in the navigation region was determined by the high-pressure area (1030 hPa), situated to the western part over the southern tip of Africa and by the low-pressure area (984–990 hPa) with its centre over Marion Island. The cold atmospheric front was moving north-east over the area, where the ship was sailing. As can be seen from the facsimile weather maps the isobars were elongated south-westward to the south of the African continent. This indicates the presence of wind with a force of 9 knots.

Judging by the synoptic situation, wind waves near the south-eastern coast of South Africa must consist of two wave systems. The first one was the wind waves connected with a local wind force of 6–7 knots. The local wind waves presented a system of short steep waves. The second system

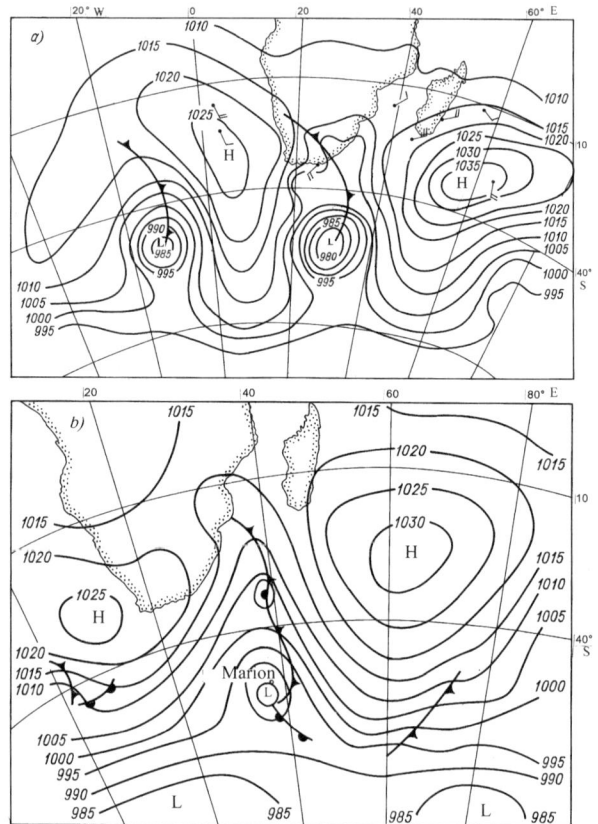

Fig. 5.30. Synoptic charts of the southern Indian Ocean on April 26 (**a**) and April 27 (**b**), 1985

presented long gentle swell waves arriving from southern latitudes. They were generated by the south-westerly wind acting at a fetch of about 1200 miles long within not less than a 24 hour period of time. The wave height must have been increased under the influence of the opposite Agulhas Current. As a result of two superimposed wave systems, the height of individual waves could be significantly increased, especially at the time when the swell and wind wave coincided by phase. Thus, under these conditions, some waves of large height might have occurred at some moment. However, the phenomenon of abnormal waves cannot be explained from the point of view of a linear wave superposition. In particular, it is impossible to explain the wave shape, its asymmetrical profile with a long deep trough, the "hole in the sea" situated in front of the high steep crest and its sudden appearance.

At present, due to the absence of theory and field measurements, it is practically impossible to forecast abnormal waves. However, on the basis of the above described facts and the analysis of similar events presented by Mallory (1974), it is possible to specify the main hydrometeorological conditions, under which the occurrence of abnormal waves is most probable in this region. They are as follows:

1. The appearance of waves is most likely in the Agulhas Current in the area with the maximum of its velocity (3–5 knots), coinciding approximately with a 200 m isobath and deeper. The probability of the appearance of a wave rapidly decreases approaching the coast from the 200 m isobath.

2. Approximately a day before the appearance of abnormal waves, a cyclone with atmospheric pressure of 980–995 hPa is observed on the synoptic map over Marion Island. The isobaths on the map are elongated south-westward to the south of Africa. A cold atmospheric front propagates with a speed of about 15 knots to the north-east along the coast.

3. Several hours before the appearance of the abnormal wave a strong persistent north-easterly wind begins blowing along the south-eastern coast of Africa increasing the maximum current velocity to 4–5 knots.

4. There is a rapid change of wind direction while passing the cold atmospheric front. The change of the north-easterly wind with a force of 6 knots to the south-easterly wind with a force of 6–7 knots occurs within 4 hours.

5. The local south-westerly wind rapidly increases the wind wave, on which the south-west direction swell is superimposed propagating opposite to the Agulhas Current.

The spatial distribution of the speed of the Agulhas Current. At present there are no theories explaining the abnormal wave phenomenon, appearing under the above described conditions. The propagation of a cold atmospheric front results in so-called "frontal swell waves", which in combination with wind waves can generate waves of a greater amplitude (Ivanenkov et al., 1977). However, the appearance of abnormal waves cannot be explained by these factors alone.

Some other factors will be investigated in this monograph, in particular, swell transformation in the Agulhas Current, which is a necessary condition for freak wave generation. According to Smith (1976) the appearance of large waves in the Agulhas Current was explained by a local wave pattern generated by wave reflection from the current. However, the problem of large-scale wave transformation in a current was not considered.

Abnormal waves are typically observed on the ocean surface, where the depth approximately coincides with a 200 m isobath. The latter passes parallel to the coastline. It is the boundary of the continental shelf, where the depth increases sharply to 3–4 km (see Fig. 5.28). That is why it was initially assumed that this abrupt depth change was actually the cause of the abnormal wave formation (Mallory, 1974). But, in reality, the bottom relief in a 200 m depth can influence only waves with lengths more than 500 m. It should be noted that typical horizontal dimensions of abnormal waves are much shorter.

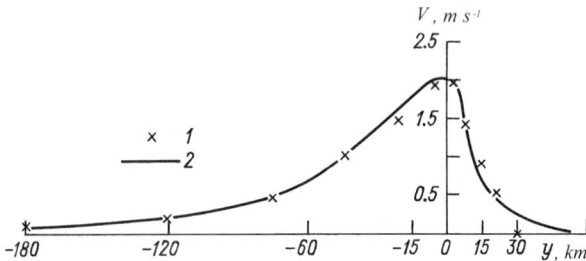

Fig. 5.31. Horizontal current velocity distribution in transversal direction. 1 – velocity values according to Schuman (1976); 2 – approximation (5.101)

The sharp depth variation at the continental shelf margin results in the maximum values of the Agulhas Current, having a jet profile directed parallel to the shoreline. The transversal profile of the Agulhas Current velocity distribution is not changed much along its entire length from the Mozambique and South Madagascar current confluence at latitude 30 degrees south (see Fig. 5.28). A typical horizontal profile of the current velocity distribution in the transversal direction (Schuman, 1976) is shown in Fig. 5.31. The Ox axis of the coordinate system is chosen along the current velocity direction, and the Oy axis is in the perpendicular direction.

The transverse profile of the current velocity $V = \{V_x(y), 0\}$ in this area can be quite accurately approximated by the relation:

$$V_x = \frac{b_1}{1 + b_2 y^2}, \qquad (5.101)$$

where $b_1 = 2.2 \, \mathrm{m\,s^{-1}}$; $b_2 = 6.25 \times 10^{-10} \, \mathrm{m^{-2}}$ at $y < 0$; $b_2 = 10^{-8} \, \mathrm{m^{-2}}$ at $y > 0$.

The current velocity distribution (5.101) is assumed to be valid at $x < 0$ in the chosen coordinate system. The point $\{x = 0, y = 0\}$ is assumed to correspond to the current velocity midstream at the point with coordinates $\{27°\mathrm{E}, 34°\mathrm{S}\}$.

The Agulhas Current diverges in a fan-shaped manner at more southern latitudes, becoming weaker.

In accordance with the Atlas (1977), it can be assumed that in the case $x > 0$, the horizontal current velocity components $V = \{V_x(x,y), V_y(x,y)\}$, satisfying the continuity equation, are approximated by the equations:

$$V_x = \frac{b_1}{(1 + b_2 y^2)(1 + b_3 x^2)};$$

$$V_y = \frac{2 b_1 b_3 x}{(1 + b_3 x^2)\sqrt{b_2}} \arctan\left(\sqrt{b_2}\, y\right), \qquad (5.102)$$

where $b_3 = 0.6 \times 10^{-11} \, \mathrm{m^{-2}}$.

Spectral solution of swell transformation in the Agulhas Current.
The swell propagation from southern latitudes, with actually no current, towards the increasing Agulhas Current will be considered. The geometric optics approximation can be applied in this case, because typical horizontal scales of current velocity variations significantly exceed the horizontal wave dimensions. Thus, the results of Sect. 5.2 can also be applied.

At a greater distance from the investigated area (see Fig. 5.28), the initial spectrum $S_0(\omega, \beta_0)$ is assumed to be prescribed at $x > 1.5 \times 10^6$ m, where the current speed can be neglected: $\sqrt{V_x^2 + V_y^2} \approx 0$. The spectrum can be described with the help of the swell spectrum approximation (5.16) with $n = 5$, $\omega_{\max} = 0.5 \,\mathrm{rad\,s^{-1}}$. It should be noted that the initial value of the angle β_0, included in $S_0(\omega, \beta_0)$, is a function of the values ω, β and the coordinates $\{x, y\}$.

In order to find β_0, it is necessary to solve the set of Hamiltonian equations (5.2), which can be written using new variables:

$$\frac{\mathrm{d}y}{\mathrm{d}x} = \frac{V_y + \frac{1}{2} f \sin(\beta)}{V_x - \frac{1}{2} f \cos(\beta)};$$

$$\frac{\mathrm{d}\beta}{\mathrm{d}x} = \frac{\sin(2\beta)\frac{\partial V_x}{\partial x} - \sin^2(\beta)\frac{\partial V_y}{\partial x} + \cos^2(\beta)\frac{\partial V_x}{\partial y}}{V - \frac{1}{2} f \cos(\beta)};$$

$$\frac{\mathrm{d}t}{\mathrm{d}x} = \frac{1}{V - \frac{1}{2} f \cos(\beta)}, \qquad (5.103)$$

where

$$f = \frac{1}{2}\frac{g}{\omega}\left\{1 + \left[4 - \frac{\sqrt{V_x^2 + V_y^2}}{g}\omega\cos\left(\beta + \arcsin\left(\frac{V_y}{\sqrt{V_x^2 + V_y^2}}\right)\right)\right]\right\}.$$

The Runge–Kutta method is used for a discrete set of frequencies ω_1 (from $\omega_1 = 0.37\,\mathrm{rad\,s^{-1}}$ to $\omega_{12} = 0.981\,\mathrm{rad\,s^{-1}}$) and the angles β_j (from -75 to $75°$ with a step size of $15°$) to obtain the values β_{0i}. They correspond to the rays arriving at the given points $\{x = 0, y - y_n\}$ from the region with practically no current $x > 1.5 \times 10^6$ m.

Typical rays $y = y(x, \omega, \beta)$, arriving at the point $\{x = 0, y = 0\}$, are depicted in Fig. 5.32. They are characterized by oscillations relative to the Ox axis. The oscillation frequencies are increased with rays approaching the intense current area. A peculiar wave channel (i.e. a waveguide) appears in the current with the wave rays being alternately reflected from one or other caustics situated on different sides of the current midstream.

The velocity distribution in the Agulhas Current is not symmetric relative to the maximum velocity line. The velocity gradients at the eastern midstream side are less than those at the western side (see Fig. 5.31). The current gradient at the western side is additionally increased due to the coastal

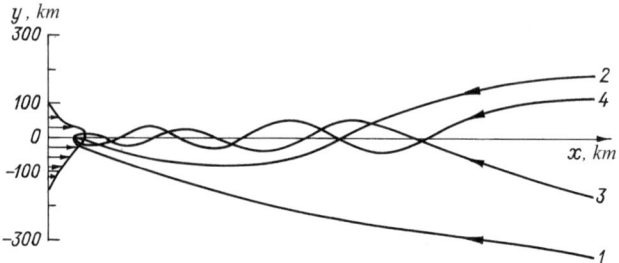

Fig. 5.32. Wave rays arriving at a given point in the Agulhas Current with frequency ω: 1 – 0.20 rad s^{-1}; 2 – 0.38 rad s^{-1} (at angle $\beta_0 = -30°$); 3 – 0.76 rad s^{-1}; 4 – 0.93 rad s^{-1} (at angle $\beta_0 = 30°$)

countercurrent (its velocity comprising $V_x \approx 0.5$ m s^{-1}) taking place during passing of the cold atmospheric front (Mallory, 1974). This transversal current velocity results in greater frequencies of caustic distribution (for waves of different lengths and directions) at the western side than at the eastern side. The wave channel is narrowed with decreasing value of x, whereas the channel width is not practically changed at $x < 0$. This wave ray behaviour becomes more pronounced with increasing frequency ω and decreasing angle β. Thus, the unusual nature of the current velocity distribution results in capturing wave rays. That is why it re-distributes the wave energy in space.

The wave spectrum in the current $S(\omega_i, \beta_j, y_n)$ is obtained at $x = 0$ as a result of numerical simulation of (5.103), using the spectral relation (5.9). The spectrum values with rays not contributing any energy (for example, going from the shore, etc.) are assumed to be zero. In order to obtain the total wave energy, the spectrum is integrated numerically over the frequency ω and the angle β. The estimation results (see Fig. 5.33) are obtained for the relative mean wave height h/h_0 in the current (where h_0 is the initial wave height without any current), whereas values of the y coordinate are plotted along the horizontal axis. The maximum value h/h_0 equal to 2.19 is estimated as being in the midstream. The ratio h/h_0 is sharply decreased along the distance to the shoreline. The relative excess of the wave height is not greater than 10 per cent at a distance of $y = 19 \times 10^3$ m from the maximum current velocity line.

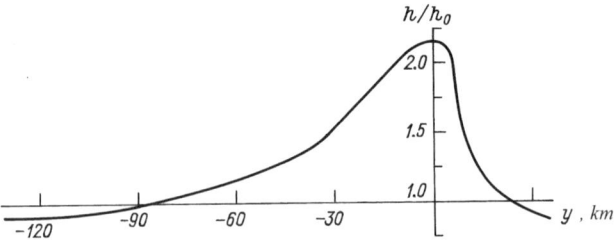

Fig. 5.33. Mean wave height distribution across current

Seaward to the midstream, the relative height h/h_0 is decreased to a lesser extent with a ten-percent wave height excess, occurring at $y = -70$ km. But the reversed case (i.e., $h < h_0$) is even observed at $|y| > 90$ km.

Later on, our theoretical estimation of wave height enlargement in the Agulhas Current was proved by satellite observations, obtained by Grundligh and Rossouw (1995).

Thus, the spatial distribution of the wave height in the Agulhas Current permits us to come to the following conclusion. The probability of observing abnormal waves is much less in going towards the shore at some distance from the line of maximum velocity with the 200 m isobath, than going the same distance from the maximum velocity current in the direction of the open sea.

Conclusion. The obtained results indicate that the peculiar velocity distribution of the Agulhas Current induces refraction of the south-west swell propagating over a wide area of the south-western Indian Ocean. This leads to the fact that the swell turns towards the current velocity maximum. The swell is trapped, intensified by the countercurrent and localized in the vicinity of the maximum velocity, propagating along the south-east coast of South Africa. As a result, there is a significant concentration of wave energy density, i.e. a focusing takes place in the given area resulting in abnormal wave generation.

Navigators sailing along the south-east coast of South Africa should be recommended to remember that freak waves can suddenly appear in that area. It may happen in the case of the south-west swell propagating against the current. If the atmospheric pressure decreases while the wind increases, changing its direction from north-east to south-west, the ship should leave the 200-m depth region with the strongest current, turning either towards the coast, with the depth being less than 100 m or seaward to deeper than the 200-m isobath areas.

5.9 Influence of Vertical Non-Uniform Current on Wind Wave Transformation

Till now we have considered the transformation of wind waves in a current that do not varying vertically. But this assumption is not always fulfilled in natural conditions as long as the currents are usually vertically non-uniform. Thus, in most cases the wave estimation error appears due to ignoring the current's vertical non-uniformity in the bottom and near-surface layers. That is why it is important to estimate the flow kinematics not only in the near-bottom layer in sediment transportation problems, but also in wind wave generation in the surface layer.

In order to describe the wave evolution in a current in general, the adiabatic invariant conservation law should be applied (see Chap. 1). This is valid for a current that is non-uniform both along the horizontal and the vertical direction. Generally there are some additional difficulties in solving of this problem. In order to determine the dispersion ratio (1.37), the marginal problem (1.35), (1.36), including the current velocity dependence on the vertical coordinate, must be solved.

In the case of the current velocity being changed only along one horizontal direction, the problem is connected with solving the equation of motion of Orr–Sommerfeld type. Its precise analytical solution can be obtained only for the simplest cases. The dispersion ratios for surface waves in water with a simple depth profile (linear, parabolic, logarithmic, etc.) can be obtained in explicit form (Hidy & Plate, 1966). The analytical solution for the piecewise-linear velocity profile approximation has also been obtained (Thomas, 1981). The solutions can be found numerically or by asymptotic methods for more general profiles (Goncharov & Leykin, 1983). Brink-Kjer & Jonsson (1975) derived the conservation laws for waves in a current with a linear depth velocity shear profile.

Investigations of waves in a current with a vertically non-uniform velocity profile have been published in several papers (Taylor, 1955; Peregrine & Smith, 1975; Goncharov & Leykin, 1983; Dreyzis et al., 1986; Kantargi et al., 1989 etc.).

In this section an attempt is undertaken to solve the problem of wave transformation in a current with velocity depth shear. It should be noted that the current is considered to be non-uniform along its direction.

Wave transformation in a current with linear depth velocity shear.
A wave transition from calm water to a current with linear depth velocity profile is considered as follows:

$$V(z) = V_{\mathrm{m}} + \Omega\left(z - H/2\right), \qquad (5.104)$$

where V_{m} is the mean depth current velocity; $\Omega = (V_{\mathrm{S}} - V_{\mathrm{B}})/H$ is the current vorticity; and V_{S} and V_{B} are the surface and bottom current velocities, respectively (see Fig. 5.34).

The wave evolution will be described in a non-uniform current within the framework of the geometrical optics approximation. The parameters $V_{\mathrm{m}}, V_{\mathrm{S}}, V_{\mathrm{B}}, \Omega$ are assumed to be changed sufficiently slowly horizontally compared with the wavelength. The choice of the simple current profile (5.104) is defined by the possibility of obtaining the dispersion ratio and the adiabatic invariant in analytical form (Brink-Kjer & Jonsson, 1975). Only one parameter characterizing the velocity depth shear is added to the current (5.104) compared with the uniform case.

In order to describe the wave motion in the vertically non-uniform current $V(z)$, the Rayleigh equation relative to the vertical component amplitude $W(z)$ can be applied (Peregrine, 1976):

$$\frac{\mathrm{d}^2 W}{\mathrm{d}z^2} - \left(k^2 + \frac{k}{Vk - \omega}\frac{\mathrm{d}^2 V}{\mathrm{d}z^2}\right)W = 0; \qquad -H \le z \le 0 \qquad (5.105)$$

5.9 Vertical Non-Uniform Current Effect

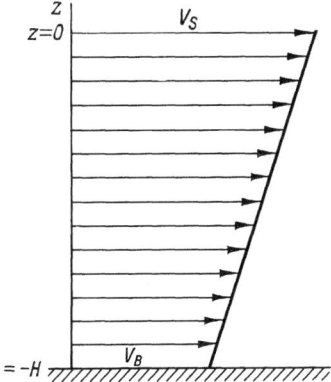

Fig. 5.34. Velocity profile of vertically non-uniform current

The equation (5.105) is obtained directly from the equations of motion presented in Chap. 1. The boundary conditions at the free water surface can be written as follows:

$$\frac{dW}{dz}(V-\omega/k)^2 = \left[g + (V-\omega/k)\frac{dV}{dz}\right]W\,; \quad z = 0\,. \tag{5.106}$$

The boundary conditions at the bottom are written in the traditional form:

$$W = 0\,, \quad z = -H\,. \tag{5.107}$$

The dispersion ratio is easily determined by solving (5.105)–(5.107) for the current velocity linear profile (5.104). Its expression was first obtained by Biesel (1950) and can be written as:

$$\omega = kV_S - \frac{\Omega}{2}\text{th}(kH) + \frac{1}{2}\sqrt{\Omega^2\text{th}^2(kH) + 4gk\,\text{th}(kH)} \tag{5.108}$$

Using the relation (5.108) and the conditions of frequency conservation along the trajectory of wave propagation, the solution for the wave number can be presented in the form:

$$\frac{V_S}{c_0}\frac{k}{k_0} - \frac{1}{2}\frac{\Omega}{k_0 c_0}\text{th}(kH)$$
$$+ \frac{1}{2}\left[\left(\frac{\Omega}{k_0 c_0}\right)^2\text{th}^2(kH) + 4\frac{k}{k_0}\text{th}(kH)\text{th}^{-1}(k_0 H)\right]^{\frac{1}{2}} - 1 = 0\,, \tag{5.109}$$

where the index "0" is related to wave elements in the area without any current. The relative wave number is obtained as a function of the Frude number V_S/c_0, the non-dimensional water depth $k_0 H$ and the non-dimensional vorticity $\Omega/(k_0 c_0)$.

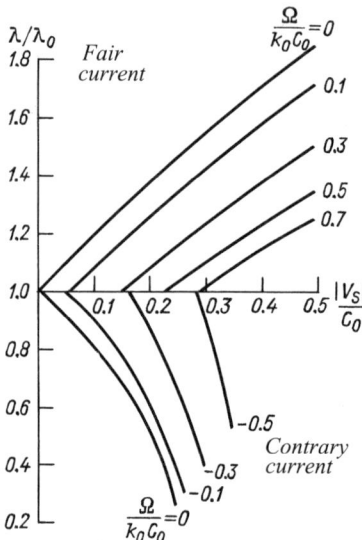

Fig. 5.35. Wavelength variation in current for different vorticity

After the equation (5.109) is derived in its transcendental form, it should be analysed in its ultimate cases, obtaining the solution to the problem in explicit form.

In the case of deep water the solution of (5.109) can be written in the form:

$$\frac{k}{k_0} = \frac{2\left(1 + \frac{\Omega}{k_0 C_0}\right)}{2\frac{V_S}{C_0} + \frac{V_S}{C_0}\frac{\Omega}{k_0 C_0} + 1 \pm \left[\left(\frac{V_S}{C_0}\frac{\Omega}{k_0 C_0} + 1\right)^2 + 4\frac{V_S}{C_0}\right]^{1/2}}, \qquad (5.110)$$

where the ($\pm$) sign in the denominator indicates the ambiguity of determining the wave number in a current, dependent on its initial value k_0 and the current velocity parameters. Similarly to the case of the vertically uniform current, the (+) sign can be referred to direct waves and the (−) sign to reverse waves reflected by the horizontal non-uniform current.

The direct wave dependence is presented in Fig. 5.35 in accordance with (5.110) for the different values of non-dimensional vorticity $\Omega/(k_0 C_0)$. The effect of current vorticity on the wavelength transformation is observed.

As for the shallow water case the following solution of (5.109) can be obtained:

$$\frac{k}{k_0} = \left\{\frac{V_S}{C_0} - \frac{1}{2}\frac{\Omega H}{C_0} + \frac{1}{2}\left[\left(\frac{\Omega H}{C_0}\right)^2 + 4\right]^{1/2}\right\}^{-1}. \qquad (5.111)$$

5.9 Vertical Non-Uniform Current Effect

Considering the ratios (5.110) and (5.111), one can expect a less significant effect of velocity depth shear in shallow water compared to the deep water case.

It is interesting to compare solutions of (5.109)–(5.111) with the case of wave transformation in a vertically uniform current. The velocity of the latter is assumed to be equal to the averaged (by depth) non-uniform current velocity. Thus, this ratio can be written as:

$$\frac{V_S}{C_0} = \frac{V_m}{C_0} + \frac{1}{2}\frac{\Omega H}{C_0} \qquad (5.112)$$

It can then be derived from (5.111):

$$\frac{k}{k_0} = \left\{\frac{V_m}{C_0} + \frac{1}{2}\left[\left(\frac{\Omega H}{C_0}\right)^2 + 4\right]^{1/2}\right\}^{-1}. \qquad (5.113)$$

It can be seen from (5.113) that the increase of the modulus of the vorticity (with a constant averaged current velocity V_m) both in a fair and in a countercurrent leads to a wavelength increase.

Now the determination of the wave height in a current will be considered. The adiabatic invariant conservation (see Chap. 3) for the current with a linear velocity depth shear in the absence of wave energy dissipation can be written in the following form:

$$\frac{\partial}{\partial x}\left\{\frac{E}{\sigma_{rm}}C_g\right\} = 0, \qquad (5.114)$$

where $E = (1/16)\rho h^2 (2g - \Omega c_S)(\sigma_S/\sigma_m)$ is the wave energy density; c_S is the phase velocity of waves relative to the surface current velocity; σ_m, σ_S are the wave frequencies relative to the mean and surface velocities, respectively; $\sigma_m = \omega - kV_m$, $\sigma_S = \omega - kV_S$, and C_g is the absolute value of the group wave velocity. Unlike the equation of ordinary wave action conservation for the case of a vertically uniform current, the equation (5.114) contains the dependence of the vertical current vorticity.

The expression for the group velocity is obtained with the help of differentiation of the dispersion dependence (5.108) with respect to the wave number:

$$C_g = V_S + \frac{g(1+E) - \Omega E c_S}{2g - \Omega c_S} c_S,$$

$$E = \frac{2kH}{\text{sh}(2kH)}. \qquad (5.115)$$

According to (5.114), the wave height in the area with a current is determined by the ratio:

$$\left(\frac{h}{h_0}\right)^2 = \left(\frac{\sigma_S}{\sigma_0}\right)\left(\frac{C_{g0}}{C_g}\right)\frac{1}{1-\frac{1}{2}\frac{\Omega}{g}c_S} . \tag{5.116}$$

The expression (5.116) for ultimate cases will be analysed. The following equation is obtained for the deep water case:

$$\left(\frac{h}{h_0}\right)^2 = \frac{1}{2}\left\{2\frac{V_S}{c_0} + 2\left[\left(\frac{\Omega}{k_0 c_0}\right)^2 + 4\frac{k}{k_0}\right]^{-1/2}\right\}^{-1}$$

$$\times\left[\sqrt{\left(\frac{\Omega}{k_0 c_0}\right)^2 + 4\frac{k}{k_0}} - \frac{\Omega}{k_0 c_0}\right]$$

$$\times\left\{1 - \frac{1}{4}\frac{k_0}{k}\frac{\Omega}{k_0 c_0}\left[\sqrt{\left(\frac{\Omega}{k_0 c_0}\right)^2 + 4\frac{k}{k_0}} - \frac{\Omega}{k_0 c_0}\right]\right\}^{-1} . \tag{5.117}$$

There is the following solution for shallow water:

$$\left(\frac{h}{h_0}\right)^2 = \frac{1}{2}\left(\frac{k}{k_0}\right)^2\left[\sqrt{\left(\frac{\Omega H}{c_0}\right)^2 + 4} - \frac{\Omega H}{c_0}\right]$$

$$\times\left\{1 - \frac{1}{4}\left(\frac{k_0}{k}\right)^2\frac{\Omega H}{c_0}\left[\sqrt{\left(\frac{\Omega H}{c_0}\right)^2 + 4} - \frac{\Omega H}{c_0}\right]\right\}^{-1} . \tag{5.118}$$

It should be noted that in the case of a current velocity with zero shear $\Omega = 0$, the well-known solutions for wave transformation by a depth uniform current are derived from (5.117) and (5.118).

Wave blocking in a current with linear depth velocity profile. The wave blocking kinematics by a countercurrent with a non-uniform depth velocity profile will be considered only for the one-dimensional case. The blocking condition can be written in the form:

$$C_g = \frac{\partial \omega}{\partial k} = 0 . \tag{5.119}$$

The wave blocking condition for the finite depth case can be derived as:

$$V_S = \frac{H}{2}\Omega\,\text{ch}^{-2}(kH)$$

$$- \left[\frac{1}{2}\Omega^2 H\,\text{th}(kH)\,\text{ch}^{-2}(kH) + g\,\text{th}(kH) + gkH\,\text{ch}^{-2}(kH)\right]$$

$$\times\left[\Omega^2\,\text{th}^2(kH) + 4gk\,\text{th}(kH)\right]^{-1/2} . \tag{5.120}$$

5.9 Vertical Non-Uniform Current Effect

In the deep water case, the following condition can be obtained:

$$V_S^* = -\frac{g}{\sqrt{\Omega^2 + 4gk}}, \qquad (5.121)$$

In the shallow water case it is as follows:

$$V_S^* = \frac{H}{2}\Omega - \frac{1}{2}\sqrt{\Omega^2 H^2 + 4gH}. \qquad (5.122)$$

The mean velocities of the countercurrent at the blocking point for the uniform profile $V = V_{\text{me}}$ and the linear-shear profile $V = V_{\text{ml}} + \Omega(z - H/2)$ in shallow water will be compared. Using the expression (5.122) the mean velocity of a linear-shear current, creating wave blocking, is found to be equal to:

$$V_{\text{ml}} = -\frac{1}{2}\sqrt{\Omega^2 H^2 + 4gH}. \qquad (5.123)$$

The ratio of blocking current velocity can be obtained dividing the expression (5.123) by the uniform current velocity $V_{\text{me}} = \sqrt{gH}$:

$$\frac{V_{\text{ml}}}{V_{\text{me}}} = \frac{1}{2}\sqrt{\frac{\Omega^2 H}{g} + 4}. \qquad (5.124)$$

It can be seen from (5.124) that the current speed is slightly different for shallow water mean velocities of the uniform and linear-shear current at the blocking point. Thus, having the values of $\Omega = 0.1\,\text{s}^{-1}$ and $H = 10\,\text{m}$, the estimation $V_{\text{ml}}/V_{\text{me}} \approx 1.001$ can be obtained.

For the deep water case, the blocking wave parameters in the linear-shear current can be compared with the ones in the uniform current equal to the surface velocity of a linear-shear current. For the latter the maximum vorticity value will be assumed to correspond to $V_B = 0$. Designating the wave number at the blocking point in a linear-shear current as k_1 and the same for the uniform current as k_2, the following expression can be obtained with the help of (5.121):

$$\frac{k_1}{k_2} = 1 - \frac{\Omega^2}{4gk_2}. \qquad (5.125)$$

It is obvious that k_1 is always less than k_2, i.e. the wavelengths at the blocking point by a linear-shear current are always greater than the wavelengths at the blocking point by a uniform current with $V = V_S$. This coincides with the graphs (see Fig. 5.35), where the curves in a countercurrent finish at the blocking points.

Now the blocking point position for linear-shear and a uniform current with equal mean velocities can be compared for deep water. The wave number in a uniform current is designated as k_3. Using (5.123), the following can be obtained:

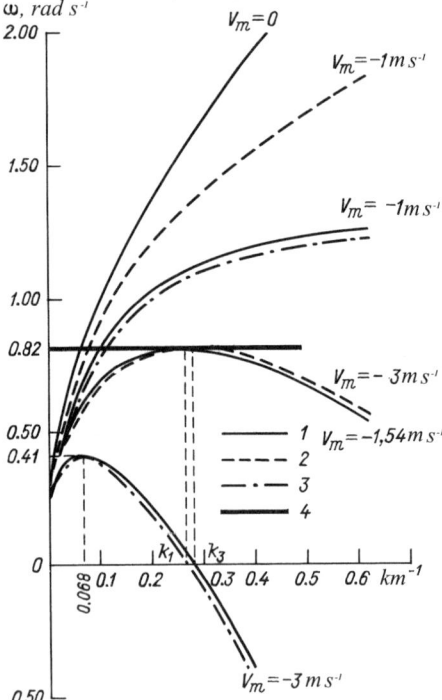

Fig. 5.36. Dispersion curves for waves in countercurrent: 1 – linear-shear current with $V_S = 2V_m$, $\Omega = -0.05\ \text{s}^{-1}$; 2 – uniform current with velocity V_m; 3 – depth uniform current with velocity equal to surface velocity of linear-shear current; 4 – wave packet with $0.82\ \text{rad s}^{-1}$ frequency

$$\frac{k_1}{k_3} = \frac{1}{4} - \frac{\Omega^2}{4gk_3}. \tag{5.126}$$

Thus, the wave numbers k_1 and k_3 can be a factor of four different at equal mean velocities, whereas k_1 is always less than k_3.

For illustration, the dispersion ratio (5.109) can be written in the coordinates ω, k in the deep water case with different countercurrent velocities (see Fig. 5.36). The dispersion curves corresponding to a linear-shear current with $V_S = 2V_m$, $\Omega = -0.05\ \text{s}^{-1}$ are marked by a solid line. The dashed line refers to a uniform current with velocity V_m, i.e. the velocity is equal to the mean velocity of a linear-shear current. The dash-dotted lines denote the dispersion curves for a depth uniform current with a velocity equal to surface velocity of a linear-shear current. The horizontal line corresponds to the wave packet with frequency $\omega = 0.82\ \text{rad s}^{-1}$.

The dispersion curves denoted by the solid and dash-dotted lines are slightly different according to the aforementioned analysis. But the dispersion curves designated by dotted lines are significantly different from them.

The blocking of waves with frequency $\omega = 0.82\,\text{rad}\,\text{s}^{-1}$ takes place for a linear-shear current at $V_m \approx -1.5\,\text{m}\,\text{s}^{-1}$, and for a uniform current at $V_m \approx -3\,\text{m}\,\text{s}^{-1}$, i.e. twice as great as the absolute value. But the wave numbers at the blocking point are approximately equal. The linear-shear current with mean velocity $V_m \approx -3\,\text{m}\,\text{s}^{-1}$ blocks a wave with half the frequency, whereas the wave number at the blocking point is approximately a third of its value in a uniform current.

As shown, the current velocity depth shear significantly affects the wave transformation in deep water. It is not enough to characterize the current only by the surface velocity in solving problems of wave–current interaction in deep water. It is also necessary to estimate the velocity depth shear. In shallow water, the effect of velocity shear can be neglected in comparison with the current effect. That is why it is admissible in this case to describe the current using only the surface velocity or the average depth velocity.

Approximate wave dispersion ratio for an arbitrary vertical current velocity profile. The simplest case of a vertical current profile is considered above. But, in most cases, it is rather difficult to find an exact solution.

That is why it is interesting to discuss a method for obtaining approximated dispersion formulas for arbitrary vertical current velocity profiles. It was proposed by Stewart and Joy (1974) and elaborated by Skop (1987) and Kirby and Chen (1989). The value $V/c \leq O(1)$ is taken as a small parameter for solving the initial problem (5.105)–(5.107) using the perturbation method.

In the general case of an arbitrary current velocity $\boldsymbol{V} = \{V_i(x,y,z,t)\}$ ($i = 1, 2$) the approximate dispersion ratio can be written as:

$$\omega = \sigma(k) + k_i V_{\text{eff}_i}. \tag{5.127}$$

The velocity V_{eff_i} is an "effective value" of the current speed. It is averaged over the depth of wave motion penetration:

$$V_{\text{eff}_i}(x,y,t) = \frac{2k}{\text{sh}(2kH)} \int_{-H}^{0} V_i(x,y,z,t)\,\text{ch}(2k(H+z))\,dz. \tag{5.128}$$

The expression (5.127) is obtained for the case $\sigma(k)/k \gg (V_1^2 + V_2^2)^{1/2}$, with the wave phase velocities being much greater than the current speed.

Integrating (5.128) partially and substituting the result into the expression (5.127), the following approximate relation can be obtained:

$$\omega = \sigma(k,h) + k_i V_{S_i} - \frac{k_i}{2k}\frac{\partial V_{S_i}}{\partial z} + O\left(\frac{1}{kH}\right), \tag{5.129}$$

where the index "S" denotes currents and their derivatives on a free water surface $z = 0$. The expression (5.129) is identical to the asymptotic relation obtained by Peregrine and Smith (1975) for short waves (i.e. $kz_V \gg 1$, where z_V is the scale of the vertical change of current velocity) propagating in currents with velocity depending on the depth.

As pointed out by Skop (1987), in the case of the value V_{eff_i} being a continuous function of the wave number k, (5.128) provides a satisfactory approximation for the dispersion ratio between large and small wave numbers. The applicability of (5.129) for intermediate wave number values requires empirical testing or verification by the existing accurate solutions for depth dependent currents.

The effective current velocity is determined with the help of (5.128) for the case of a current velocity profile with constant vorticity (5.104) as:

$$V_{\text{eff}} = V_s - \frac{\Omega}{2k}\text{th}(kH) . \qquad (5.130)$$

Comparing the approximate formulas (5.127), (5.130) and the accurate solution of the problem (5.108), it can be concluded that their agreement is quite satisfactory. There are some differences decreasing with the relative depth. The maximum deviation in the phase velocity value is achieved at $kH \to 0$ and is equal to $\frac{1}{2}\sqrt{(\Omega H)^2 + 4gH}$.

A comparison of the other approximate solution, namely, for the surface jet current and laboratory measurements, is made by Kantargi (1995). The current velocity profile is given as a piecewise-linear approximation:

$$V(z) = \begin{cases} -V_0(1 + z/D) & -D \leq z \leq 0 ; \\ 0 & z > D . \end{cases} \qquad (5.131)$$

Substituting the current velocity (5.131) into (5.127), (5.128) gives an approximate expression for the dispersion relation as:

$$\omega = (gk\,\text{th}\,(kD))^{1/2} - kV_0 + \frac{V_0}{2H}\text{th}(kD) . \qquad (5.132)$$

Comparison of the wave elements (5.132) with the obtained data for different conditions including wave blocking in a countercurrent reveals good agreement.

Some peculiarities of the influence of a vertical non-uniform current on the wind wave spectrum. The presence of a non-uniform current with velocity depending not only on the horizontal, but also on the vertical coordinates, influences the waves in a more complex and diverse way than a current independent of the vertical coordinate. The vertical non-uniformity of the current velocity can significantly influence the wave blocking, redistributing the spatial location of the wave turning points and caustics, where the wave heights reach anomalously large values. As shown above, the use of the spectral approach removes the singularity occurring in the classical solution. The statistical averaging used in the spectral approach helps smooth

5.9 Vertical Non-Uniform Current Effect

too large wave heights. As shown in Sect. 5.5, the singularity arising in the classical solution for waves propagating in the increasing countercurrent is eliminated around the caustic and the calculation results are in sufficiently good agreement with full-scale observations. At the same time the vertical non-uniformity of the flow seems to change significantly the solution.

As an example, the wave propagation in a countercurrent with its increasing value dependent not only on the horizontal coordinate x, but also on the vertical z coordinate: $V = \{-V(x,z), 0\}$ will be considered. The wave packet is prevented from arriving at the point where the wave group velocity is equal to the current speed and opposite to its direction. In this case wave blocking takes place. The waves start propagating in the reverse direction. The current speed can be determined at this point with the help of the effective velocity value (5.127), (5.128):

$$C_{gx} = \frac{\partial \omega}{\partial k} = \frac{\partial \sigma}{\partial k} - \frac{\partial}{\partial k}(kV_{\text{eff}}(x,k))$$

$$= \frac{1}{2}\sqrt{\frac{g}{k}} - V_{\text{eff}}(x,k) - k\frac{\partial}{\partial k}V_{\text{eff}}(x,k), \qquad (5.133)$$

where $V_{\text{eff}}(x,k) = 2k \int_{-\infty}^{0} V(x,z)e^{2kz}\, dz$.

Using this condition an integral relation is obtained to determine the current velocity at the blocking point (when $C_{gx} = 0$):

$$\frac{1}{2}\sqrt{\frac{g}{k}} = 2k \int_{-\infty}^{0} V(x,z)e^{2kz} + \frac{\partial}{\partial k}(kV(x,z)e^{2kz})\, dz$$

$$= 4k \int_{-\infty}^{0} V(x,z)(1+kz)e^{2kz}\, dz\ .$$

Integrated by parts, this expression can be transformed into the following form:

$$\frac{1}{2}\sqrt{\frac{g}{k}} = 2k \int_{0}^{\infty}\left(V - z\frac{\partial V}{\partial z}\right)e^{-2kz}\, dz\ . \qquad (5.134)$$

Applying the reverse Laplacian transformation (i.e. the so-called Mellin transformation) to (5.134), an ordinary differential equation relative to the current velocity can be obtained:

$$\frac{\partial V}{\partial z}z - V + qz^{(1/2)} = 0\ , \qquad (5.135)$$

where $q = \sqrt{2g/\pi}$.

It is easy to find a solution for (5.135) in analytical form as:

$$V = 2q\sqrt{-z} + z \text{ const} \qquad (z \leq 0), \qquad (5.136)$$

where "const" is the constant of integration, prescribed by the problem formulation.

The profile of the current velocity (5.136) is shown in Fig. 5.37. It should be noted that the value $V(z)$ is not dependent on the wave number k. In the case of waves propagating in increasing countercurrent and its value with the vertical profile (5.136) being achieved, the blocking of all wave spectrum components will take place at one spatial point. This is the point where the location of caustics of all spectral components coincides. The spectral solution does not eliminate these singularities, as was done in case of a vertically uniform current. A more intensive increase of the wave height and total energy is observed in the vicinity of this point.

It should be noted that the solution (5.136) is formally obtained using the approximate dispersion relation (5.127), (5.128), and its correct use is limited by the condition of the small vorticity parameter $\Omega^2/gk \leq 1$ (Skop, 1987). This means that the vorticity of the current velocity profile should be less than the wave frequency. Thus, the applicability condition of the solution (5.136) can be written in the form:

$$\frac{\Omega^2}{gk} = \frac{1}{gk}\left(\frac{\partial V}{\partial z}\right)^2 = \frac{1}{gk}\left(\sqrt{\frac{2g}{\pi z}} + \text{const}\right)^2 \leq 1. \qquad (5.137)$$

In the case with const = 0, the condition (5.137) can be written as $\Omega^2/gk = 2/(zk\pi) \leq 1$. Due to the value z becoming zero this condition becomes unfulfilled in the uppermost layer of the surface current.

It can be expected that the considered solution is realized locally, i.e. neither in entire surface layer, nor for all wave spectrum components. Even in the case when the blocking is simultaneously observed within some range of wave numbers, a more intensive increase of the total wave height and, correspondingly, wave breaking could be expected. This situation (see Fig. 5.37) might be observed in the presence of a background current entraining a significant water layer with the surface countercurrent, probably of drift origin, being superimposed on it. The described wave blocking appears in the case with the wave propagation direction coinciding with the drift current direction and being opposite to the background flow.

In conclusion it should be noted that the vertical non-uniform current influences the waves in a sufficiently diverse way. Another aspect of wave interaction with a vertical non-uniform current, generated by wind, is discussed in the paper by Zakharov & Shrira (1990), explaining the narrowness of the angular spectrum. The evolution of a random field of weak non-linear surface waves is investigated with consideration of the induced dispersion mechanism in the wind-driven near-surface current. This makes a contribution to

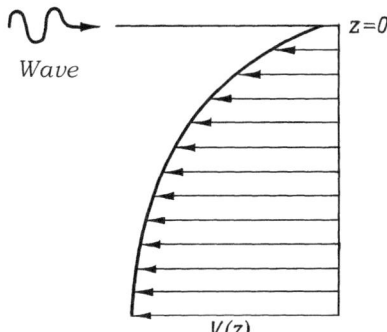

Fig. 5.37. Current velocity profile increasing with depth (5.136) blocking wave spectrum range

the kinetic equation for the wave action spectral density $N(\mathbf{k})$. Its parameterization is proportional to $N^2(\mathbf{k})$ and to a small parameter – the ratio of the drift current velocity to the wind wave phase velocity and to some large parameter determined by the Reynolds number. The suggested mechanism determines the intense redistribution of the wave action angle. Practically it does not affect the wave number spectrum. The induced dispersion action leads to formation of narrow bimodal angular distributions.

5.10 Wind Wave Generation in a Current

Statement of the problem. Among different aspects of the current effect on waves, special attention is usually paid to two of them: wave transformation in a current and the current effect on wind wave generation. In natural conditions the processes of generation and transformation are usually interrelated. The character of this distinction is a conventional one. Transformation means a change of wave parameters as a result of wave interaction with a spatially non-uniform current. The previous sections of this monograph are devoted to investigation of this mechanism.

The fact that the current can significantly influence wind wave generation is well known. River navigators take into consideration the fact that waves are low in a strong wind directed downstream in the river, and it is easier to make ship routing. But the waves are high in a wind of the same speed upstream in the flow, preventing ship routing. Similar effects are observed in strong tidal currents, especially at maximum current velocity values. It cannot be ascertained in general that wave transformation effects in a current are insignificant for such cases, although influence of a current on wind wave generation can be suggested to prevail.

Special investigations of the influence of a current on the processes of wind wave development in a wind–water basin were held in Russia (Kononkova et al., 1977; Sudolsky & Teplov, 1982; Pokazeyev et al., 1986) and in other countries (Francis & Dudgeon, 1967; Kato et al., 1976, 1978, 1981; Plate & Trawle, 1970; Tsuruya et al., 1987).

The detailed measurements of wind waves in a current under laboratory conditions have been made by Japanese researchers in the wind–water flume at the Hydrodynamic Laboratory of the Institute of Ports in Simonoseki (Kato et al., 1976, 1978, 1981; Tsuruya et al., 1987). The wind–water flume has an operating length of 28.5 m, a width of 1.5 m and a depth of 1.3 m. Waves are measured at seven stations along the fetch. Both mean wave parameters and spectral densities of wave energy are determined. The experiments were performed at wind speeds from 5.6 to 11.3 m s^{-1} and current velocities from 0 to 29.9 cm s^{-1} for a fair current and from 0 to 14.4 cm s^{-1} for a countercurrent. Later, the corresponding current velocity range was increased to 67.2 cm s^{-1} (Tsuruya et al., 1987). The ratio of the current velocity V to the wind speed U is up to 0.6, corresponding to strong countercurrents.

The Japanese researchers used an effective fetch notion in analysing the experimental data. This notion is different from the traditional one, meaning the usual distance from the shore, along which the wind blows. The time of wind interaction with the water surface is considered to be a decisive wave development characteristic. It is quite clear that the interaction time is different within one fetch, depending on the presence or absence of a current. The effective wave fetch in a current X_{eff} is defined as the distance at which waves are developed to the same age as in the absence of current. Equal time periods of wave interaction with wind in the absence and in the presence of a current are defined as:

$$\frac{X_{\text{eff}}}{C_{\text{g0}}} = \frac{x}{c_{\text{g}} + V}, \qquad (5.138)$$

where X_{eff} is the effective fetch length; C_{g0} is the wave group velocity in the absence of a current; and $C_{\text{g}} = c_{\text{g}} + V$ is the absolute wave velocity in a current. When the wave parameters are changed due to wind effect along the fetch, the formula for estimating the effective fetch length can be written in the integral form:

$$\frac{X_{\text{eff}}}{x} = \frac{\int_0^x \mathrm{d}x / C_{\text{g}}(x)}{\int_0^x \mathrm{d}x / C_{\text{g0}}(x)}, \qquad (5.139)$$

where the group velocities are referred to the wave frequency $f_{\text{p}} = \omega_{\max}/2\pi$ of the spectral maximum and are dependent on the fetch x.

Using data from the measurements of Mitsuyasu & Rikiishi (1975), the following formulas for waves in a current have been proposed (Kato et al., 1981):

$$\frac{U_* f_\text{p}}{g} = 0.939 \left(\frac{gX_\text{eff}}{U_*^2}\right)^{-0.354} ; \qquad (5.140)$$

$$\frac{gh_\text{S}}{U_*^2} = 0.0222 \left(\frac{gX_\text{eff}}{U_*^2}\right)^{0.669} , \qquad (5.141)$$

where h_S are the significant wave heights and U_* is the wind friction velocity in the immovable coordinate system.

It should be pointed out that the results obtained under full-scale conditions are not numerous. They usually correspond to shallow water areas with sufficiently limited fetches. For example, full-scale measurements are carried out at straight large soil reclamation canals (Mass et al., 1988). Unlike the laboratory experiments it is more difficult to control natural wave-generation conditions. Besides, it is necessary to take into account varying dimensions of the water basin sizes, especially in depth and width. Probably for this reason, there is a great scatter in the experimental data. However, the conclusions of these experiments are mainly the same. The wave heights and lengths are smaller in a fair current, while they are greater in a countercurrent than in calm water for the same fetch and under other equal conditions. Thus, the effective fetch value can be used to estimate the wave parameters.

Problem formulation and its formal solution. In spite of the fact that principal empirical formulas obtained under laboratory conditions can be considered to be established, it is not clear whether they can be applied for the description of sea waves. An attempt should be undertaken to solve this problem from the theoretical point of view. There are some disputable questions in this issue besides the well-known unsolved problems in wind wave theory. For example, there is the viewpoint (Kononkova & Pokazeyev, 1985) that the waves can be generated by wind in a countercurrent only with the period being greater than a critical period ($\tau \geq 8\pi V/g$). In reality, the waves in a current can be generated by wind over a wider range. However, the wave frequency ω measured in the immovable coordinate system cannot be greater than the value $gk/(4kV)$, due to the dispersion dependence ($\sigma^2 = \omega - k\mathbf{V}$), valid for gravity waves in deep water ($\sigma^2 = gk$).

The problem of wave generation is also rather questionable. It is known that wind waves are not always formed in rapid flows. That is why it is suggested (Labzovskiy, 1973) that there is some critical current velocity and wind speed ratio at which the wind cannot generate waves in a water surface. But this viewpoint cannot be accepted fully for the following reasons. If the wind speed relative to the water surface is greater than the value necessary for the instability occurring at the water–air interface, then waves can be generated. The situation is different for shallow rapid streams (Labzovskiy, 1973), where probably intense fluid turbulence appears, naturally preventing wave generation.

In this section an attempt is undertaken to describe the evolution of wave parameters along the fetch in the presence of a spatially uniform current. The wave development in a current is assumed to occur according to the same laws as in its absence. The role of a current is attributed to a change of the relative wind speed generating the waves, i.e. the speed measured in the coordinate system connected with the current velocity. Furthermore, the current velocity is added to the wave propagation velocity relative to a fixed fetch changing the time of the wave interaction with the air flow.

In accordance with the general formulation of the problem (5.1), (5.2), a constant wind speed U directed along the Ox axis is assumed to be prescribed at the horizon: $z = 10$ m. The current velocity is equal to the value V.

Wind wave evolution along the fetch x in the established regime will be investigated with the initial energy spectral density value $S = S(k, \beta)$, being prescribed at the boundary $x = x_0$ (see Fig. 5.38).

In order to describe wind wave development a coordinate system moving with the current V will be used. The coordinates of a mobile system x' and y' are connected with the initial coordinate system $x' = x - Vt$; $y' = y$. The current is absent in the new coordinate system and the waves develop according to similar laws as in calm water with the exception that the wind speed relative to the water surface is equal to the value $U - V$.

The evolution of spectral density of the wind wave energy is assumed to be described by (5.1) which can be written as follows:

$$\frac{dS}{dt} = \frac{\partial S}{\partial t} + \frac{\partial S}{\partial x'}\frac{dx'}{dt} + \frac{\partial S}{\partial y'}\frac{dy'}{dt} + \frac{\partial S}{\partial k}\frac{dk}{dt} + \frac{\partial S}{\partial \beta}\frac{d\beta}{dt} = G', \qquad (5.142)$$

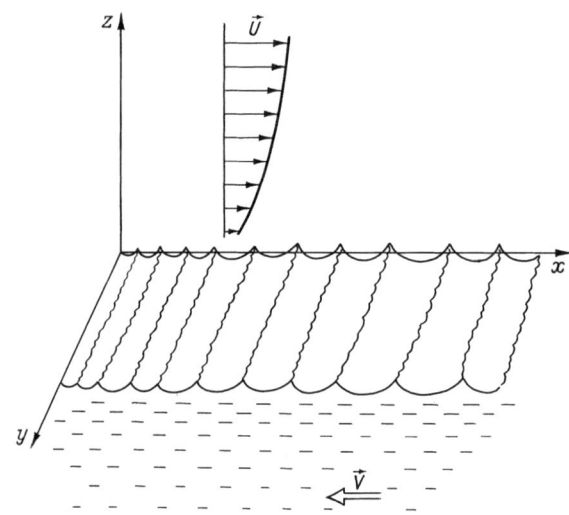

Fig. 5.38. Schematic presentation of wind wave generation in current: U – wind speed; V – current speed

5.10 Wind Wave Generation in a Current

where $G' = G(k, \beta, S, \boldsymbol{U} - \boldsymbol{V})$ is a source function forming the wind wave spectrum. The equation characteristics are written in the form:

$$\frac{dx'}{dt} = c_{gx} \,; \quad \frac{dy'}{dt} = c_{gy} \,; \quad \frac{dk}{dt} = 0 \,; \quad \frac{d\beta}{dt} = 0 \,, \qquad (5.143)$$

where c_{gx} and c_{gy} are the group velocity components equal to the values $0.5\sqrt{g/k}\cos(\beta)$ and $0.5\sqrt{g/k}\sin(\beta)$, respectively.

As long as the current speed is uniform and stationary, the wave number k and the angle β are constant along the characteristic. The boundary conditions for the spectrum $S = S_0$ can be prescribed at the movable boundary: $x'_0 = x_0 - Vt$. Using the method of characteristics and separating the variables, the solution of (5.142) can be written as follows:

$$\int_{S_0}^{S} \frac{dS}{G(k, \beta, S, \boldsymbol{U} - \boldsymbol{V})} = t - t_0 = \frac{x' - x'_0}{c_{gx}} \,. \qquad (5.144)$$

The left-hand part of the formula is a function of the spatial spectrum S, not depending on the choice of the coordinate system, i.e. it is an invariant. The values k, β, $\boldsymbol{U}$ and $\boldsymbol{V}$ are constant. Hence, proceeding to the immovable coordinate system, this part of the formula is not changed. Passing to the immovable coordinates, the right-hand part of (5.144) is transformed into the form $t - t_0 = (x - x_0)/(c_{gx} - V)$. The final implicit form of the solution for the initial coordinate system can be written as:

$$\frac{1}{2}\sqrt{g/k}\cos(\beta) \int_{S_0}^{S} \frac{dS}{G(k, \beta, S, \boldsymbol{U} - \boldsymbol{V})} = \frac{x - x_0}{1 - 2V\sqrt{k/g}\cos^{-1}(\beta)} \,. \qquad (5.145)$$

The spectral density value can be determined by transforming the expression (5.145) as follows:

$$S = F\left(\frac{k, \beta, \boldsymbol{U} - \boldsymbol{V}, (x - x_0)}{1 - 2V\sqrt{k/g}\cos^{-1}(\beta)}\right) \qquad (5.146)$$

This solution reveals that the presence of a constant current leads to the additional wind speed value (the left-hand part of (5.145)) and to a change in the wind fetch. This is connected with the variation of group velocity due to the presence of the current speed (the right-hand part of (5.145)).

It should be noted that the obtained solution (5.146) is a formal one, due to the absence of an explicit form of the source function G. Its determination presents the main difficulty in the wind wave evolution problem. However, an attempt is undertaken to solve the problem in a roundabout way.

If the spectrum approximation describing its evolution along the fetch under constant wind speed in the absence of a current is known as

$S(k, \beta, \boldsymbol{U}, x - x_0)$, then the expression (5.145) indicates that in the presence of a constant current the spectrum expression is transformed according to the substituted variables as follows:

$$x - x_0 \to \frac{x - x_0}{1 - 2V\sqrt{k/g}\cos^{-1}(\beta)}; \qquad \boldsymbol{U} \to \boldsymbol{U} - \boldsymbol{V}. \tag{5.147}$$

In should be noted that after replacing the expression (5.147), the fetch value can be formally extremely large and it becomes negative at $1 - 2V\sqrt{k/g}/\cos(\beta) \leq 0$. This is due to the fact that the group velocity projection $C_{gx} = c_{gx} - V$ can attain zero or negative values for some spectral components. These waves cannot propagate along the positive Ox axis due to large current velocities, i.e. from the initial boundary x_0, with the initial spectral density S_0 being set, to the calculated point x. However, these spectral components can still arrive at the calculated point as they are driven backwards by the current (i.e. from right to left, see Fig. 5.38).

Since the fetch value is not limited to the right of the calculated point (see Fig. 5.38), the spectral components drifting backwards can be considered to travel an infinitely large distance. The time period of their interaction with the air flow is not limited either. That is why the fetch value can be assumed to be extremely large at $C_{gx} < 0$. In fact this means that the energy of these spectral components should have reached the saturation condition.

The dependence (5.16) is used as the spectrum approximation S in the absence of current. There are the following values: m_0 is the zero moment, and $\sigma_{\max}$ is the spectral maximum frequency. They are functions of the fetch x and the wind speed U. The value n is assumed to be equal to 5.5, corresponding to the mean parameter value at the main stage of wave development (Davidan et al., 1985).

The following relationship between the non-dimensional values $\tilde{m}_0 = m_0 g^2/U^4$ and $\tilde{\sigma}_{\max} = \sigma_{\max} g/U^2$ in the absence of a current is used (Davidan et al., 1985):

$$\tilde{\sigma}_{\max} = 0.11\, \tilde{m}_0^{-0.34}. \tag{5.148a}$$

The dependence of the spectral zero moment m_0 of the non-dimensional fetch $\tilde{X} = (x - x_0)\, g/U^2$ is assumed to be equal to:

$$\tilde{m}_0 = \tilde{m}_\infty \left\{ \text{th}\left(2.727 \times 10^{-5} \tilde{X} \right)^\delta \right\}^{\frac{0.84}{\delta}}, \tag{5.148b}$$

where $\tilde{m}_\infty = 3.03 \times 10^3$. This is a non-dimensional zero moment of the spectrum for completely developed waves, with the spectral maximum frequency being equal to $\tilde{\sigma}_{\max} = 0.8$.

The expression (5.148b) is a known generalized power dependence (Davidan et al., 1985) for $\tilde{X} > 3.6 \times 10^4$. The value $\tilde{m}_0$ is asymptotically smoothly transferred into $\tilde{m}_\infty$ for large values of the argument to the hyperbolic

5.10 Wind Wave Generation in a Current   243

tangent. The character of this transfer is determined by the parameter δ. The value δ is estimated for developed wave stages as 2.0 according to the JONSWAP experiment.

Estimation of wind wave elements in a current. The obtained solution is used for estimating the wave element evolution along the fetch in a current. In order to emphasize the current effect on waves the variables are changed in the spectral expression $S\left(\tilde{k}, \beta, \tilde{X}, \tilde{\gamma}\right)$ as follows:

$$S = \frac{1}{2}U^4(1-\tilde{\gamma})^4 g^{-2} Q(\beta)(n+1) m_\infty f_1 f_2^n \tilde{k}^{-1} \exp\left(-\frac{n+1}{n}f_2^n\right), \quad (5.149)$$

where:

$$f_1\left(\tilde{k}, \beta, \tilde{X}, \tilde{\gamma}\right)$$

$$= \begin{cases} \text{th}^{\frac{0.84}{\delta}} \left(\dfrac{2.72 \times 10^{-5} \tilde{X}}{(1-\tilde{\gamma})^2 \left(1 - 2\tilde{\gamma}\cos^{-1}(\beta)\sqrt{\tilde{k}}\right)}\right)^{\delta} & \text{when } 1 - 2\tilde{\gamma}\sqrt{\tilde{k}}/\cos(\beta) \geq 0 \, ; \\ 1.0 & \text{when } 1 - 2\tilde{\gamma}\sqrt{\tilde{k}}/\cos(\beta) < 0 \, ; \end{cases}$$

$$f_2\left(\tilde{k}, \beta, \tilde{X}, \tilde{\gamma}\right) = \frac{\tilde{\sigma}_\infty}{\sqrt{\tilde{k}(1-\tilde{\gamma})}} f_1^{-0.34}\,.$$

Now the spectrum S is a function of the non-dimensional wave number $\tilde{k} = k\sqrt{V/g}$, the angle β, the non-dimensional fetch $\tilde{X}$ and the relative current velocity parameter $\tilde{\gamma} = V/U$. At the value $\tilde{\gamma} = 0$, the spectrum (5.149) is transformed to the wind wave spectrum in the absence of current. The ratio of the mean wave height $\bar{h}$ in a current to the height $\bar{h}_0$ in the absence of a current can be expressed by the ratio of corresponding statistical moments:

$$\frac{\bar{h}}{\bar{h}_0} = \left[\frac{2\pi}{m_0}\int_0^\infty \int_{-\pi}^{\pi} S\left(\tilde{k}, \beta, \tilde{X}, \tilde{\gamma}\right) d\beta\, d\tilde{k}\right]^{\frac{1}{2}} = F_1\left(\tilde{X}, \tilde{\gamma}\right), \quad (5.150)$$

where $F_1\left(\tilde{X}, \tilde{\gamma}\right)$ is a function of two non-dimensional parameters: the fetch $\tilde{X}$ and the relative current velocity $\tilde{\gamma}$.

The ratio of the mean wavelengths is written similarly:

$$\frac{\bar{\lambda}}{\bar{\lambda}_0} = \frac{\sqrt{\frac{5}{6}\sigma_m^2\left(\tilde{X}\right)}\left(\frac{n+1}{n}\right)^{\frac{2}{n}} \sqrt{\Gamma(1-4/n)}\sqrt{m_0\left(\tilde{X}\right)}}{\left[\int_{-\pi}^{\pi}\int_0^\infty \tilde{k}^2 \cos^2(\beta) S\left(\tilde{k}, \beta, \tilde{X}, \tilde{\gamma}\right) d\tilde{k}\, d\beta\right]^{\frac{1}{2}}} F_1\left(\tilde{X}, \tilde{\gamma}\right)$$

$$= F_2\left(\tilde{X}, \tilde{\gamma}\right). \quad (5.151)$$

The mean period can be determined differently depending on the chosen coordinate system. It can be calculated with the help of formulas similar to (5.150), (5.151) in the coordinate system moving with the current:

$$\frac{\bar{\tau}_1}{\bar{\tau}_0} = \frac{\sigma_m\left(\tilde{X}\right)\left(\frac{n+1}{n}\right)^{\frac{1}{n}}\sqrt{\Gamma\left(1-2/n\right)}\sqrt{m_0\left(\tilde{X}\right)}}{\left[\int\limits_{-\pi}^{\pi}\int\limits_{0}^{\infty}\tilde{k}S\left(\tilde{k},\beta,\tilde{X},\tilde{\gamma}\right)\,d\tilde{k}\,d\beta\right]^{\frac{1}{2}}} F_1\left(\tilde{X},\tilde{\gamma}\right)$$

$$= F_3\left(\tilde{X},\tilde{\gamma}\right). \qquad (5.152)$$

Determining the mean period in the immovable coordinate system $\bar{\tau}_2$, it is necessary to take into account the Doppler shift connected with the current. The frequency ω used to determine the mean period $\bar{\tau}_2$ can attain both positive and negative values (Peregrine, 1976). The latter holds for waves with their own phase velocity $\boldsymbol{c} = \sigma\boldsymbol{k}/k^2$, directed opposite to the current and being less in value. In order to calculate the mean period $\bar{\tau}_2$, the expression of the mean value should be used rather than for the RMS value, as usually used in (5.150)–(5.152). Otherwise, the contribution of harmonics with a negative frequency to the mean period $\bar{\tau}_2$ could be taken into account with the wrong sign. The expression for the mean wave period in the immovable coordinate system can be written as follows:

$$\frac{\bar{\tau}_2}{\bar{\tau}_0} = \frac{\sigma_m\left(\tilde{X}\right)\left(\frac{n+1}{n}\right)^{\frac{1}{n}}\Gamma\left(1-4/n\right)m_0\left(\tilde{X}\right)}{\left[\int\limits_{-\pi}^{\pi}\int\limits_{0}^{\infty}\left(\sqrt{\tilde{k}}-\tilde{\gamma}\tilde{k}\cos(\beta)\right)S\left(\tilde{k},\beta,\tilde{X},\tilde{\gamma}\right)\,d\tilde{k}\,d\beta\right]^{\frac{1}{2}}} F_1^2\left(\tilde{X},\tilde{\gamma}\right)$$

$$= F_4\left(\tilde{X},\tilde{\gamma}\right). \qquad (5.153)$$

Thus, the obtained integral formulas (5.150)–(5.153), describing the change of the mean wave elements in a current, are functions of the non-dimensional fetch $\tilde{X}$ and the relative current velocity $\tilde{\gamma}$. The non-dimensional form of the determining parameters $\tilde{X}$ and $\tilde{\gamma}$ makes the obtained expressions more universal.

The obtained integral expressions F_1, F_2, F_3, F_4 are calculated numerically. The integrands are presented in a discrete way as eight angular directions β_i (within the range from $\pi/2$ to π) and 50 components by the wave numbers k_j, taken in geometrical progression. Thus, it provides accuracy of integral calculation with not more than 1–2 per cent numerical error.

The non-dimensional fetch $\tilde{X}$ is varied from 10 to 10^5, and the parameter of the relative velocity $\tilde{\gamma}$ is assumed to be within the range from -0.5 to $+0.5$. It covers a reasonable range of these parameter variations under full-scale conditions.

5.10 Wind Wave Generation in a Current   245

The results of numerical calculations are given in Fig. 5.39. The fetch $\tilde{X}$ is plotted along the horizontal axis. The ratios $\bar{h}/\bar{h}_0$, $\bar{\lambda}/\bar{\lambda}_0$, $\bar{\tau}_1/\bar{\tau}_0$, and $\bar{\tau}_2/\bar{\tau}_0$ for different values of the current velocities $\tilde{\gamma}$ are plotted along the vertical axis. As can be seen (see Fig. 5.39a) there is a significant influence of the current on the wave elements. The wave height is increased by a current opposite to the wind, while it is decreased by a fair current. The influence of the current is of great importance in the initial stages of wave development. The influence

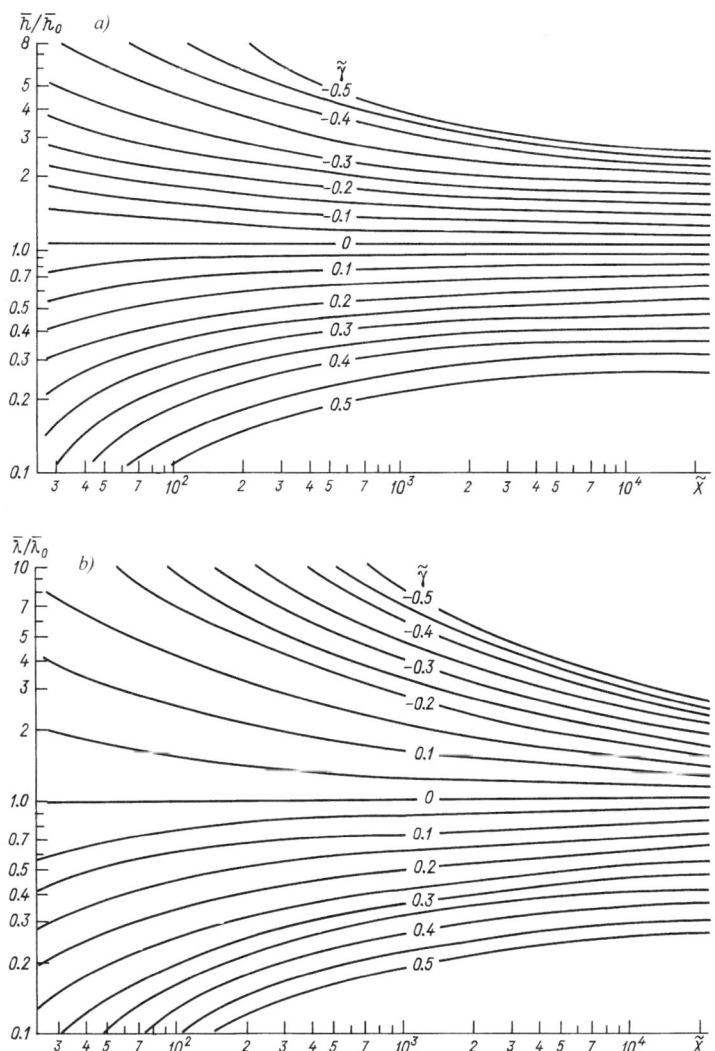

Fig. 5.39. Relative values of wave height $\bar{h}/\bar{h}_0$ (**a**) and length $\bar{\lambda}/\bar{\lambda}_0$ (**b**) along fetch for different relative velocity $\tilde{\gamma}$ (Fig. 5.39c,d see next page)

246   5 Wave Evolution in Non-uniform Currents in Deep Water

of the current decreases with further increase of the wave development, i.e. with an increasing fetch.

For sufficiently large fetches ($\tilde{X} > 10^4$), the ratio of heights $\bar{h}/\bar{h}_0$ tends to the value $(1 - \tilde{\gamma})^2$, corresponding to the ratio of wave heights for completely developed waves at the wind speed U with and without added current velocity. This means that the influence of the fetch length distortion effect due to

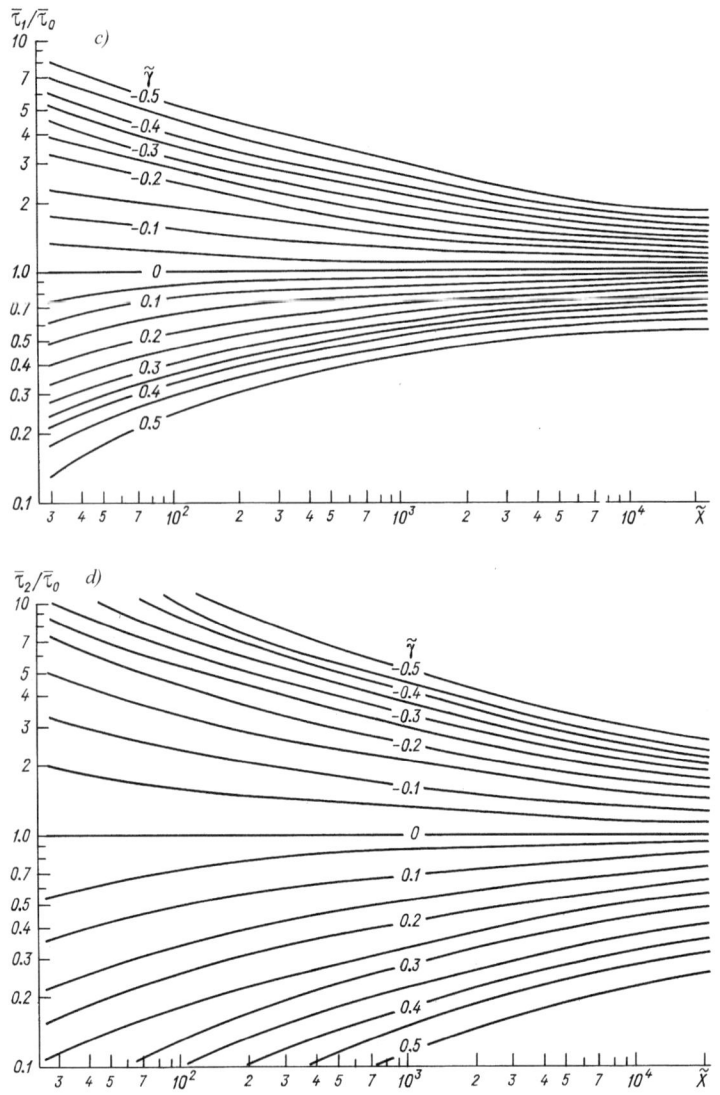

Fig. 5.39 (continued). Relative values of wave period $\bar{\tau}_1/\bar{\tau}_0$ (**c**) in movable coordinate system and wave period $\bar{\tau}_2/\bar{\tau}_0$ (**d**) in immovable coordinate system for different relative velocity $\tilde{\gamma}$

the presence of a current ("effective fetch") is decreased with wave development. In this case the wind speed value relative to the water surface becomes decisive.

The wavelength dependence $\bar{\lambda}/\bar{\lambda}_0$ for different values $\tilde{\gamma}$ as a function of the non-dimensional fetch $\tilde{X}$ is shown in Fig. 5.39b. The character of this dependence is similar to the height ratio $\bar{h}/\bar{h}_0$. The wavelength is increased in a countercurrent, while it is decreased in a fair current, whereas $\bar{\lambda}/\bar{\lambda}_0 \approx \bar{h}/\bar{h}_0$.

The wave period is similarly influenced by the current (see Fig. 5.39c,d). However, it is interesting to note that the mean wave period measured in the movable coordinate system (see Fig. 5.39c), for example, on a ship drifting with a current, is significantly less influenced by it (as long as $\bar{\tau}_1/\bar{\tau}_0 \approx \sqrt{h\bar{h}_0}$) in comparison with measurements in the immovable coordinate system (see Fig. 5.39d), for example, made on a stationary platform.

The influence of current on waves is increased with increasing parameter of the relative current velocity $\tilde{\gamma}$. It is necessary to point out that the current velocity value V is not so important, as its value compared to the wind speed U.

5.11 Some Practical Recommendations for Estimation of Current Influence on Waves

Difference of the current influence on wave generation and transformation. There is a necessity to estimate wind wave development in the presence of a current. However the accurate corresponding practical recommendations are still absent (Building Standards and Rules, 1983). In this case numerical models can be applied.

The results of this monograph can be used to obtain an estimation of current influence on waves. In order to do so, the nature of the aforementioned prevailing mechanisms should be determined, whether it is a wave transformation or wave generation in a current. In the case of wave generation the height and length are increased by the countercurrent due to increasing relative wind speed and wave interaction time with air flow at a fixed fetch. But, in the other case, the wavelength is diminished with waves propagating in a countercurrent being increased along the stream direction (see Sect. 5.4). This means that the waves become "compressed" by length at a practically constant mean wave period measured in the immovable coordinate system.

The processes of wave generation and transformation in a current are interconnected in natural conditions. It should be noted that the separation of these two mechanisms is rather conventional. Nevertheless, an attempt should be undertaken to estimate the contribution of these two mechanisms to the source function in the spectral density balance equation (5.1).

The wind energy input G_{in} is estimated by the formula (4.34) according to the Miles mechanism (1960). The wind input occurs mainly in the high-frequency spectral range. Its energy is partially lost due to wave breaking and partially transferred to the low-frequency spectral range by non-linear wave interaction. The contribution of wave interaction with a horizontal non-uniform current to the source function (i.e. the radiation stresses) can be estimated as $G_{\text{cur}} \approx \frac{d\sigma}{\sigma \, dt} S = 2 \frac{\partial V}{\partial x} S$. In this case the energy flux is distributed to the wave spectrum in all frequencies. The non-linear interaction contribution can be of less importance.

Thus, a quantitative comparison of these two mechanisms can be used as a rough estimation of its predominant influence on wave development. It can be suggested that wave generation is predominant over the wave transformation mechanism when $G_{\text{in}} > G_{\text{cur}}$. As a result the wave transformation mechanism prevails in limited water basins (for example, sea straits, river mouths, etc.) with significant current velocity gradients being observed at small wind speeds. And, on the contrary, wave generation becomes the determining factor for extensive water areas with small current velocity gradients.

Estimation of wave element transformation in a current changing along its direction. In the case of a large gradient area it is necessary to use the wave transformation results (see Sect. 5.4). In order to determine the wind wave elements for deep water in a counter or fair current, the mean wave period $\bar{\tau}_0$ must be found in the area without any current. Then the parameter ν is estimated according to the formula:

$$\nu = 5.2 \frac{V}{g\bar{\tau}_0}, \qquad (5.154)$$

where V (m s^{-1}) is the current velocity and $g = 9.81$ m s^{-2}.

A positive value of the parameter ν is used for a wave propagating opposite to the current, and negative value for a fair current. The wave components in a current are determined with the help of Fig. 5.40 by the nomograms given for different initial directions of wave propagation. The ratio of mean wave height in a current $\bar{h}$ to the mean height $\bar{h}_0$ in the absence of current is determined by curves I (see Fig. 5.40) according to the parameter ν, depending on the initial angle of wave propagation relative to the current velocity direction.

The relative value for the mean wavelength $\bar{\lambda}/\bar{\lambda}_0$ is determined by curves II, similarly.[4] The mean relative period is determined by curve III.

The following estimations should be fulfilled. Wind waves with a mean period $\bar{\tau}_0 = 5.3$ are assumed to propagate from the open sea area to the strait. The countercurrent's velocity V is 2 m s^{-1}. The parameter ν is equal to 0.2 according to the formula (5.154). The mean wave height is increased by 218 per cent, the length is decreased by 210 per cent and the period is increased by 8 per cent according to the nomogram (see Fig. 5.40).

---

[4] If the calculated wave steepness is larger than the ultimate value equal to 1/7, the wave height corresponding to the ultimate steepness for estimated wavelength should be used.

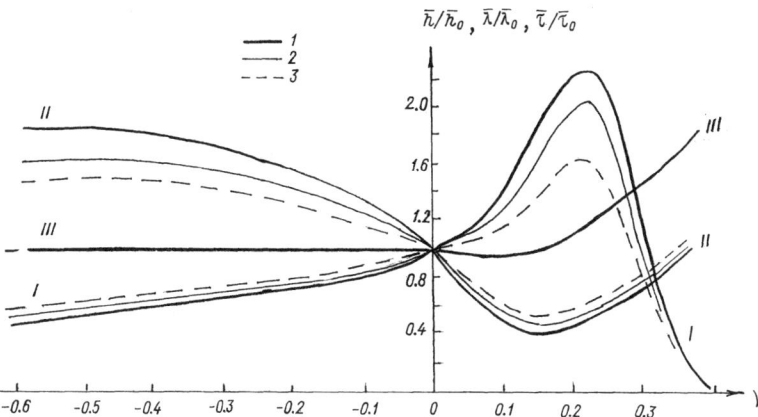

Fig. 5.40. Variation of wave mean heights $\bar{h}/\bar{h}_0$ (I), lengths $\bar{\lambda}/\bar{\lambda}_0$ (II) and periods $\bar{\tau}/\bar{\tau}_0$ (III) in counter- ($\nu > 0$) and fair ($\nu < 0$) current: 1 – initial direction of wave propagation and current being collinear (at $\nu < 0$) or opposite (at $\nu > 0$); 2 – direction different by 30°; 3 – direction different by 60°

Estimation of current influence on wave generation. The nomograms obtained in Sect. 5.10, can be recommended to be used in the case of the current influence on wave generation by wind prevailing over the wave transformation by a non-uniform current. It should be noted that the results are obtained using the empirical dependencies of the wave generation conditions (5.148) for the wave development at the main stage ($\tilde{X} > 100$). That is why the obtained results can be considered to be valid for the same age of wave development.

An example of estimating the wind wave parameters in a current can be presented using the above-obtained nomograms. Suppose that the wind blows with a speed $U = 10 \text{ m s}^{-1}$ along the 50 km fetch. The current velocity is assumed to be $V = -2 \text{ m s}^{-1}$, with the current direction being opposite to the wind direction.

The mean height $\bar{h}_0$ and the period $\bar{\tau}_0$ in the absence of current are estimated to be equal to 0.62 m and 3.8 s, respectively (Building Standards and Rules, 1983). The non-dimensional fetch $\tilde{X} = xg/U^2$ is estimated as 4900, and the parameter of the relative velocity is $\tilde{\gamma} = -0.2$. The wave height is determined using Fig. 5.39a as $\bar{h}/\bar{h}_0 = 1.3$, i.e. $\bar{h} = 0.81$ m. The wavelength is equal to $\bar{\lambda}/\bar{\lambda}_0 \approx \bar{h}/\bar{h}_0 = 1.3$, where $\bar{\lambda}_0$ is determined by the mean wave period $\bar{\lambda}_0 = 1.56$, $\bar{\tau}_0^2 = 22.5$, i.e. $\bar{\lambda} \approx 29$ m.

The mean wave period measured in the movable coordinate system (for example, on a ship drifting with the current) is determined by the ratio $\bar{\tau}_1/\bar{\tau}_0 \approx \sqrt{\bar{h}/\bar{h}_0} = 1.44$, i.e. $\bar{\tau}_1 = 4.3$ s. The mean wave period measured on a stationary platform is estimated as $\bar{\tau}_2/\bar{\tau}_0 = 1.2$, i.e. $\bar{\tau}_2 = 4.6$ s according to Fig. 5.39b.

Estimation of current influence on wind wave distribution functions. The problem of the influence of current on the regime-climatic functions of the wave parameters is still incompletely investigated. However, it is of great practical importance. The regime functions allow obtaining the wave parameters used directly in engineering practice.

Using the aforementioned results, an attempt is undertaken to estimate the influence of strong tidal currents in the White Sea on the regime functions of mean wave heights in the small probability range (Korobov & Lavrenov, 1989). It should be noted that there is a regular semi-diurnal tide with current velocities up to 250–270 cm s^{-1}. In order to do this it is divided into regions according to the prevailing role of one of two possible current influence mechanisms on the wave generation or transformation. As the estimation shows, the diagrams of regime functions, with currents taken into account, are displaced relative to the diagrams for calm water conditions towards larger wave heights. The estimates of the calculated wave heights appear to increase by 14–16 per cent in sea regions with strong tides.

As far as the problem of current influence on the distribution of wave elements within a quasi-stationary range is concerned, this still remains open. At least it is well known that currents can significantly change the distribution functions of wave heights and lengths. The proportion of wave heights and lengths greater than the mean values is decreased according to the experimental data (Krivoshey, 1976). In addition, the variation and asymmetry coefficients are considerably changed (Mass et al., 1988). Observations in natural conditions show that the currents significantly change the distribution function of wave heights and periods (Matushevskiy, 1985).

An estimation of wave elements of prescribed probability is performed on the basis of the mean wave height using a distribution function. The wave height distribution function is known to be subordinate to the Rayleigh law for all stages of wave development. The distribution function of wave elements in deep water is dependent on the non-dimensional fetch (Krylov et al., 1976). This assumption is confirmed by field and laboratory experiments.

Using these results, the distribution function for the case of the influence of current on the wave generation was estimated by Mass et al. (1990). The current and the fetch restriction by a coastline are taken into account using the effective fetch length. It is assumed that the wave height distribution function preserves its form and is determined by the effective fetch:

$$F(h/\bar{h}) = \exp\left[-\phi_1\left(\frac{gX_{\text{eff}}}{U^2}\right)(h/\bar{h})^{\phi_2\left(\frac{gX_{\text{eff}}}{U^2}\right)}\right], \quad (5.155)$$

where

$$\phi_2 = \frac{10^3}{(gX_{\text{eff}})^{1/2}/U + 3.9 \times 10^2}; \quad \phi_1 = \Gamma^{\phi_2}\left[\frac{1}{\phi_2} + 1\right];$$

Γ is the gamma function, and X_{eff} is the effective fetch.

5.11 Recommendations for Current Influence Estimation

As shown above, the effective fetch in a fair current is less than that without any current. The ratio (5.154) shows that the wave heights of small probability are decreased and those of large probability are increased. The ratio of effective fetches is inverse in a countercurrent. The wave heights of small probability are increased. But, starting from some definite or greater probability, they are decreased compared to the situation of the absence of current. This phenomenon of the wave height distribution function corresponds qualitatively to available observations. In particular, the greatest extreme waves are observed in a countercurrent wind. A similar transformation of the wave height distribution function under the influence of a current is revealed in the Karakumy Channel (Mass et al., 1988; Rzhanitsyn et al., 1988). Since the greatest changes occur with wave heights of small probability, this transformation should be taken into account to determine the wave elements in designing hydrotechnical construction.

6 Wave Transformation in Shallow Water

6.1 Spectrum Transformation
Due to Wave Refraction in Shallow Water

Wave spectrum refraction a in coastal area. The wave description given in Chap. 1 allows easy analysis of wind wave propagation in a coastal area,[1] i.e. when relatively short sea waves propagate from deep to shallow water, approaching a coastline. In this case refraction plays a special role among the different factors affecting wave behavior. Due to depth variation it results in the evolution of wave parameters, namely, propagation direction, length, amplitude and wave profile. As noted in the Introduction, there are many papers devoted to the problem of wave transformation in a coastal area. An application of the spectral approach to this problem following from the general formulation is considered in this chapter.

Sticking to physical terminology, wave refraction in shallow water can be considered in the framework dispersive wave propagation in a spatially non-uniform isotropic medium. It follows from Sect. 1.6 that the propagation of a wave packet is described with the help of the ray equations on a spherical surface (1.86)–(1.89) and in a plane (5.2), with the frequency ω being constant along the ray:

$$\omega^2 = \sigma^2 = f^2\left(\boldsymbol{k}, H\left(\boldsymbol{r}\right)\right) = gk \operatorname{th}\left(kH\right) = \text{const} . \tag{6.1}$$

This condition yields a useful ratio for determining the wave number $|\boldsymbol{k}|$ and phase velocity c along the trajectory, depending on the slowly changing depth H.

A second simple consequence of the general kinematic ratios can be obtained in the "cylindrical" case, i.e. with the depth H changing only in one direction, for example $H = H(x)$. It follows from (5.2) that a component of the wave vector k_y should be constant during wave packet propagation

$$k_y = k\sin\left(\beta\right) = \text{const} , \tag{6.2}$$

where $\pi/2 - \beta$ is the angle between the direction of the wave vector $\boldsymbol{k}$ and the isobath. The dependence between the angle β and the wave number k or

---

[1] In this case the wave spectrum transformation apart from the surf zone is considered.

the phase velocity c (taking into account (6.1)) at the initial and subsequent time moments follows directly from (6.2):

$$\frac{\sin(\beta_0)}{\sin(\beta)} = \frac{k}{k_0} = \frac{c_0}{c}. \tag{6.3}$$

The formula (6.3) is known in optics as the Snellius law. The formula is not dependent on bottom relief changes between the initial and final ray points, and is determined only by the depths at these points.

If waves were not destroyed in shallow water near a shore line (with $H \to 0$), they would approach perpendicular to it (with the angle approaching zero: $\beta \to 0$), regardless of their previous movement. In reality, waves are usually destroyed before reaching a shore line. That is why it is possible to apply the formula (6.3) for determining the angle of a wave approaching the breaking zone up to the depths where strongly non-linear effects begin playing their role. These non-linear effects are manifested in continuous wave profile deformation resulting in wave overturn.

The wave spectrum transformation in shallow water will now be considered. At the same time the area of application of the spectral approach based on the equations for a sphere (1.84), (1.86)–(1.89) or for a plane (local) coordinate system (5.1), (5.2) will be defined. The equation of wave energy spectral density evolution is obtained using the assumption of weak non-linearity and independence of separate wave phases. This assumption cannot be fulfilled due to strong non-linear effects in coastal shallow water. Thus, the spectral approach is assumed to be used outside the zone of non-linear wave transformations and wave break-down.

The simplest case of spectral transformation will be considered, when it is relatively easy to obtain a solution analytically. The initial spectrum is assumed to be uniform and stationary $S_0 = S_0(\omega, \beta)$. It is given at the boundary $Q = Q(\mathbf{r})$, where the depth is equal to $H = H(\mathbf{r})$. The spectral value for the whole area can be obtained using (5.1) or (5.4), neglecting the source function effect, i.e. assuming that $G = 0$. The frequency-angular spectrum along the characteristics (6.2) can be presented as follows:

$$S(\omega, \beta, \mathbf{r}) = \frac{\partial \kappa^2}{\partial \omega} \left(\frac{\partial \kappa_0^2}{\partial \omega}\right)^{-1} S_0(\omega, \beta_0), \tag{6.4}$$

where $\kappa = \kappa(\omega, \mathbf{r})$.

The angle $\beta_0 = \beta_0(\omega, \beta, \mathbf{r})$ can be easily obtained when $H(\mathbf{r})$ changes only in one direction, for example, $H = H(x)$. According to the ratio (6.2) $k_y = k_{y0}$, and so:

$$\beta_0 = \arcsin\left(\frac{\kappa}{\kappa_0} \sin(\beta)\right). \tag{6.5}$$

The arcsine argument in (6.5) must not be greater than 1. Hence, the kinematic condition limiting the area of determining the wave number and

6.1 Spectrum Transformation Due to Wave Refraction in Shallow Water

angle is as follows:

$$\left|\frac{\kappa}{\kappa_0} \sin(\beta)\right| \leq 1 . \tag{6.6}$$

If this condition is not fulfilled, it means that with decreasing depth, when κ/κ_0 is increased, the corresponding spectral component is absent at the point with coordinate $x = x'$. Constructing a complete spectral solution for this combination of parameters: ω, β, x', the following holds:

$$S(\omega, \beta, x) \equiv 0 . \tag{6.7}$$

It follows from the solution (6.4) that the spectrum $S(\omega, \beta, x)$ does not depend on the bottom slope, and is determined only by depth. The wave spectrum in shallow water can be easily obtained using this relation, if the wave spectrum in deep water is known. Thus, the expression for a frequency-angular spectrum is written as:

$$S(\omega, \beta, H) = \frac{(g\kappa)^2}{\omega^4 \left(1 + \frac{2\kappa H}{\sh(2\kappa H)}\right)} S(\omega, \beta_0) . \tag{6.8}$$

As can be seen, this expression coincides with the analogous ratio obtained earlier (Krasitskii, 1974; Krylov et al., 1976).

The initial wave spectrum in deep water can be prescribed in the following form:

$$S_0(\omega, \beta_0) = S_0(\omega) \cos^{n_0}(\beta_0) Q(n_0) , \tag{6.9}$$

where $Q(n_0)$ is the normalizing function of the angular distribution:

$$Q(n_0) = \begin{cases} \frac{2^{n_0} \Gamma^2\left(\frac{n_0}{2}+1\right)}{\pi \Gamma(n+1)} & \text{at } |\beta_0| \leq \pi/2 \\ 0 & \text{at } |\beta_0| \geq \pi/2 , \end{cases}$$

where $\Gamma(n)$ is the gamma function.

According to the ratio (6.6), the value $\kappa/\kappa_0 = [\th(\kappa H)]^{-1} = 1$ increases with decreasing depth, and the direction distribution of the wave energy flux becomes narrower with waves approaching a shore. If the ratio $|\beta| \leq \pi/2$ is fulfilled in deep water, then in shallow water: $|\beta| \leq \arcsin(\th(\kappa H))$. Thus, the following is obtained:

$$\cos^{n_0}(\beta_0) = \left[1 - \left(\frac{\kappa}{\kappa_0} \sin(\beta)\right)^2\right]^{\frac{n}{2}} \approx \cos^n(\beta) , \tag{6.10}$$

where $n = n_0 (\kappa/\kappa_0)^2$.

So, with decreasing depth, the index of the degree of angular direction $n = n_0 \left(\kappa/\kappa_0\right)^2$ increases, which is also confirmed by observations (Krylov et al., 1976). The frequency spectrum value:

$$S(\omega, H) = \int_{-\pi/2}^{\pi/2} S(\omega, \beta, H) \, d\beta$$

changes with the depth as:

$$S(\omega, H) \approx \frac{(g\kappa)^2}{\omega^4 \left(1 + \frac{2\kappa H}{\text{sh}(2\kappa H)}\right)} S_0(\omega) \, 2^{n_0} \left[1 - \left(\frac{\kappa}{\kappa_0}\right)\right]^2$$

$$\times \frac{\Gamma^2\left(\frac{n_0}{2} + 1\right) \Gamma(n+1)}{\Gamma^2\left(\frac{n}{2} + 1\right) \Gamma(n_0 + 1)} . \quad (6.11)$$

The spectral parameter variations in shallow water are shown in Fig. 6.1. Relative values of the wave number κ/κ_0, power index n/n_0, phase velocity of the waves c/c_0, as well as the ratio of the spectral densities $S(\omega, H)/S(\omega)$, obtained by the formula (6.11) depending on the non-dimensional depth $H\omega^2/g$, are shown. If the wave number and the power index are increased and the phase velocity is decreased monotonically with decreasing depth, then at first the ratio of spectral densities is slightly decreased and after that begins

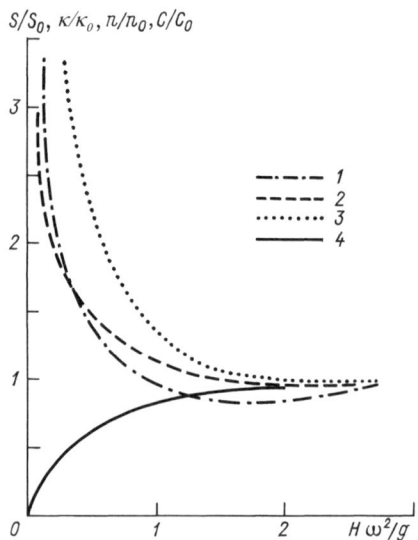

Fig. 6.1. Transformation of wave spectrum ratio S/S_0 (1), wave number κ/κ_0 (2) and power index $n = n_0 \left(\kappa/\kappa_0\right)^2$ (3) of angular direction distribution and phase velocity c/c_0 (4) at their normal propagation towards a coast

6.1 Spectrum Transformation Due to Wave Refraction in Shallow Water   257

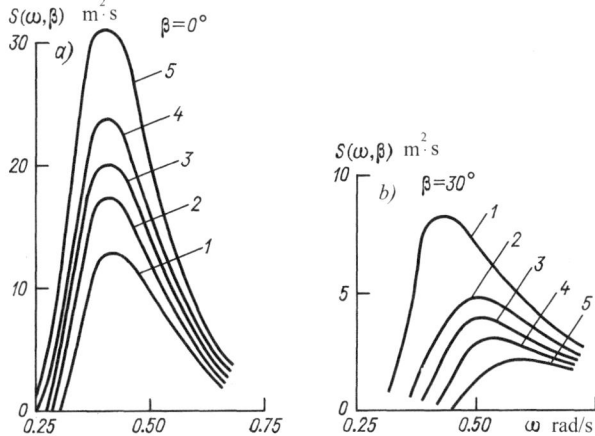

Fig. 6.2. Example of frequency-angular spectrum transformation in shallow water for $\beta = 0°$ (**a**) and $\beta = 30°$ (**b**): 1 – initial spectrum in deep water; 2 – in water with 30 m depth, 3 – in water with 25 m depth; 4 – in water with 20 m depth; 5 – in water with 15 m depth

increasing. This is connected with non-monotonic group velocity variations with depth.

The transformation of the frequency-angular spectrum in shallow water will be considered as an example. Thus, the frequency-angular spectrum $S(\omega, \beta, H)$ for $\beta = 0$ and $\beta = 30°$ at different depths is shown in Fig. 6.2. The group velocity is decreased for the selected depth variations. The change of spectral parameters results in increasing the spectrum at average frequencies and its decreasing at small and large frequencies. As for large angles (see Fig. 6.2b), the spectral density is decreased at all frequencies. A sharp decrease of the curves 2, 3, 4 and 5 to zero at the small depth H is connected with violation of the condition (6.6), when the corresponding spectral parameters are absent.

The expression (6.8) allows calculating the change of the main wave parameters in shallow water. In particular, it can be shown that the average wave period is practically not changed for rectilinear isobaths, although in a more complicated isobath pattern, the average wave period measured at different coastal points can be different due to redistribution of the component wave energy. It should be generally noted that the theoretical calculations based on a linear model of regular and irregular waves in the coastal shallow zone are proved by field observations (Krylov et al., 1976). However, the area of applicability of the aforementioned ratios is limited by the case of neglecting wave breaking in shallow water.

There are numerous measurements of wind waves in shallow coastal sea areas obtained from stable platforms and scaffold bridges, whereas such measurements in open sea are absent. That is why it is quite natural that researchers use these measurements for analysing wind wave properties in deep sea areas (Kostichkova et al., 1990). The frequency spectra obtained using these measurements are often identified with deep-water spectra without sufficient justification. However, even in the same wind the spectra and other statistical characteristics far from the coast (in deep water) and near it (in the coastal area) are different. The results of a linear wave refraction theory, showing the connection of frequency-angular spectra at two different depth points, can be used to solve the problem of recalculating the wind wave spectra in the coastal to deep-water areas. However, the validity of this approach is admissible if the isobaths are rectilinear and the wave fetch between the points in deep and shallow water is so small that wave generation by wind, dissipation due to bottom friction and non-linear wave interaction do not essentially affect the wave spectrum variation.

Wave parameter evolution in a coastal area. The solution (6.8) allows easily obtaining the ratios describing the evolution of mean wave components in shallow water. In the general case the spectral expression (6.8) is integrated over the frequencies ω and the directions β.

The wave amplitude change can be written in explicit form for the ultimate narrow spectrum $S_0(\omega, \beta) = m_0 \delta(\omega - \omega_0) \delta(\beta - \beta_0)$ as:

$$a = a_0 \sqrt{\frac{\cos(\beta_0) \kappa}{\cos(\beta) \kappa_0 \left(1 + \frac{2\kappa H}{\text{sh}(2\kappa H)}\right)}}, \tag{6.12}$$

where β and κ are determined by (6.1) and (6.3) as functions of their initial values and the depth H.

It can be seen that the expression under the square root in (6.12) coincides with the ratio of group velocity components c_{gx_0}/c_{gx} in (5.49), indicating the conservation of energy flow directed perpendicular to the shore.

As for long waves ($kH \ll 1$), the ratio (6.12) becomes simpler and it can be written as:

$$a = a_0 \left(\frac{H_0}{H}\right)^{\frac{1}{4}} \left[\frac{\cos(\beta_0)}{\sqrt{1 - \sin^2(\beta_0) H/H_0}}\right]^{\frac{1}{2}}. \tag{6.13}$$

The ratio (6.13) when $\beta_0 = 0$ is known as the Green formula.

In accordance with (6.12) the wave height is slightly decreased with waves propagating from deep to shallow water. This value comprises $a/a_0 = 0.95$ for the general direction $\beta_0 = 0$, and it is equal to 0.80 for $\beta_0 = 60°$ (see Fig. 6.3).

Infinitely large values of wave amplitudes at a shoreline, following from the solution (6.12) as $H \to 0$, are physically unreal. The spectral approach to the solution does not "spread" this singularity as in the case of wave

6.1 Spectrum Transformation Due to Wave Refraction in Shallow Water

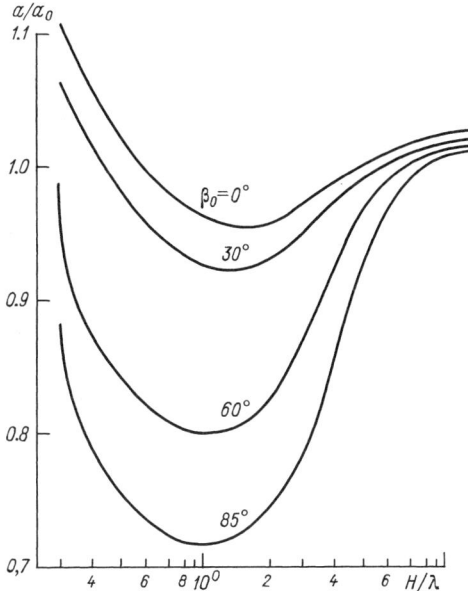

Fig. 6.3. Wave amplitude variations due to approaching a shore for different initial angles

refraction in a horizontal non-uniform countercurrent. This singularity exists simultaneously for all spectral components at the shoreline (at $H = 0$).

This solution behaviour is connected with the fact that it does not only take into account non-linear effects and the possibility of wave breaking, but also does not consider wave reflection from a sloping bottom. The linear theory taking into account the possibility of reflection is considered both in the framework of long wave equations (Mazova and Pelinovskiy, 1982; Shuleykin, 1959) and potential fluid movement equations (Sretenskiy, 1977). In particular, it is shown that an accurate solution is described by the Bessel function in the case of shallow water equations for a basin with constant bottom slope $H - \alpha x$. At a distance from the shoreline (as $x \to \infty$) this solution describes a standing wave with amplitude changing according to the linear Green formula. At the shoreline (at $H = 0$), the wave amplitude is finite and equal to $a = a_0\, 2\sqrt{\pi \omega H_0 / \alpha \sqrt{g}}$.

The increasing amplitude and decreasing wavelength in its propagation to shallow water result in the increased role of non-linear effects causing some increase of the average water level and generation of a countercurrent compensating the mass flow induced by waves (Matushevskiy, 1975; Whitham, 1974; Longuet-Higgins and Stewart, 1962, 1964).

Non-linear effects are also manifested in continuous deformation of the wave profile, resulting in its overturn. In shallow water, the Stokes wave theory is invalid for finite amplitude values. In this case the Bussinesq equations should be used as the basic ones. They are valid for rather small and smoothly changing depth. The Bussinesq non-linear dispersion equations describe knoidal and solitary waves propagating from sea to shallow water. This problem is investigated with the help of analyses of finite amplitude wave refraction above a rough bottom. A method similar to non-linear geometrical acoustics, but generalized for the dispersed medium, is used by Ostrovskii & Pelinovskii (1975).

In the case of waves propagating to deep water, there is the possibility of the appearance of caustics at some distance from the shoreline (at $x = x^*$). After turning at the point $x = x^*$, the wave begins propagating in the opposite direction, i.e. to the shore. The angle of wave propagation direction is determined by the ratio (6.3), from which the location of the caustic as $\sin(\beta^*) = 1$ or $c(x^*) = c_0/\sin(\beta_0)$ can be determined. The basin depth at this point is equal to:

$$H = H^* = \frac{1}{k_y}\operatorname{arcth}\left(\frac{k_y c_0^2}{g \sin^2(\beta_0)}\right), \qquad (6.14)$$

which can be written as $H^* = H/\sin^2(\beta_0)$ in the shallow water case.

The wave amplitude at a caustic tends to infinity, as follows, for example, from (6.12) when $\cos(\beta) = 0$ in the frame of the geometrical optics approximation. The asymptotic method presented in Sect. 5.6 can be easily used for estimating a wave field near such caustics. The free water surface can be presented in the form (5.47), where $\sigma = \sigma_0 = \text{const}$ and $F = \sqrt{gk\,\text{th}(kH(x))}$.

Proceeding from (5.52), the formula for the maximum wave amplitude near caustics can be written as:

$$a_{\max} = 1.69 a_0 \sqrt{|C_{gx}|}\left(\frac{\partial F}{\partial H}\frac{\partial H}{\partial x}\right)^{-\frac{1}{6}}\left(\frac{\partial^2 F}{\partial k_x^2}\right)^{-\frac{1}{3}}\bigg|_{x=x^*}, \qquad (6.15)$$

where

$$\frac{\partial^2 F}{\partial k_x^2} = \frac{C_g}{k}; \qquad \frac{\partial F}{\partial H} = \frac{1}{2}\frac{gk}{c\,\text{ch}^2(kH)}.$$

For waves in shallow water, the ratio (6.15) can be written, taking into account (6.14), as:

$$a_{\max} = 1.90 a_0 \left(\frac{H_0 k_0 \sin^2(\beta_0)}{\partial H/\partial x}\right)^{\frac{1}{6}}\sqrt{\cos(\beta_0)}. \qquad (6.16)$$

The amplitude $a_{\max}$ is a function of the initial angle β_0. The maximum value $a_{\max}$ is achieved at:

$$\beta_0 = 35.26°; \qquad a_{\max} = 1.50\left(\frac{H_0 k_0}{\partial H/\partial x}\right)^{\frac{1}{6}}\bigg|_{x=x^*}.$$

It is obvious that there is no need for the isobath to be rectilinear for turning rays. Entrapment of waves near the shore can occur at any frequency, for which the corresponding conditions exist for the appearance of a caustic at some distance from the shore. The waves can be trapped in a similar way in the absence of a shore by submerged topographical bottom features, for example, by underwater ridges under conditions that caustics exist on both sides of this feature. A number of similar cases for geophysical waves of a different nature is given in the papers by Dobrokhotov & Zhevandrov, 1988a,b; Le Blond & Maysek, 1981; Rabinovich, 1993. They describe the entrapment, resonance and radiation of waves in the shelf ocean zone, resulting in the appearance of seiche oscillations in gulfs, bays and harbours.

6.2 Weak Non-Linear Wave Interaction in Shallow Water

Statement of the problem. Weak non-linear interaction of waves is one of the main physical mechanisms determining wind wave formation and evolution. That is why the estimation of its effect on the wave spectrum transformation in shallow water is of some definite interest. The numerical calculation of the collision integral describing weak non-linear energy transfer is very complicated and demands much computing time (see Sect. 4.1) even for the condition of infinitely deep water. For finite depth water, calculation of the collision integral is even more complicated.

At the same time, it was shown (Hasselmann & Hasselmann, 1981) that the effectiveness of non-linear wave interaction is increased with diminishing water depth. The applicability of the weak turbulent description of this mechanism is disturbed at sufficiently small depths. The wave movements in a free water surface are usually described within the framework of non-linear dispersion equations of Boussinesq or Korteveg de Vriz, whose studies are often quoted in scientific papers. In this connection the study of non-linear evolution of the wave spectrum in the intermediate case is of some interest. This means that, on the one hand, the water is not deep enough for the considered wavelengths and, on the other, it is not so shallow as to disturb the applicability of the weak turbulent theory. A similar situation can be seen in shallow water or in the case of wave propagation from deep sea to a coastal area.

Formulation of the problem. In the spatially uniform case, with the right-hand side of the kinetic equation (5.1), describing the non-linear interaction, the formulation of the spectrum evolution problem can be presented in the form:

$$\frac{dN}{dt} = G_{nl} . \tag{6.17}$$

The wave frequency ω is assumed to be connected with the wave number k using the dispersion ratio $\omega^2 = gk\,\text{th}\,(kH)$.

A numerical solution of (6.17) for finite depth water is quite a difficult problem demanding much computing time. The recommendations of Herterich & Hasselmann (1980) can simplify the solution. The results of numerical calculation of non-linear energy transfer for the frequency-angular spectrum $S(\omega, \beta)$ in the case of finite depth are shown to be similar to those obtained for infinitely deep water ($\omega^2 = gk$) and differ only by the multiplier R:

$$G_{\text{NL}}(\omega, \beta)|_{\omega^2 = gk\,\text{th}(kH)} = R(kH)\,G_{\text{nl}}(\omega, \beta)|_{\omega^2 = gk}, \qquad (6.18)$$

where G_{NL} is a function of the non-linear energy transfer for the wave spectrum in finite depth water and G_{nl} is a similar function for the infinitely deep water case. The expression for the function R can determined as follows (Komen et al. 1988, 1994):

$$R(\tilde{z}) = 1 + \frac{5.5}{\tilde{z}}\left(1 - \frac{5}{6}\tilde{z}\right)\exp\left(-\frac{5}{4}\tilde{z}\right). \qquad (6.19)$$

where the value R is a function of the parameter $\tilde{z} = kH$.

Thus, the solution of the spectrum evolution (6.17) is reduced to solving the equation where the right-hand side is calculated using the formula for infinitely deep water multiplying it by the function R.

Algorithm for solving the evolution problem. In order to solve numerically the evolution problem, the collision integral is calculated using the method described in Chap. 4.

It should be taken into account that the time required for calculation of the integral is sufficiently large. It has to be calculated numerically many times, i.e. at every step of the spectrum evolution. To save calculation time, the spectrum and the collision integral can be presented in the non-dimensional form:

$$S(\omega, \beta) = S_{\max}\tilde{S}(\tilde{\omega}, \beta); \qquad (6.20)$$

$$G_{\text{nl}}(\omega, \beta) = S_{\max}^3 \omega_{\max}^{11}\tilde{G}_{\text{nl}}(\tilde{\omega}, \beta)/g^4, \qquad (6.21)$$

where $S_{\max}$ is the maximum spectrum value; $\tilde{S}$ is its non-dimensional value; and $\tilde{G}_{\text{nl}}$ is the non-dimensional collision integral value. In a similar way, the core function T in (4.4) of the integrand (4.3), not depending on the spectrum, is reduced to non-dimensional form.

In order to calculate the non-dimensional values, a grid with the relative frequencies ω_i ($i = 1.30$) and directions β_j ($j = 1.36$) is chosen. A set of core values is calculated for these grid points. The numerical solution of the equation is obtained by stages. At the first stage, using the given spectrum $S(\omega, \beta, t_n)$, the spectral maximum $S_{\max}$ and the corresponding frequency

6.2 Weak Non-Linear Wave Interaction in Shallow Water

$\omega_{\max}$ are determined for the time moment t_n. At the second stage, non-dimensional values for the right-hand side of (6.17) are calculated. The values of the core function T are not calculated at every time step, because they are calculated at the preprocessing stage. The dimensional value is obtained by multiplying the calculated value $\tilde{G}$ by $S_{\max}^3 \omega_{\max}^{11}/g^4$. Then the evolution equation (6.17) is solved with the help of a first-order numerical scheme. The spectrum maximum is chosen using spectral values obtained at this step $S(\omega_i, \beta_j, t_n)$, and the next step of solving the problem is made.

It should be noted that the arguments ω, β occurring in the integrands do not coincide with the grid point values ω_i, β_j in the numerical solution changing constantly. The spectrum at these points is determined by interpolating its value using the four nearest grid points.

A significant increase of the processing speed is obtained by choosing the most considerable area from the full three-dimensional calculation space, making the main contribution. The sizes of the area are determined by the calculation accuracy prescribed in advance. This reduces the number of calculations. For example, the calculation is decreased by an order of magnitude, prescribing a 5 per cent calculation error.

The integrating step Δt is chosen automatically. It is taken into account that the spectrum is displaced to the low frequency area due to non-linear evolution, while the energy transfer intensity is decreased. Meeting the requirement of approximate equality of the relative spectrum variations at each step, the following uniformity ratio can be obtained:

$$\Delta t_n = \Delta t_{n-1} \cdot \left(S_{\max}^{(n)}/S_{\max}^{(n-1)}\right)^2 \cdot \left(\omega_{\max}^{(n)}/\omega_{\max}^{(n-1)}\right)^{11} \cdot R^{(n)}/R^{(n-1)}, \quad (6.22)$$

where Δt_n is the time increment at step n; $S_{\max}^{(n)}$ is the corresponding maximum of the spectrum; and $\omega_{\max}^{(n)}$ is its frequency.

As a result of this approach, a sufficiently optimal algorithm is developed, yielding numerically stable results with required accuracy using comparatively little computing time.

Results of numerical calculations. The purpose of numerical calculations is to study the basin depth effect on non-linear energy transfer and spectrum evolution. A typical wind wave spectrum approximation (the JONSWAP spectrum) with $\gamma = 3.3$ and the angular energy distribution $\sim \cos^4(\beta)$ are chosen. The results of calculating the non-linear transfer function and the frequency spectrum evolution in deep and shallow water at different time moments are presented in Figs. 6.4 and 6.5. The non-linear transfer function values are normalized by the function maximum calculated for the initial spectrum (at $t = 0$) in deep water.

The initial non-linear transfer function (see Fig. 6.4) is of a typical form with a positive maximum located at the frequency $\tilde{\omega} = \omega/\omega_{\max}^0 = 0.95$ and a negative minimum at frequency $\tilde{\omega} = 1.08$. The second minimum, being equal to 75 per cent of the first value, occurs at the frequency $\tilde{\omega} = 1.40$. This

264   6 Wave Transformation in Shallow Water

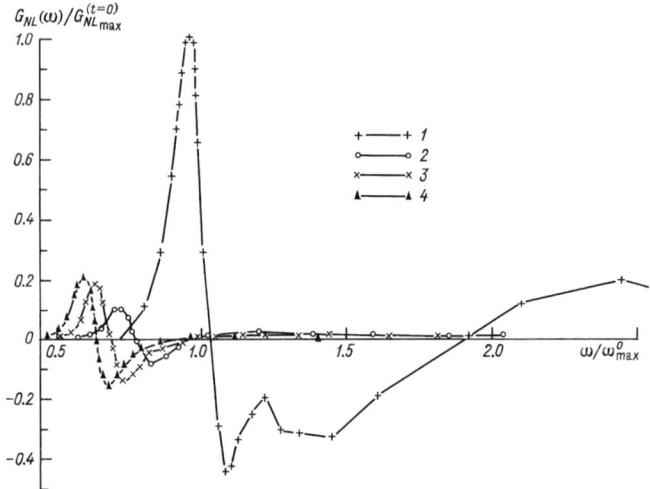

Fig. 6.4. Functions of non-linear energy transfer: 1 – initial non-linear energy transfer in deep water; 2 – similar value at $t = 5 \times 10^4$ s; 3 – non-linear transfer in the case of finite depth with $\omega_{Hm} = 0.7$ at $t = 5 \times 10^4$ s; 4 – similar value with $\omega_{Hm} = 0.6$

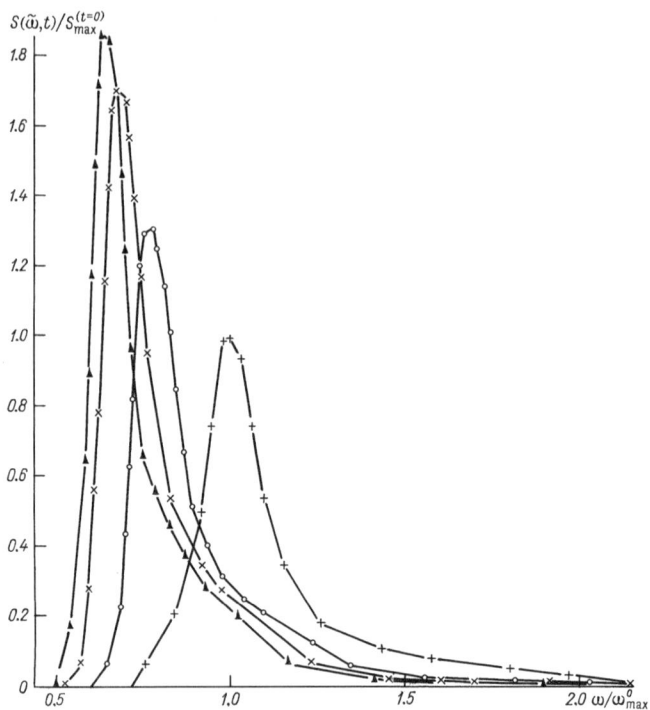

Fig. 6.5. Non-linear frequency spectrum evolution (for designations, see Fig. 6.4)

figure also shows values of the non-linear transfer function in the deep water case at $t = 5 \times 10^4$ s after the evolution begins. The function maximum is displaced to the left at the frequency $\tilde{\omega} = 0.725$, and the minimum at the frequency $\tilde{\omega} = 0.83$. The maximum of the non-linear energy transfer value is decreased by an order of magnitude and the minimum by a factor of 5. The second minimum disappears at the same time. The results reveal an intensive decrease of the non-linear energy transfer and its influence on the spectrum and displacement of the latter to the low-frequency area. Due to the spectrum maximum frequency decrease, the value of the non-linear transfer becomes smaller and the spectrum shape is stabilized. This is also in agreement with the estimates obtained from the dimensional ratio $G_{nl} \sim S^3_{max}\omega^{11}_{max}/g^4$, showing a strong dependence of the non-linear mechanism on the spectrum maximum frequency. The waves become longer and smoother with the same mean height, i.e. their steepness and non-linearity are decreased.

The behaviour of the non-linear transfer in the wave spectrum (see Fig. 6.5) in a finite depth basin is partly analogous to the case of infinite depth. However, the value of the function G_{NL} is larger in the finite depth case with its maximum displaced to the left. The frequency spectrum becomes narrower and its maximum is greater compared to the deep water case. This tendency is intensified with decreasing depth. The angular distribution function (see Fig. 6.6) shows its narrowing near the spectral maximum. It indicates the increased intensity of the non-linear wave interaction in shallow water. In this case an increase of the energy-carrying harmonics occurs. Due to this a narrow spectrum maximum area is singled out from the entire wave spectrum, creating a tendency to the wave monochromatization.

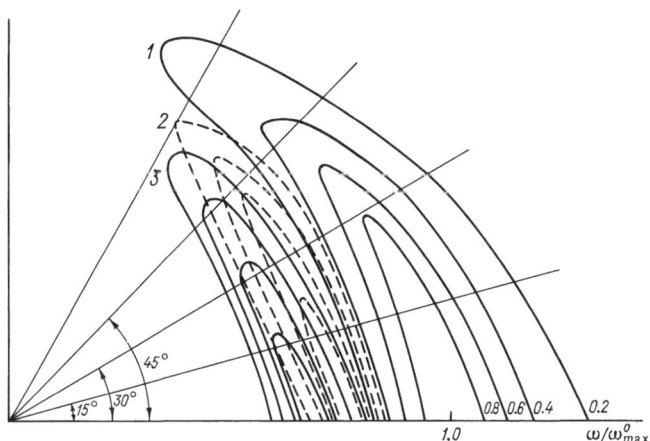

Fig. 6.6. Functions of energy angular distribution: 1 – initial angular energy distribution; 2 – angular distribution at $t = 5 \times 10^4$ s in deep water; 3 – distribution at the same time in shallow water at $\omega_{Hm} = 0.7$.

Continuation of this process disturbs the applicability of the initial assumptions introduced in the model under consideration. According to the estimation of Zakharov (1968), the possibility of using the kinetic equation (6.17) can be defined by the inequality: $\omega''(k)|\Delta k|^2 \gg T\tilde{N}\sqrt{R}$, where Δk is the spectrum width over the wave number; and $\tilde{N}$ is the integral wave action. Using the estimation $\tilde{N} \sim m_0 g/k\sqrt{gH}$, where m_0 is the spectrum zero moment, the following inequality can be obtained:

$$\left(\frac{\Delta k}{k}\right)^2 \gg \sqrt{\frac{5.5}{(kH)^3}\frac{m_0}{H^2}}. \tag{6.23}$$

This condition is disturbed at small depths and the waves cease to be "spectral-wide" for implementing the kinetic equation.

Estimation of the spatial–temporal scale of non-linear wave interaction in shallow water. For the manifestation of non-linear effects it is necessary to allow some time for their accumulation. The typical time of wave interaction allows estimating whether the non-linear effects are important or not for some wave propagation area. This gives the possibility of estimating the sphere of application of linear wave transformation models in shallow water.

According to Zakharov (1968), the typical time of the effectiveness of the non-linear interaction can be estimated as:

$$T \sim \left(\frac{\Delta k}{k}\right)^2 \left(\frac{H}{h}\right)^4 kH\tau, \tag{6.24}$$

where h is the mean wave height and τ is the mean wave period.

It is found that $T \sim 10^2 \tau$ for $\left(\frac{\Delta k}{k}\right)^2 \sim 10^{-1}$, $H/h \sim 10$. With a two-fold depth decrease, the typical interaction time is decreased by more than an order of magnitude.

Assuming the typical horizontal scale of wave propagation is L and its velocity is $C_g \approx \sqrt{gH}$, the condition of the applicability of the linear approach can be written as:

$$\frac{L}{h} < (2\pi)^{-1}\left(\frac{\Delta k}{k}\right)^2 \left(\frac{H}{h}\right)^4. \tag{6.25}$$

If the ration $L/h = 10^3$, a linear approximation can be used.

Three-wave interactions in shallow water. The non-linear effects become of increasingly greater importance with further wave propagation shoreward and, correspondingly, with depth decrease. Now they should be described by a three-wave interaction rather than by four-wave interactions. According to measurements (Beij & Battjes, 1993), the generation of divisible harmonics in the frequency wave spectrum occurs at the same time.

One of the first attempts to describe this effect theoretically was undertaken in the paper by Abreu et al. (1992). A solution for the shallow-water approximation is obtained, although wave dispersion is excluded, limiting the use of this result in the models. In the paper by Zaslavskii (1998) a formula describing three-wave interactions in finite depth water was derived. The so-called "quasi-kinetic" equation is obtained, where the conservatism condition is not fulfilled for a non-linear wave interaction in shallow water.

A simplified energy formula for three-wave interactions, which can be used for coastal wind wave models, has been proposed (Eldeberky & Battjes, 1995). The model is one-dimensional and called the Discrete Trial Approximations (DTA). It is tested using data from laboratory experiments displaying good agreement for describing the principal features of energy transfer from the main maximum to the spectrum of divisible harmonics. The formula describing this mechanism has been extended for the case of spectral angular directions (Ris, 1997). It is used in the SWAN model in the form of:

$$G_{\text{nl}\,3}(\omega,\beta) = G^{(+)}_{\text{nl}\,3}(\omega,\beta) + G^{(-)}_{\text{nl}\,3}(\omega,\beta) \qquad (6.26)$$

where:

$$G^{(-)}_{\text{nl}\,3}(\omega,\beta) = -2 G^{(+)}_{\text{nl}\,3}(\omega,\beta)\,,$$

$$G^{(+)}_{\text{nl}\,3} = \max\left\{0,\, 2\pi \alpha_{\text{EB}} c_{g,\omega} J^2 \left|\sin\left(-\pi/4(\log \text{Ur}+1)\right)\right|\right.$$

$$\left. \times \left(\frac{\omega}{k_\omega} S^2(\omega/2,\beta) - \frac{\omega}{k_{\omega/2}} S(\omega/2,\beta) S(\omega,\beta)\right)\right\}$$

α_{EB} is the coefficient of interaction, assumed to be equal to 1.0; $k_{\omega/2}$ is the value of the wave number for a spectral component, its frequency being equal to $\omega/2$; and Ur is the Ursell number, equal to $\text{Ur} = \frac{g}{2\sqrt{2}} \frac{h_s}{H^2 \omega_{\max}^2}$ (h_s is the significant wave height).

The three-wave interactions are calculated in case of $\text{Ur} > 1$ (in the opposite case the interactions are not taken into account). The interaction coefficient $J = g \alpha_{\text{nl}\,3} y \alpha_{\text{nl}\,3} / \mu_{\text{nl}\,3}$ was determined by Madsen & Sorensen (1993):

$$\begin{cases} \alpha_{\text{nl}\,3} = \left(2 k_{\omega/2}\right)^2 \left[\frac{1}{2} + \frac{c_{\omega/2}}{gH}\right]\,, \\ \mu_{\text{nl}\,3} = -2 k_\omega \left[gH + 2 B g H^3 k_\omega^2 - \left(B + \frac{1}{3}\right) \omega^2 H^2\right]\,, \end{cases}$$

where $B = 1/15$.

As mentioned above, the three-wave interactions resulting in the generation of divisible harmonics are well supported by the laboratory experiments and under field conditions in the absence of wind. But this effect is obscured by wind action due to dissipation in the equilibrium spectral range.

In conclusion, it should be noted that the results of this section explain the observed wave transformation pattern in a coastal area. Approaching the surf area, quasi-regular, almost monochromatic waves with a narrow frequency spectrum and maxima at divisible frequencies are distinguished in the wave field. If the angular energy distribution function is predominantly narrowed due to wave refraction with decreasing depth, then the behaviour of the frequency spectrum can be explained with the help of the non-linear wave interaction.

6.3 Joint Influence of Bottom and a Non-Uniform Current on Wave Transformation

Wave transformation in a non-uniform current in deep water or separately in shallow water without any current have been studied so far. A more general solution describing the combined effect of an uneven depth and a horizontal non-uniform current on waves is considered in this section. Among different variants of the joint variations of water depth and current velocity the case of depth and current, changing monotonically in the same direction, is chosen. This situation can take place, for example, in channels. In particular, the case of the deep water wave transformation in a current considered above follows from the general problem solution. The second particular solution following also from the general one and presenting another ultimate case describes shallow water wave evolution without any current.

Spectral solution of the problem. The problem formulation should be considered in the local rectangular coordinate system $\{x, y, z\}$. Let the Oz axis be directed vertically upwards, and the non-disturbed water surface coincide with the plane xOy. Waves from an area with depth H_0 and current velocity V_0 (at $x < 0$) are assumed to propagate to the area with current V and depth H (at $x > 0$). The velocity V is assumed to be directed along the Ox axis and to change monotonically in this direction: $V = \{V(x); 0\}$. In accordance with the mass conservation law a relationship is considered to exist between the variation of the horizontal current velocity V and the depth H. This relationship is due to the water flow conservation law: $V(x) H(x) = V_0 H_0$. Thus, the depth H is changed in the same direction as the current velocity (see Fig. 6.7).

The wave evolution problem will be considered within the framework of the spectral equation of the wave action density balance (5.19)–(5.22). Although an accurate solution for this equation set is obtained for deep water only, a similar solution can be assumed to be valid also for the finite depth water case. The solution can be considered to have the following form at least for large values of the parameter q ($q \gg 1$):

$$N(\boldsymbol{k}, \boldsymbol{r}, t) = \begin{cases} N_0(\boldsymbol{k}_0, \boldsymbol{r}, t_0) & \text{when } N_0 < N_\infty \\ N_\infty(\boldsymbol{k}) & \text{when } N_0 > N_\infty \end{cases} \tag{6.27a}$$

6.3 Bottom and a Non-Uniform Current Influence on Waves

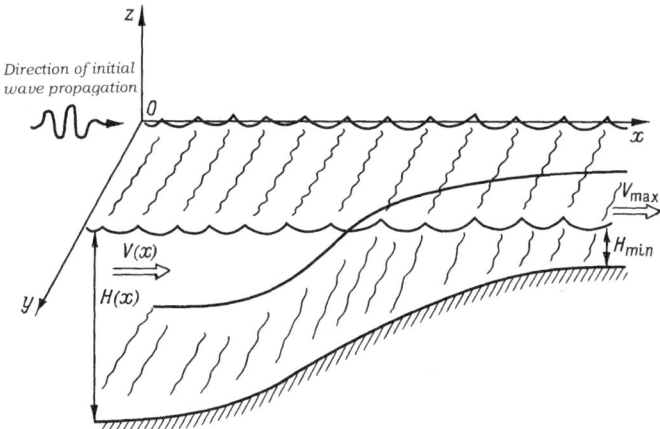

Fig. 6.7. Scheme of wave transformation due to the joint effect of uneven bottom and horizontal non-uniform current: $H(x)$ is depth; $V(x)$ is current velocity

where N_0 and N_∞ are the initial values of the wave action density and the equilibrium range correspondingly, given at the initial time t_0. The arguments N_0 and N_∞ depend on the wave vector $\boldsymbol{k}$ considered at time t and point $\boldsymbol{r}$. The latter dependencies are derived using the set of the equations (5.20)–(5.22).

Using the spectral density of the wave action N, the energy spectral density F, depending on the wave number $|\boldsymbol{k}|$ and the angle between its components $\beta = \arctan(k_y/k_x)$, can easily be found:

$$F(k, \beta, \boldsymbol{r}, t) = \min\left\{\frac{\sigma k}{\sigma_0 k_0} F_0(k_0, \beta_0, \boldsymbol{r}_0, t) \, ; \, F_\infty(k, \beta)\right\}, \qquad (6.27\text{b})$$

where the least value of the two functions in braces is assigned to the function F; F_0 is the initial spectral energy density; F_∞ is the spectral density of the equilibrium range; and $k_0, \beta_0, \boldsymbol{r}_0$ are derived with the help of the equation set (5.20)–(5.22). Using the dispersion ratio $\sigma^2 = gk \operatorname{th}(kH)$, the expression $\sigma k/\sigma_0 k_0$ can be written as: $\sqrt{\frac{k^3 \operatorname{th}(kH)}{k_0^3 \operatorname{th}(k_0 H_0)}}$.

The spectral energy density, prescribed for the initial boundary at $x = 0$, is assumed to be the uniform and stationary function: $F_0(k_0, \beta_0)$, determined using the spectrum (5.16).

Wave kinematic investigation in finite depth water. In order to determine the unknown values $k_0, \beta_0, \boldsymbol{r}_0, t_0$ in (6.27), it is not necessary to solve directly the set the equations (5.20)–(5.22) numerically, but an attempt is undertaken instead to derive their solution analytically. The motion integrals of wave packet propagation, following from this set, will be used. They are: the preservation of the wave vector component k_y and the frequency w, as soon as the current velocity V and depth H are only dependent on the x coordinate (without depending on the y coordinate and time t). These terms can be rewritten in the following form:

$$k_0 \sin(\beta_0) = k \sin(\beta) \; ; \qquad (6.28a)$$

$$\sqrt{gk_0 \operatorname{th}(k_0 H)} - k_0 V_0 \cos(\beta_0) = \sqrt{gk \operatorname{th}(kH)} - kV \cos(\beta) \; . \qquad (6.28b)$$

It seems that the initial problem solution can be obtained by substituting the values $k_0 = k_0(k, \beta, V)$ and $\beta_0 = \beta_0(k, \beta, V)$ from (6.28) and (6.27). However, this is more a necessary than a sufficient condition, so long as only two movement integrals (6.28) are used instead of solving the complete set of five equations. That is why some additional conditions, following from the kinematics of propagating wave packets, should be imposed on the formal solution obtained in this way. Thus, the shortcoming connected with non-solving the complete set of equations describing the wave packet propagation can be compensated. A similar method is used earlier for waves in deep water (see Sect. 5.3).

The situation is more complicated for the case of finite depth water, connected with two circumstances. Firstly, the current should not be considered to be absent at the initial point, where the initial wave spectrum value is given. Proceeding from the problem formulation, the initial non-zero current velocity should be taken into account. Secondly, there appears a question about wave blocking and the creation of reverse waves in a countercurrent under conditions of finite water depth.

The problem of wave blocking in a horizontal non-uniform current was thoroughly described earlier for waves in deep water. It follows formally from the ambiguity of determining the wave number using the dispersion ratio, that two different wave numbers can correspond to the same frequency in a current at the same point in space. However, if the current is absent, this ambiguity is also absent. The explanation is as follows: the dispersion ratio, valid for the absence of current in the deep water case ($\omega = \sqrt{gk}$, for $kH \gg 1$), is principally changed in water with a current ($\omega - Vk = \sqrt{gk}$). It is transformed to a quadratic equation, having two root values. Actually it formally describes the possibility of the existence of a reverse wave appearing at the point of blocking straight waves in a current.

In shallow water ($kH \ll 1$) the presence of a current velocity does not principally change the dispersion ratio, because it was and remains linear: $\omega - Vk = k\sqrt{gH}$. Thus, reverse waves should not exist in shallow water.

A natural question arises: What would happen in the intermediate case, i.e. for waves in finite depth water ($kH \sim 1$)? In order to investigate it, the approximation for waves in shallow water can be used. But the additional terms of dispersion corrections should be taken into account. Considering only the second term of an expansion in the dispersion ratio:

$$\sigma = \sqrt{gk \operatorname{th}(kH)} \approx k\sqrt{gH} \cdot \left(1 - (kH)^2/6 + O\left((kH)^3\right)\right) ,$$

6.3 Bottom and a Non-Uniform Current Influence on Waves

the expression for the frequency can be written as follows:

$$\omega = k\sqrt{gH} - \frac{(kH)^2}{6} k\sqrt{gH} - Vk\cos(\beta) . \qquad (6.29)$$

The designation $\tilde{k} = kH$, $\tilde{V} = V\cos(\beta)/\sqrt{gH}$, $\tilde{\omega} = \omega\sqrt{H/g}$ will be introduced and the ratio for the non-dimensional frequency $\tilde{\omega}$ can be presented as:

$$\tilde{\omega} = \tilde{k}\left(1 - \tilde{V}\right) - \frac{1}{6}\tilde{k}^3 . \qquad (6.30)$$

The dispersion curves $\tilde{\omega}\left(\tilde{k}, \tilde{V}\right)$ for different values of the parameter $\tilde{V}$ (i.e. current velocities for a fixed depth) are shown in Fig. 6.8. The curves $\tilde{\omega}\left(\tilde{k}, \tilde{V}\right)$ can be interpreted as the evolution of wave packet parameters with its propagation in a horizontal non-uniform current. As can be seen (see Fig. 6.8), several values of the wave number $\tilde{k}$ can correspond to the same frequency. Thus, there are three values of the wave number $\tilde{k}$ (for $\tilde{V} < 1$) having the frequency $\tilde{\omega}$ with two of them (for $\tilde{\omega} > 0$) being positive and one negative. The latter should not be considered, as it makes no physical sense and appears as a root of the dispersion equation expansion. Two other values can be determined using a solution of the cubic equation (6.30), being equal to:

$$\tilde{k}_1 = -\frac{u+v}{2} + \frac{u-v}{2}i\sqrt{3}; \quad \tilde{k}_2 = -\frac{u+v}{2} - \frac{u-v}{2}i\sqrt{3}, \qquad (6.31)$$

where

$$u = \sqrt[3]{-3\tilde{\omega} + \sqrt{(3\tilde{\omega})^2 - \left[2\left(1-\tilde{V}\right)\right]^3}};$$

$$v = \sqrt[3]{-3\tilde{\omega} - \sqrt{(3\tilde{\omega})^2 - \left[2\left(1-\tilde{V}\right)\right]^3}}.$$

The values $\tilde{k}_1$ and $\tilde{k}_2$ are merged into one at the point B, in which a local extreme value for the function $\tilde{\omega}\left(\tilde{k}\right)$ is achieved at $\tilde{k} = \sqrt{2\left(1-\tilde{V}\right)}$. At this point, the following ratio is valid: $\tilde{\omega} = \tilde{k}^3/3$; $V_B = \sqrt{gH}\left[1 - \frac{1}{2}\sqrt[3]{(3\tilde{\omega})^2}\right]$. The group wave velocity $C_{gx} = \sqrt{gH}\left[1 - \frac{1}{2}(kH)^2 - V\right]$ becomes zero. If the current velocity were greater than the value V_B, there would be no waves at this point (see Fig. 6.8, curve for $\tilde{V} = 0.99$).

The point B is a blocking point. If the wave packet propagating opposite to the current and arriving at the blocking point (see Fig. 6.8, curve $\tilde{V} < 0.5$)

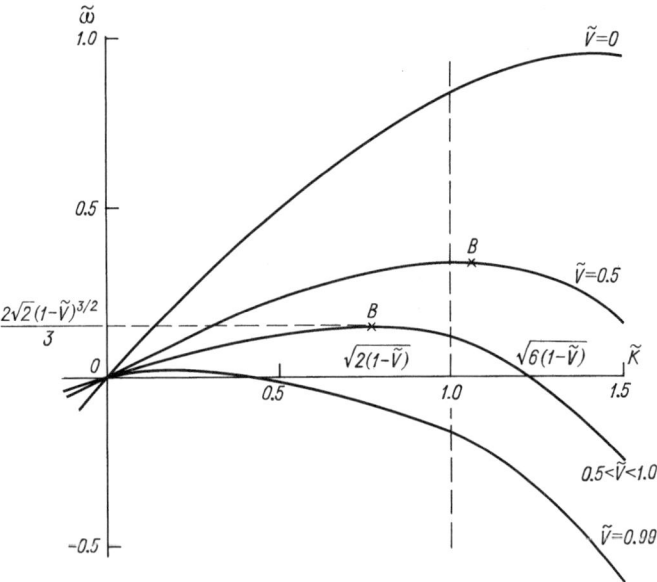

Fig. 6.8. Variation of wave dispersion dependence in shallow water with different current velocities $\tilde{V}$ (the *dotted line* is the boundary of the applicability of the dispersion ratio at $\tilde{k} = 1$)

stops at this point, it is carried backward by the current. Its wave number is increased. Thus, the wave ceases being long in comparison with the depth, violating the application area of the description, because the dispersion ratio expansion is true only for small values of $kH < 1$. The vertical dotted line $\tilde{k} = 1$ (see Fig. 6.8) denotes the boundary of the applicability area of this approach. It is necessary to use a more accurate expression for the dispersion ratio for $\tilde{k} > 1$, giving the possibility of describing waves in deep water.

The dispersion curves (see Fig. 6.8) show that the wave blocking and appearance of reverse waves are also possible in relatively shallow water. In order to describe these effects it is necessary to take into account the wave dispersion.

A more accurate notion of the wave number variation within a wide range $(0 < k < \infty)$, occurring with a wave propagating in a countercurrent increasing along the Ox axis and with depth variations, can be obtained. In order to do it, the set of equations (6.28) can be solved numerically. The relative value of the wave number $\tilde{y} = \sqrt{k/k_0}$, depending on the current velocity V/V_0 (or the depth H_0/H), is shown in Fig. 6.9 for two values of the parameter α. At first the initial relative depth $k_0 H_0 = \alpha$ is assumed to be equal to 1.0, corresponding to the case of "almost" deep water. The initial current velocity is assumed to be of such a small value that its influence on the wave could be neglected at the initial moment ($v = V_0\sqrt{k_0/g} = 10^{-3}$). As the wave packet

6.3 Bottom and a Non-Uniform Current Influence on Waves

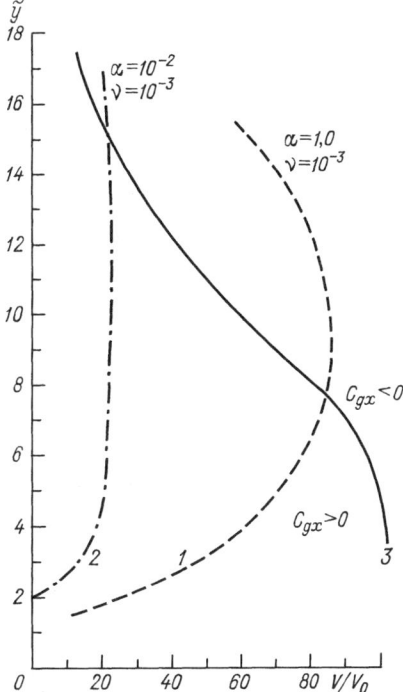

Fig. 6.9. The function $\tilde{y} = \sqrt{k/k_0}$ of wave number variation in a countercurrent: 1 – at $\alpha = 1.0$ and $\nu = 10^{-3}$; 2 – at $\alpha = 10^{-2}$ and $\nu = 10^{-3}$; 3 – boundary at which the wave group velocity becomes zero, $V < 0$

is propagating, its wave number is monotonically increased up to the value $k/k_0 = 60$ at $V/V_0 = 87$. After that, the wave packet is returned (i.e. it is carried away to the area with lower current velocities) and the wave number continues increasing. The point where the derivative $\partial V / \partial k$ becomes zero, is the blocking point of the wave packet (i.e. it is the point B in Fig. 6.8). In this case the group velocity component becomes zero $C'_{gx} = 0$. At this point the value $\tilde{k}$ is equal to 0.65, with the wave blocking, which takes place in non-deep water conditions.

A similar graph for the value $\alpha = 10^{-2}$ is shown in Fig. 6.9, corresponding to the wave behaviour in shallow water. Wave blocking also takes place at $\tilde{k} = 0.15$ (see Fig. 6.8), which can be considered as the shallow water case. On passing the blocking point, the wave packet is carried away backwards. Its wave number is quickly increased and the wave packet becomes "a wave in deep water". Nevertheless, it is important to underline once more that wave blocking is also possible in relatively shallow water. Under other equal conditions, the current velocity, with blocking taking place in shallow water, is less than in deep water.

274   6 Wave Transformation in Shallow Water

Now, the problem of calculating the wave spectrum in a current will be considered, taking into account the presence of straight and reverse waves. The straight waves, with a positive value of the group velocity component C_{gx}, are assumed to propagate the direction of the countercurrent:

$$C_{gx} = c_g \cos(\beta) - V \geq 0, \qquad (6.32)$$

where $c_g = \frac{1}{2}\sqrt{\frac{g\,\text{th}(kH)}{k}}\left(1 + \frac{2kH}{\text{th}(2kH)}\right)$.

There are also reverse waves ($C_{gx} < 0$), existing at the point considered in addition to the straight wave. The reverse waves are reflected at the point $x_B > x$ and carried away backwards by the current. In other words, the reverse waves can be observed at the point x, if the blocking current velocity V_B (see Fig. 6.7) is larger than the current velocity V at the point x, but smaller than the maximum current velocity $V \leq V_B \leq V_{\max}$.

The value of the velocity V_B at which blocking occurs, can be determined as a function of the values k, β, V, H, using the following set of transcendental equations:

$$k \sin(\beta) = k_0 \sin(\beta_0) \; ;$$

$$\sqrt{gk_0\,\text{th}(k_0 H)} - k_0 V_0 \cos(\beta_0) = \sqrt{gk\,\text{th}(kH)} - kV \cos(\beta) \; ;$$

$$\frac{1}{2}\sqrt{\frac{g\,\text{th}(k_B H_B)}{k_B}}\left(1 + \frac{2k_B H_B}{\text{th}(k_B H_B)}\right) \cos(\beta_B) - V_B = 0 \; ;$$

$$VH = V_B H_B . \qquad (6.33)$$

Furthermore, it should be taken into account that as a result of wave kinematic transformation in a current, the function of the angular energy distribution is changed (see Sect. 5.3). Thus, the following inequality should be fulfilled:

$$|\sin(\beta_0)| = \left|\frac{k}{k_0}\sin(\beta)\right| \leq 1 . \qquad (6.34)$$

A non-zero solution for the spectrum is determined for a specific set of wave numbers k and angles β. The variation range of the wave number k and the angle β can be essentially different from their typical determination area in the deep water case and in the absence of spatial non-uniformity of the wave medium, when the usual range $0 < k < \infty$ and $0 \leq \beta \leq 2\pi$ is considered. Determination of the boundary changes of the values k and β in (6.28) are due to the above described kinematic conditions connected with the spatial non-uniformity of the wave medium.

The presence of reverse waves imposes some specific limitations on the prescribed initial spectrum (at $x = 0$). As the initial current velocity (at $x = 0$) is not equal to zero, the reverse waves can be carried away to the initial boundary and can exist within this area. Their spectral density cannot be prescribed in a random way at $x = 0$. This is connected with the expression (6.27) of the corresponding straight wave spectral density.

Reducing the problem solution to a non-dimensional form. The obtained spectrum solution (6.27) will be presented in a more universal form. In order to do so, it is expressed as a function of non-dimensional arguments and parameters. The following non-dimensional parameters are introduced: the initial depth $\tilde{\alpha} = k_m H_0$; the initial current velocity $\tilde{v} = V_0\sqrt{k_m/g}$; the parameter characterizing the maximum current velocity relative to its initial value $\mu = V_0/V_{\max}$ ($0 < \mu < 1$) and the parameter of the relative present current velocity $\gamma = V/V_0$ (or $\gamma = H/H_0$; $1 \leq \gamma \leq 1/\mu$); and the wave number of the spectrum maximum k_m.

The relative wave number ratio can be designated as $k_0/k = f$, and the value of the function f can be determined numerically by using the set of equations (6.33). The initial angle β_0 is determined as $\arcsin\left(\frac{1}{f}\sin(\beta)\right)$. Using the replacement of the variables $y^2 = k/k_m$ and omitting intermediate calculations, the wave spectrum (6.27) can be written as a function of the non-dimensional variables y and β:

$$F\left(y, \beta, \beta_0^0, \tilde{\alpha}, \tilde{v}, \mu, \gamma\right)$$
$$= \min\left\{(n+1)\, m_0 \frac{8}{3\pi} \cos^4\left(\beta_0 - \beta_0^0\right) \sqrt{\frac{\operatorname{th}^n(\tilde{\alpha})\,\operatorname{th}(fy^2/\gamma)}{f^{n+5}\operatorname{th}^{n+1}(fy^2\tilde{\alpha})}}\right.$$
$$\times \exp\left[-\frac{n+1}{n}\left(\frac{1}{fy^2}\frac{\operatorname{th}(\tilde{\alpha})}{\operatorname{th}(fy^2\tilde{\alpha})}\right)^{\frac{n}{2}}\right] \frac{1}{y^{n+1}}\left[1 + \frac{4fy^2\tilde{\alpha}\exp\left(2fy^2\tilde{\alpha}\right)}{\exp\left(4fy^2\tilde{\alpha}\right) - 1}\right];$$

$$Q(\beta)\frac{\alpha_F}{y^5 k_{\max}^2}\right\}$$

$$\times \left\{\Theta\left[\frac{1}{2}\sqrt{\operatorname{th}(y^2\tilde{\alpha}/\gamma)}\left(1 + \frac{4\tilde{\alpha}y^2/\gamma\exp\left(2\tilde{\alpha}y^2/\gamma\right)}{\exp\left(4\tilde{\alpha}y^2/\gamma\right) - 1}\right)\cos(\beta) - \tilde{v}\gamma y\right]\right.$$
$$+ \Theta\left[\tilde{v}\gamma y - \frac{1}{2}\sqrt{\operatorname{th}(y^2\tilde{\alpha}/\gamma)}\left(1 + \frac{4\tilde{\alpha}y^2/\gamma\exp\left(2\tilde{\alpha}y^2/\gamma\right)}{\exp\left(4\tilde{\alpha}y^2/\gamma\right) - 1}\right)\cos(\beta)\right]$$
$$\left.\times \left[\Theta(\eta - 1) - \Theta(\eta\mu\gamma - 1)\right]\Theta\left[1 - (\sin(\beta)/f)^2\right]\right\}, \qquad (6.35)$$

where $\Theta(y, \tilde{\alpha}, \gamma, f, \mu)$ is the Heaviside function describing the area boundaries of changing arguments y and β. This is obtained using the kinematic considerations described above; $\eta = \eta(y, \beta, \tilde{\alpha}, \gamma, f) = V_B/V$ is a function obtained by solving the set of equations (6.33); $Q(\beta)$ is the angular energy distribution within the equilibrium range; α_F is the Phillips constant; and β_0^0 is the initial general direction of wave propagation.

The constructed spectrum is a function of the arguments y and β, whereas its defining parameters $\beta_0^0, \tilde{\alpha}, \tilde{v}, \gamma, \mu$ fix the initial spectrum state: the initial depth H_0 and the current velocity V_0, as well as the maximum current velocity value $V_{\max}$. Using the given initial values of these parameters, the spectral evolution at different basin points is determined only by the parameter γ.

The mean values for wave elements changing along the current are obtained with the help of the spectral expression (6.35). By integrating the spectrum (6.35) over y and β, similar to (5.40)–(5.42), the following values can be found:

the ratio of the mean height $\bar{h}$ to the mean height of the initial wave $\bar{h}_0$:

$$\tilde{h} = \bar{h}/\bar{h}_0 = f_1\left(\beta_0^0, \tilde{\alpha}, \tilde{v}, \gamma, \mu\right),$$

the ratio of mean wavelengths:

$$\tilde{\lambda} = \bar{\lambda}/\bar{\lambda}_0 = f_2\left(\beta_0^0, \tilde{\alpha}, \tilde{v}, \gamma, \mu\right)$$

and the ratio of mean periods:

$$\tilde{\tau} = \bar{\tau}/\bar{\tau}_0 = f_3\left(\beta_0^0, \tilde{\alpha}, \tilde{v}, \gamma, \mu\right).$$

The values f_1, f_2, f_3 are functions dependent on the non-dimensional parameters β_0^0, $\tilde{\alpha}$, $\tilde{v}$, γ, μ. This makes the obtained formulas more universal.

The ratios f_1, f_2, f_3, presented as integrals, are numerically calculated using the Chebyshev quadrature method.

In order to simplify the physical interpretation of the results, the parameters $\tilde{\alpha}$ and $\tilde{v}$ are expressed through the mean values of the initial wave elements. With the help of the initial spectrum (5.16), the parameter $\tilde{\alpha}$ can be expressed using the mean length $\bar{\lambda}_0$:

$$\tilde{\alpha} = \frac{2\pi \frac{H_0}{\bar{\lambda}_0}\left(\frac{n+1}{n}\right)^{-\frac{2}{n}}}{\Gamma\left(1 - \frac{2}{n}\right)}. \tag{6.36}$$

Analogously, the parameter $\tilde{v}$ can be written as:

$$\tilde{v} = \frac{V_0}{\sqrt{g\bar{\lambda}_0}}\sqrt{2\pi\Gamma\left(1 - \frac{2}{n}\right)}\left(\frac{n+1}{n}\right)^{-\frac{1}{n}}. \tag{6.37}$$

It should be noted that the ratios (6.36) and (6.37) are accurate for wave propagation cases in a fair current, when the reverse waves are absent and all spectral components propagate from the initial boundary to the calculated point. In such a case the following is obtained for $n = 5.5$:

$$\tilde{\alpha} = 4.19 H_0/\bar{\lambda}_0; \qquad \tilde{v} = 2.05 V_0/\sqrt{g\bar{\lambda}_0}. \tag{6.38}$$

If waves propagate in a countercurrent, an accurate determination of the value of the parameters $\tilde{\alpha}$ and $\tilde{v}$ makes some additional difficulties, because the spectrum is different from (5.16) at the initial point due to the influence of the current. It cuts off a high-frequency spectrum range, where components cannot propagate countercurrent. The expression (6.38) can be used for estimation in the cases of no reverse waves, for example, as in a current not increasing along the Ox axis. If there are reverse waves (at the current increasing along the Ox axis), then their spectral density can be approximated by the equilibrium range due to wave breaking. In order to estimate the values $\tilde{\alpha}$ and $\tilde{v}$, the parameter $n = 4$ will be taken instead of $n = 5.5$. After that the values $\tilde{\alpha} = 3.17 H_0/\overline{\lambda}_0$; $\tilde{v} \approx 1.78 V_0/\sqrt{g\overline{\lambda}_0}$ can be estimated.

Results of mean wave element calculations. Numerical calculations of the obtained solution will be made. Firstly, the calculation results of the expressions f_1, f_2, f_3 are presented for the cases when their values can be compared with the known measurement data. A transformation of waves, propagating without any current from deep water to the shallow coastal area, is referred to such cases. The current velocity is assumed to be very small at the initial boundary ($\tilde{v} = 10^{-5}$) and the non-dimensional initial depth is equal to $\tilde{\alpha} = 10$ (i.e. $H_0/\overline{\lambda}_0 \approx 2.5$), corresponding to the deep-water case.

Variations of mean wave elements, with waves approaching a coast, are shown in Fig. 6.10. A comparison of calculations and observations (Krylov et al., 1976) is also presented. As the waves propagate, their heights are smoothly decreased to $\tilde{h} \approx 0.91$ and then begin increasing. The wavelength decreases and the mean period is changed. The wave height change coincides accurately enough with the known results of wave transformation in a coastal area. It should be noted that the evolution of the mean wave elements calculated using the spectral solution is quantitatively different from the monochromatic wave solution presented in Fig. 6.3 for different angles of waves approaching a shore. The spectral approach slightly smoothes the solution, averaging it over angle directions. The quantitative difference of two solutions is determined by the angular distribution width for the spectrum of waves directed to the shore.

The second control case of testing the general solution is wave evolution in deep water for their propagation from a small current area against the current increasing along its direction. This case has already been considered in Sect. 5.5. Calculations of wave element transformation, made in this section, coincide practically with those obtained earlier (see Fig. 5.15).

Not only quantitative, but also qualitative agreement of calculations and observations given above indicates that the proposed mathematical model could describe the wave parameter transformation both in the presence of a horizontal non-uniform current and under basin depth variation. Now it would be interesting to investigate the problem of the depth and current velocity changing simultaneously.

278   6 Wave Transformation in Shallow Water

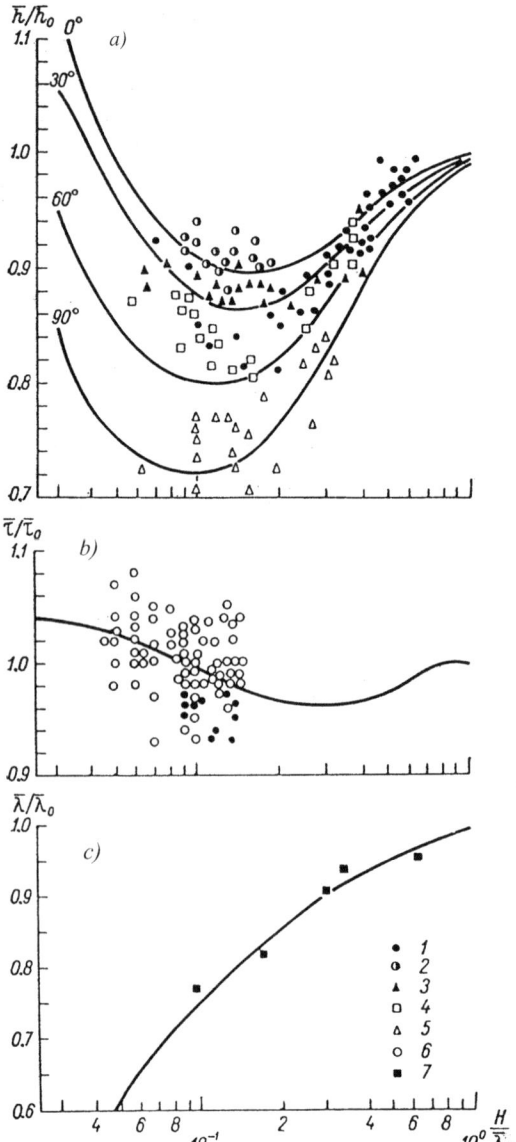

Fig. 6.10. Comparison of calculated elements with observations for different angles β_0^0 of waves approaching a shore (Krylov et al., 1976): data of Tsyplukhin for $\beta_0^0 = 20\text{–}40°$ (1); Vapnyar's for $\beta_0^0 = 0°$ (2); $\beta_0^0 = 30°$ (3); $\beta_0^0 = 60°$ (4); $\beta_0^0 = 90°$ (5); $\beta_0^0 = 0\text{–}90°$ (6); "Soyuzmorniiproject" Institute (7)

As the next step the calculations of wave element transformation are made for different combinations of defining parameters. The situations and

6.3 Bottom and a Non-Uniform Current Influence on Waves

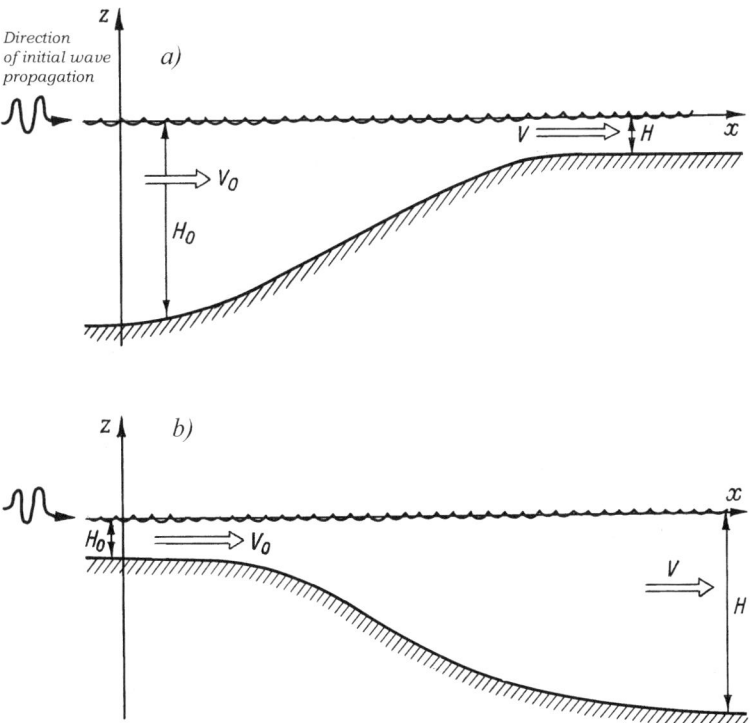

Fig. 6.11. Wave propagation in fair increasing (**a**) and decreasing (**b**) currents

the corresponding relative calculated values of the mean wave elements $\tilde{h}, \tilde{\lambda}, \tilde{\tau}$ as functions of the parameter γ, characterizing relative variations of current velocity (or depth), are presented schematically in Figs. 6.11–6.15. The value $\gamma = 1$ corresponds to wave elements at the initial boundary. The value $\gamma > 1$ describes wave variations taking place with the wave propagating in the area with increasing current velocity (or decreasing depth). The value $\gamma < 1$ describes the situation of wave propagation in a decreasing current. The use of a logarithmic scale presents the results in a more compact form.

Wave propagation in a deep water fair current (see Fig. 6.11) will be considered in detail (i.e. when the effect of depth variations is of no importance). However, unlike the second test results, the initial current velocity will be considered to be significantly different from zero, assuming that $\tilde{v} = 0.1$. Wave element variations in this case are shown in Fig. 6.12a.

The wave heights are monotonically decreased, and their lengths are increased in the fair increasing current ($\gamma > 1$). The wave periods are constant and later on they are decreased at $\gamma > 6.0$. In a fair decreasing current ($\gamma < 1$), the wave periods are practically constant, the heights are increased and the lengths are decreased. The wave parameter variations become more intensive with increasing initial current velocity.

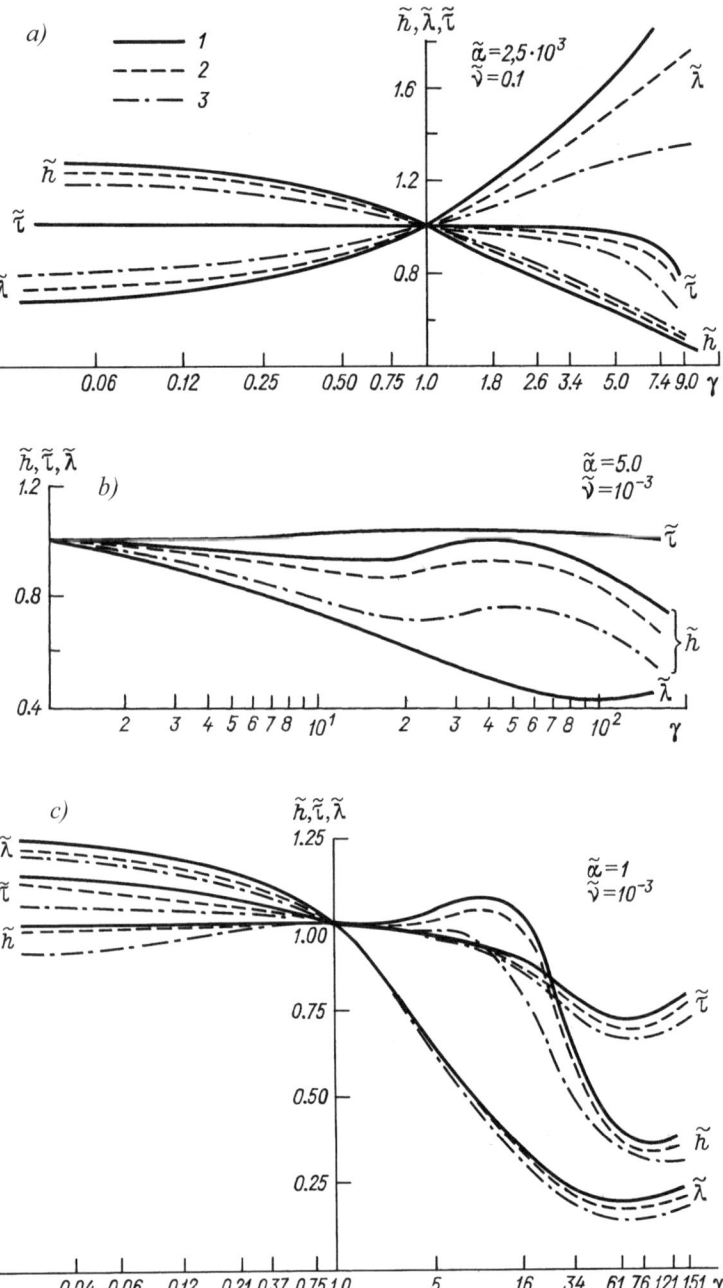

Fig. 6.12. Variations of mean wave element in deep water with fair current, $\tilde{\nu} = 0,1$ (**a**); passing from deep to shallow water (**b**) and for shallow water (**c**): 1 – initial general direction 0°; 2 – direction 30°; 3 – direction 60°; $\gamma > 1$ – increasing current; $\gamma < 1$ – decreasing current

6.3 Bottom and a Non-Uniform Current Influence on Waves

The values of wave elements are shown in Fig. 6.12b for the parameter $\tilde{\alpha} = 5$. It should be noted that waves at the initial point can be considered to be short in comparison with the depth, but the situation can be changed with wave propagation. This situation occurs in the case of a wave propagating from deep to shallow water in a current with $\gamma > 1$. It should be remembered that a similar case is considered in the first test calculation, but without any current. This allows revealing the effect of decreasing depth in its "pure form". At the initial stage of wave propagation, with the current velocity being small ($\tilde{v} = 10^{-3}$, see Fig. 6.12b), and the depth variations being important, the wave element transformation is in agreement with ordinary wave presentations in a coastal area. At first, the wave heights are decreased, and increased later on. At this stage the periods are not practically changed and the lengths are decreased. When the current velocity is large, and the wave velocities are small due to decreasing depth, a decrease of the wave heights takes place again. For small depths ($\gamma \approx 100$), with the prevailing current effect, a gradual increase of wavelengths is observed ($\tilde{\lambda} \sim \gamma$) and the waves become flatter. With increasing the parameter $\tilde{v}$, the wave transformation becomes more influenced by current velocity variations along the flow compared to depth variations. A monotonic height decrease and length increase are observed.

The graphs of wave element variations with initial basin depth being less than the wavelength ($\tilde{\alpha} = 1$) are presented in Fig. 6.12c. If the initial current is negligible ($\tilde{v} = 10^{-3}$), a depth decrease results at first in the wave height increase and in the wavelength decrease. There is no initial decrease of the height, as in case of a wave propagating from deep to shallow water (Fig. 6.12a). The variations of wave heights in shallow water are estimated using the Green formula: $\tilde{h} \sim \sqrt[4]{\gamma}$. However, later on ($\gamma > 15$), when the wavelength becomes shorter and the current velocity is comparable with the velocity of wave propagation, the role of the current becomes decisive. The wave height decreases with increasing current velocity. With further depth decrease and current velocity increase, the wavelength increases ($\gamma > 50$). If the initial current velocity is $\tilde{v} = 10^{-1}$, the monotonic height $\tilde{h}$ decrease and length $\tilde{\lambda}$ increase are observed with increasing current velocity.

The wave element transformations for different initial general directions show that the described tendencies become weaker with increase of the angle β_0^0 (i.e. the angle between the general wave direction and the current velocity). At the same time the opposite situation can be revealed at small depth.

Now, the wave behaviour in a countercurrent will be considered. Two situations can be identified: increasing and decreasing currents (see Fig. 6.13). It should be noted that in the case of an increasing countercurrent, reverse waves carried backwards by the flow can appear in a basin area with changing current velocity. The wave description is limited by consideration of decreasing and increasing countercurrent, excluding the part where the gradients of current velocity are different from zero.

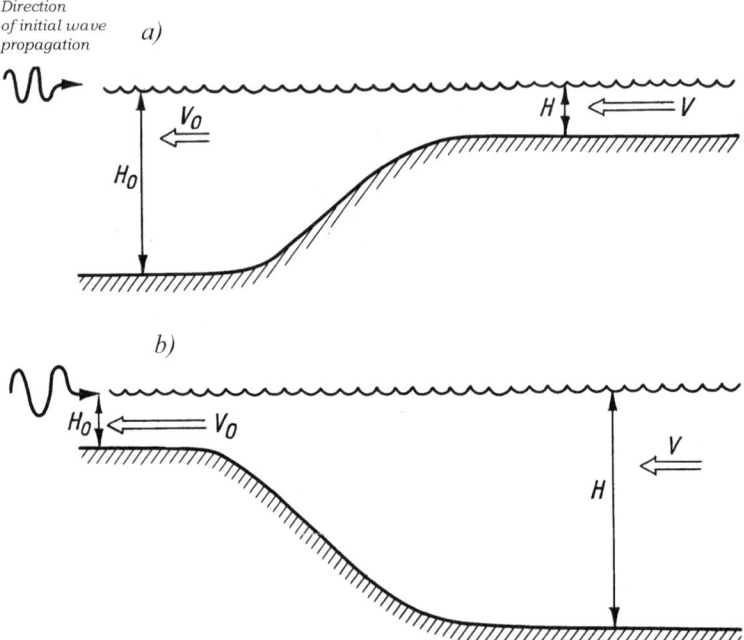

Fig. 6.13. Wave propagation in a countercurrent, increasing (**a**) and decreasing (**b**) along its direction

Omitting a detailed description of wave transformation in deep water for different initial values, it should be noted that the wave element variations are analogous to that given before (see Fig. 5.15). With increasing velocity, the maximum height is decreased (at $\tilde{v} = 0.1$; $\tilde{h} \approx 1.8$) and the minimum wavelength is increased ($\tilde{\lambda} \approx 0.5$).

Now, the wave propagation from deep to shallow water with existing countercurrent will be considered (for $\tilde{\alpha} = 5$). In the case of negligible initial current velocity ($\tilde{v} = 10^{-5}$), the wave element variations in a wide depth range actually coincide with the element transformations in a coastal area in the absence of current considered above. With increasing the initial current velocity (for $\tilde{v} = 10^3$), at first, a decrease of the wave height is observed, after which it is increased with the height achieving its maximum ($\tilde{h} \approx 1.25$ at $\gamma \approx 50$). Later on the wave height is decreased to zero. This is due to wave blocking by the countercurrent. The wavelength is monotonically decreased within the entire considered range of changing the parameter γ.

With a further increase of the initial current velocity (see Fig. 6.14a, for $\tilde{v} = 10^1$), the maximum wave height is increased ($\tilde{h} \approx 1.8$) and the behaviour of the wave parameters is very similar to that for wave evolution in deep water. In this case the depth does not influence the waves. The current influences the waves before becoming long enough in comparison with the

6.3 Bottom and a Non-Uniform Current Influence on Waves 283

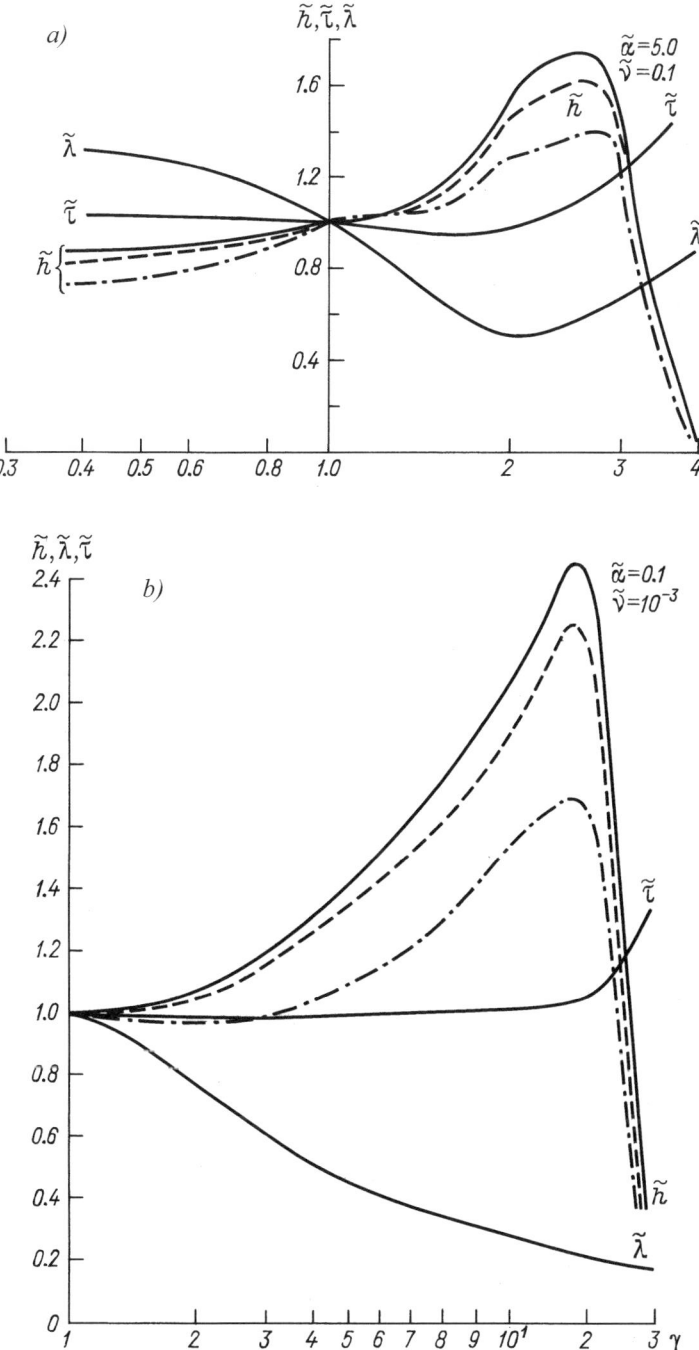

Fig. 6.14. Variations in mean wave element in increasing ($\gamma > 1$) and decreasing ($\gamma < 1$) countercurrent (**a**) and in shallow water countercurrent (**b**)

depth. The described tendency of wave element variations is preserved with decreasing initial depth. A joint decrease in depth and increase in current velocity results in increasing the maximum wave height ($\tilde{h} \approx 2.5$ at $\tilde{\alpha} \approx 0.1$ and $\tilde{v} \cong 10^{-3}$, see Fig. 6.14b).

The wave element transformation will be considered in the section of a counterflow, with the current velocity spatial variation: $\partial V/\partial x \neq 0$. The peculiarity of the wave behaviour is connected with reflecting wave packets propagating countercurrent at this precise section. After being reflected, the wave packets are carried away backwards by the current in the form of reverse waves, increasing their heights and decreasing their lengths. The total wave pattern in this case is represented by a superposition of straight and reverse waves.

The appearance of reverse waves in a flow is determined by the blocking condition (6.33). If it is fulfilled, the spectral components of reverse waves make a significant contribution to the total wave spectrum. Limitations of the spectrum value connected with possible wave breaking are introduced, using the notion of the equilibrium range. Its presence in the solution results in the appearance of the parameter characterizing the ratio of the initial spectrum value relative to the equilibrium range value. In Sect. 5.5, this value is expressed by the mean steepness $\bar{h}_0/\bar{\lambda}_0$ of the initial wave spectrum. The initial steepness value can significantly affect the wave element changes in a current. The values of wave elements along the increasing flow ($\tilde{\alpha} = 5$; $\tilde{v} = 0,1$) for waves with steepness $1/36$ (corresponding to the mean wind wave steepness) are shown in Fig. 6.15, where the straight and reverse waves as well as their breaking are taken into account. The behaviour of wave elements is similar to that described above (see Fig. 5.16). It is seen that the presence of the reverse wave and its breaking do not result in a large increase of mean heights.

In conclusion it should be noted that the evolution of mean wave elements occurs under the effect of two mechanisms. They are, firstly, a wave transformation by the action of a horizontally non-uniform current and, secondly, by changing depth of the basin. Although these mechanisms are described using the same equations (within the framework of the initial approximation), their effects are different. The current influences are greater on the short-wave spectrum components, while the changing depth affects the long wave components. They penetrate into deeper water and are subjected to depth variations.

As shown by the obtained results the effect of non-uniform depth and current velocity lead to such changes of wave elements, which are impossible to predict using the mechanisms of transformation separately. In some cases, the combined effect of the changing depth and current essentially increases the wave transformation, while decreasing it in some other cases. In shallow water areas with spatial non-uniform currents, such situations can be observed, when the wave heights and other wave elements are increased greatly (by several times). But there are also opposite situations, with waves decrease up to their almost complete disappearance.

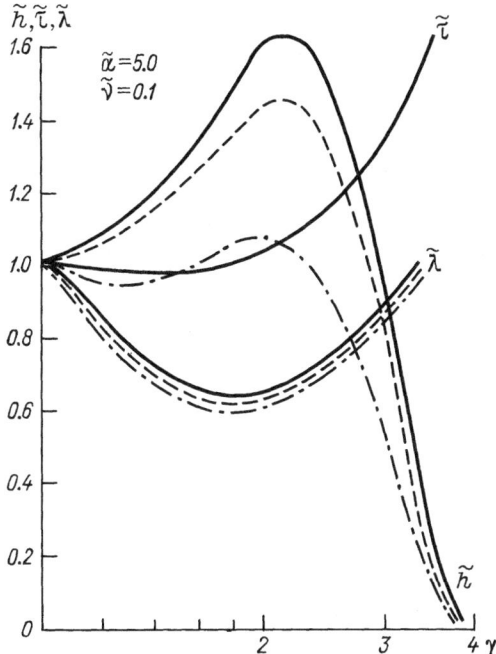

Fig. 6.15. Variation of mean wave elements in a countercurrent, taking into account reverse waves and their breaking

6.4 Wave Energy Dissipation in Shallow Water

Statement of the problem. As waves are propagating in shallow water, wave movements reach the bottom. Besides the wave refraction described above, the interaction of wave water movement and the bottom takes place. As a result, additional physical mechanisms appear, affecting the formation of the wave spectrum.

The main parameter determining the bottom effect on waves is still the non-dimensional product of the wave number k and the depth H. Due to strong non-linear effects at very small depths, the kinetic equation used for describing wave spectrum evolution is not applicable. The investigation in this section is limited to the frame of the intermediate case (i.e. when, on the one hand, the bottom affects the waves, but on the other hand, the depth is not so small for the kinetic equation to be valid). Thus, the non-dimensional parameter kH is supposed to be within the range $0.5 < kH \leq 3.0$.

It should be noted that the effect of the bottom on waves was studied in many papers (Komen et al., 1994; Massel, 1996). For example, the following mechanisms have been pointed out describing the bottom effect on waves: scattering by bottom non-uniformities, movement of the viscous boundary layer, water filtering through a porous bottom and turbulent friction in the boundary layer (Shemdin et al., 1978). The first mechanism results in a local wave energy redistribution due to wave component scattering, for example, by small-scale coral reefs or sand riffles. The last three mechanisms are dissipative. Their effectiveness is determined by concrete conditions of forming the boundary layer at the bottom. The filtration depends on the porous structure (penetrability) of the bottom material, and the bottom friction is connected with the dimensions of the bottom roughness.

Dissipation dependence on wave bottom stress. The wave energy dissipation determined by bottom friction due to a random wave field with an average current will be considered. This approach for a regular wave was suggested by K. Kajiura (1968) and generalized for monochromatic waves and current in the paper by Christoffersen & Jonsson (1985).

The linear equation of momentum for the water boundary layer can be written in the following form:

$$\frac{\partial \bm{v}}{\partial t} + \frac{1}{\rho_2} \nabla P = \frac{1}{\rho_2} \frac{\partial \tau}{\partial z} , \tag{6.39}$$

where ρ_2 is the water density; $\bm{v}$ is the average horizontal velocity; P is the average pressure; and τ is the turbulent stress in the boundary layer.

The values $\bm{v}$, P and τ can be divided into a constant value (averaged over phases) and a fluctuating remainder:

$$\tau = \tau_\mathrm{w} + \tau_\mathrm{c} , \tag{6.40}$$

where $\tau_\mathrm{c} = \langle \tau \rangle$, $\tau_\mathrm{w} = \tau - \langle \tau \rangle$, and $\langle \ \rangle$ means averaging over random wave phases. The components of velocity and pressure consist of a linear superposition of constant current and wave movement. The stresses τ_c and τ_w are non-linearly dependent on the total movement. The value τ_w is supposed to be equal to zero outside a boundary wave layer being a thin sub-layer in a current boundary layer.

The equation of wave momentum can be found by averaging the equation (6.39) over random wave phases and subtracting the average equation from the complete one (6.39):

$$\frac{\partial \bm{v}_\mathrm{w}}{\partial t} + \frac{1}{\rho_2} \nabla P_\mathrm{w} = \frac{1}{\rho_2} \frac{\partial \tau_\mathrm{w}}{\partial z} . \tag{6.41}$$

Supposing that the pressure gradient is not dependent on wave height inside the wave boundary layer, it is exchanged by the value defined at the surface and written as:

$$\frac{\partial \left(\bm{v}_\mathrm{w} - \bm{V}_\mathrm{w}^\mathrm{H} \right)}{\partial t} = \frac{1}{\rho_2} \frac{\partial \tau_\mathrm{w}}{\partial z} , \tag{6.42}$$

where $\bm{V}_\mathrm{w}^\mathrm{H}$ is the orbit velocity of free flow, and the index H designates the velocity at the bottom accepted to be equal to the value at the top of the boundary layer for non-viscous water flow.

6.4 Wave Energy Dissipation in Shallow Water

The wave energy dissipation determined by the bottom current can be found as follows. The equation (6.42) is multiplied by $v_{w\,k}$ (i.e. the orbit velocity of the wave component k) and integrated over depth and averaged over the random wave phase:

$$g\frac{\partial}{\partial t}S(k) = \left\langle \int_{-H}^{-H+\delta} v_{w\,k}\frac{\partial}{\partial z}\tau_w\,dz \right\rangle$$

$$= \left\langle \int_{-H}^{-H+\delta} \{V_{w\,k}^{\rm H} + (v_{w\,k} - V_{w\,k}^{\rm H})\}\frac{\partial}{\partial z}\tau_w\,dz \right\rangle$$

$$= \langle V_{w\,k}^{\rm H}\tau_w\rangle\Big|_{-H}^{-H+\delta} + \left\langle \int_{-H}^{-H+\delta} \tau_w\frac{\partial}{\partial z}V_{w\,k}^{\rm H}\,dz \right\rangle$$

$$+ \left\langle \int_{-H}^{-H+\delta} (v_{w\,k} - V_{w\,k}^{\rm H})\frac{\partial}{\partial t}(v_{w\,k} - V_w^{\rm H})\,dz \right\rangle,$$

where δ is the thickness of the wave boundary layer. The second term in this formula disappears, as the value $V_{w\,k}^{\rm H}$ is constant within the wave boundary layer; the third term also disappears due to movement periodicity. Finally, it can be seen that:

$$G_{\rm f}(k) = -\frac{1}{g}\langle V_{w\,k}^{\rm H}\tau_w^{\rm H}\rangle. \tag{6.43}$$

It follows from (6.43) that the wave energy dissipation is dependent on the orbital velocity of the undisturbed bottom flow (it is usually known) and turbulent stress at the bottom (its values are indefinite). The problem is that it is rather difficult to make direct measurements of bottom stress produced only by random waves or their superposition with the average flow. The available measurements are obtained over the boundary layer (Christoffersen & Jonsson, 1985) and, mainly, under laboratory conditions for monochromatic waves. That is why it is supposed that the bottom stress parameterization for random waves can be obtained, using the monochromatic wave theory, as the approximations of the bottom velocity spectrum are usually of a narrow and one-mode character (Komen et al., 1994).

Parameterization of bottom stress. There is a great number of different approximations for solving the problem of closing the turbulent boundary layer equations: from the simplest empirical formulas to the highest-order closing schemes (Monin & Yaglom, 1992). The approximation of turbulent viscosity is widely used for modelling the stationary flows in atmosphere and ocean. The stress is parameterized analogously to the case of the usual viscosity with the help of this approximation:

288   6 Wave Transformation in Shallow Water

$$\tau = \nu_e \frac{\partial \boldsymbol{v}}{\partial z}, \qquad (6.44)$$

where $\nu_e \sim \tau^{1/2}$ is the coefficient of turbulent viscosity. Proceeding from the law of resistance (6.44), the stress τ for stationary flows can be expressed in the following form:

$$\tau = C_D(z) V^2(z), \qquad (6.45)$$

where $V^2(z)$ is the velocity at the height z, and the resistance coefficient C_D is a function of the ratio of surface roughness to the given level, which is further accepted to be equal to the boundary layer thickness.

The total stress is divided into wave and current stresses:

$$\tau = \tau_c + \tau_w = \nu_e \frac{\partial \boldsymbol{v}_c}{\partial z} + \nu_e \frac{\partial \boldsymbol{v}_w}{\partial z}. \qquad (6.46)$$

Thus, the stress is calculated as the sum of two components. One of them is dependent on the current in the bottom boundary layer, where $\nu_e = \nu_{ec} \sim (\tau_c^H)^{1/2}$. The second one is determined by wave movement in the internal wave boundary sub-layer, where $\nu_e = \nu_{ew} \sim \langle (\tau_w^H)^2 \rangle^{1/4}$. It should be noted that the coefficient ν_e is determined differently in these two boundary layers.

Solving the equations for the boundary layer with proper boundary conditions for τ_c, the law of resistance (6.45), with $V = V_c^H$, can be obtained. There are two ways of determining the value τ_w. The first one is connected with the wave field spectral presentation. It provides the possibility of obtaining the law of resistance for every spectral component. The second approach is connected with calculations for some frequency spectral range, for example, for the frequency of spectrum maximum. The integral law of resistance can be obtained:

$$\tau_w^H = C_D (V_w^H)_{rms} V_w^H, \qquad (6.47)$$

where $(V_w^H)_{rms}$ is the mean square current velocity at the bottom.

It should be noted that the bottom stress caused by waves and current is connected with the proper current components above the bottom, described with the help of the formulas (6.45) and (6.47). The interaction of wave flow and stationary current is taken into consideration with the help of the parameter C_D. Its value is increased due to the total movement in comparison with the separately taken case of wave movement or constant current. In most cases the boundary layer created by a constant current is affected by the wave movement, rather than vice versa (Christoffersen & Jonsson, 1985).

Another approximation for the bottom stress, determined by the random wave field under a constant current, was suggested in the paper by Hasselmann & Collins (1968). The law of resistance (6.45) is used for the case when the total bottom stress is connected with the complete velocity of flow at some height, for example, at the height of the wave boundary layer:

$$\tau^H = C_D \left| V^H \right| V^H. \qquad (6.48)$$

It should be noted that the wave phase is included non-linearly in (6.48), while this does not happen in (6.47). In this case the resistance coefficient can be determined using experimental estimations for wave energy dissipation.

Substituting (6.47) into the formula for dissipation (6.43), it can be obtained that:

$$G_f(\mathbf{k}) = -\frac{1}{g} C_D \left(V_w^H\right)_{rms} \left\langle \left(V_{w\mathbf{k}}^H\right)^2 \right\rangle . \qquad (6.49)$$

The formula (6.49) can be transformed in terms of the spectral energy density. Thus, the dissipation is presented in the following form:

$$G_f(\mathbf{k}) = -2C_D \left(V_w^H\right)_{rms} \frac{k}{\operatorname{sh}(2kH)} S(\mathbf{k}) , \qquad (6.50)$$

where the designation $C_f = 2C_D(V_w^H)_{rms}$ is introduced. A similar expression is obtained using the formula (6.48), taking $C_f = 4C_D V_c^H$. In the two-layer boundary model the coefficient C_f depends linearly on the mean square near-bottom flow velocity, while in the one-layer model its value is proportional to the flow velocity.

A number of attempts have been undertaken to determine the coefficient value C_f as a function of bottom roughness, integral parameters of the spectrum, and frequency and direction of wave component propagation. According to the JONSWAP experiment, the mean value is estimated as $C_f = 0.038 \text{ m}^2\text{s}^{-3}$ (Hasselmann et al., 1973). The WAM model is believed to correspond to ordinary wind wave development conditions (Komen et al., 1994). However for a strong storm the coefficient is dependent on wind sea parameters, being changed during a storm. According to estimations (Weber, 1991), the coefficient dependence can be written in the form:

$$C_f = \exp\left(-8.34 + 6.34 z_H^{0.08}\right) (V_w^H)_{rms} , \qquad (6.51)$$

where $z_H = k_N \omega_{\max}/(V_w^H)_{rms}$. The roughness parameter can be equal to $k_N = 0.04$ both for wind sea and swell. Tide currents can be neglected in this case.

Wave energy dissipation due to bottom-through filtration. The above mentioned wave energy dissipation is described under the suggestion of bottom impermeability. Although this simplifies the solution, in reality it is not always true. It is possible for wave movements to penetrate through the bottom soil, in the case of its structure being porous, or permeable for liquid flows. Part of wave energy is lost by overcoming the resistance of the bottom porous structure.

The simplest formulation of the problem can be done by describing wave movements in a two-layer medium. The movements are supposed to be potential in the upper layer $(-H \leq z \leq \eta)$. A loss of wave energy due to dissipation, determined by filtration, takes place in the lower layer $(-H_n < z < -H)$. The conditions for normal stress continuity and liquid flow constancy are set at the boundary of the two media ($z = -H$).

The problem is considered for the thick-sand layer (Shemdin et al., 1978) and for sand of certain thickness (Massel, 1996). Thus, the dissipation function is obtained in the following form:

$$G_\mathrm{p}(\boldsymbol{k}) = -\frac{gKk}{\nu} \frac{\mathrm{th}\left[k\left(H_n - H\right)\right]}{\left\{\mathrm{ch}(kH) + \frac{\omega K}{\nu} \frac{k(H_n-H)\,\mathrm{sh}(kH)}{\mathrm{ch}^2[k(H_n-H)]}\right\}^2} S(\boldsymbol{k}) \qquad (6.52)$$

where K is the sand penetrability coefficient; $(H_n - H)$ is the sand layer thickness; and ν is the viscosity coefficient.

In the case of the thick-sand layer $((H_n - H) \to \infty)$, the dissipation function is simplified and can be written as:

$$G_\mathrm{p}(\boldsymbol{k}) = -C_\mathrm{p} \frac{k}{\mathrm{ch}^2(kH)} S(\boldsymbol{k})\,, \qquad (6.53)$$

where $C_\mathrm{p} = gK/\nu$ is the penetrability coefficient. It is equal to $0.0006\,\mathrm{m\,s^{-1}}$ for fine-grained sand with dimensions of particles $D_\mathrm{m} = 0.25\,\mathrm{mm}$. These values are $C_\mathrm{p} = 0.01\,\mathrm{m\,s^{-1}}$ and $D_\mathrm{m} = 1.0\,\mathrm{mm}$ for coarse-grained sand.

A comparison of the dissipation functions caused by bottom friction (6.50) and filtration (6.53) shows that their behaviour is similarly dependent on the parameter kH. The function values are only different for the low-frequency swell. Their values are similar for fine-grained sand with fraction dimensions $D_\mathrm{m} = 0.25\,\mathrm{mm}$ and coefficient $C_\mathrm{f} = 0.001$, which is typical for a plain bottom.

Wave energy dissipation due to liquid soil. The bottom is often covered with liquid soil (i.e. miry silt of different suspended consistence) at coastal areas and especially at river mouths. Wind waves, penetrating into the depth, cause liquid soil movements with soil viscosity resulting in wave energy dissipation. Due to the liquid soil effect the damping intensity can be higher than wave energy losses caused by other reasons (Tubman & Suhayda, 1976).

It is known that the river Amazon carries out much silt to the Surinam region. Experiments (Wells, 1983) show that significant wave energy losses take place there. The waves lose up to 88 per cent of their energy, propagating between two stations. The depth is 7.1 m at the first station and it is 4.7 m at the second one. The distance between them is equal to 1.5 km. Such significant energy losses within so small a distance are connected with neither wave breaking, nor their transformation.

Nowadays a number of wave dissipation models with liquid soil have been developed. Their classification is given in the paper by Massel (1996).

As a rule, a two-layer liquid is considered in the models. The movement of the upper layer $(-H \leq z \leq \eta)$ is described using the potential approximation. In order to describe the dynamics of the lower layer $(-H_n \leq z \leq -H)$, which

6.4 Wave Energy Dissipation in Shallow Water

is viscous, the flow function is used. A solution to the problem was obtained by Hsiao & Shemdin (1980). The connection between the wave number and the frequency is written in the form of:

$$k = \frac{\omega^2}{g} \frac{1 + \text{th}(kH\Omega)}{\text{th}(kH) + \Omega}, \qquad (6.54)$$

where

$$\Omega = r \frac{(m^2 - k^2)\left[s_1 - \frac{k}{m}c_1\right]}{(m^2 + k^2)\left[c_2 - \frac{k}{m}s_2\right] - 2k^2} ;$$

where

$$m = k\left[1 - \frac{\omega^2}{k^2(J - i\omega\nu)}\right]^{\frac{1}{2}},$$

$$s_1 = \text{sh}\left(k(H_n - H)\right)\text{ch}\left(m(H_n - H)\right),$$
$$s_2 = \text{sh}\left(k(H_n - H)\right)\text{sh}\left(m(H_n - H)\right),$$
$$c_1 = \text{ch}\left(k(H_n - H)\right)\text{sh}\left(m(H_n - H)\right),$$
$$c_2 = \text{ch}\left(k(H_n - H)\right)\text{ch}\left(m(H_n - H)\right),$$

ν is the coefficient of kinematic viscosity; $r = \rho_2/\rho_m$ is the ratio of water to liquid soil density; and $J = E/\rho_m$, where E is the shift elasticity coefficient.

The value k is a complex quantity in (6.54): $k = k_r + ik_i$. The imaginary part k shows the wave energy dissipation connected with the liquid soil layer. In the case of values $E \to \infty$ and $\nu \to \infty$, the formula (6.54) is transferred into an ordinary dispersion relation for waves in water of final depth H.

Using (6.54), the spectral function of wave energy dissipation is obtained as:

$$G_{ds}(\mathbf{k}) = -2k_i C_g S(\mathbf{k}) . \qquad (6.55)$$

As long as the dissipation function (6.55) is linearly connected with the spectrum, the energy evolution of every spectral component is independent of the other components.

This dissipation effect is decreased with increasing the water layer thickness and wavelength decrease. The wavelength is decreased due to the presence of liquid soil. This change usually makes up from 10 to 20 per cent, but it can reach 50 per cent. The wave energy dissipation is increased with enlarging the layer thickness of the liquid soil and its viscosity. However, after reaching a certain value for the viscosity, the effectiveness of the influence of the liquid soil starts decreasing the dissipation effect.

Wave scattering due to bottom roughness. There is another mechanism of wave energy decrease with propagation over an uneven bottom. This mechanism can be described using non-linear interaction theory. In this approach the wave movement in the water surface $\eta(\boldsymbol{r},t)$ and the bottom roughness $\varepsilon(\boldsymbol{r})$ (where the depth is written as $H = H_0(1+\varepsilon(\boldsymbol{r}))$, $|\varepsilon| \ll 1$) are presented with the help of the Fourier integral. These values are distributed into wave vectors and frequencies, whereas the frequency of spectral components describing the bottom unevenness is accepted to be equal to zero.

Due to the roughness of the bottom, wave scattering results in mean wave field damping with a resonance character (Pelinovskiy, 1979). The wave amplitude change is caused only by those spectral components of bottom roughness, whose wave vectors satisfy the synchronous conditions:

$$\boldsymbol{k} \pm \boldsymbol{k}_1 \pm \boldsymbol{k}_2 = 0 , \qquad (6.56)$$

where $\boldsymbol{k}$ is the wave vector of the incident wave; $\boldsymbol{k}_1$ is the wave vector of the scattered wave; and $\boldsymbol{k}_2$ is the wave vector of the "frozen field" of bottom roughness. In other words, the wave vectors of scattered waves satisfy the known Bragg conditions. Non-resonance components of bottom roughness can result in changing the wave frequency, but the damping is affected only by resonance components.

The formula for the mean wave field damping increment γ (i.e. the wave amplitude which can be written as $\langle a \rangle = ae^{\gamma t}$) is obtained (Pelinovskiy, 1979) using the Burre approximation and generalized (Lavrenov, 1980) for the case of long dispersed waves ($kH_0 < 1$). The dispersion ratio is taken as $\omega^2 = gH_0/(1+1/3k^2H_0^2)$. The formula for the damping increment is written as follows:

$$\mu = \operatorname{Re}\gamma = -\frac{\pi\sqrt{gH_0}k^3}{2\sqrt{1+1/3k^2H_0^2}} \int_0^{2\pi} \cos^2\theta\, \Phi\left(2k\sin\frac{\theta}{2}, \frac{3\pi+\theta}{2}\right) d\theta , \qquad (6.57)$$

where $\Phi(\boldsymbol{k})$ is the spectrum of the bottom roughness.

In the case of one-dimensional depth roughness with the angle between the isobath and the wave propagation direction being equal to $\pi/2 - \beta$, the two-dimensional spectrum of the roughness can be written as:

$$\Phi(\boldsymbol{k}_2) = F\left[2k\sin\frac{\theta}{2}\sin\left(\frac{\theta}{2}-\beta\right)\right] \delta\left[2k\sin\frac{\theta}{2}\cos\left(\frac{\theta}{2}-\beta\right)\right].$$

So, the integral (6.57) can be estimated analytically, obtaining that:

$$\mu = -\frac{\pi\sqrt{gH_0}k^2}{2\sqrt{1+1/3k^2H_0^2}} \frac{F(0) + F(2k\cos\beta)\cos^2(2\beta)}{\cos(\beta)} . \qquad (6.58)$$

In the case of normal wave propagation ($\beta = 0$), the damping is determined by the components of the bottom roughness spectrum at two wave numbers $k_2 = 0$ and $k_2 = 2k$. The first one corresponds to forward scattering, whereas the second to backward scattering. If the following ratio $F(0) = F(2k)$ is fulfilled, wave scattering is absent in spite of the bottom roughness. The spectrum $F(k)$ is considered to be a monotonic decreasing function of the wave number k. Two cases can be distinguished: the spectrum $F(k)$ is wide enough to correspond to a small correlation scale; and the spectrum $F(k)$ is narrow enough (i.e., a large correlation length). The first case is considered to be scattering in the medium with small-scale fluctuations. The second one is considered to be scattering in the medium with large-scale fluctuations. But in both cases the scattering is effected by the same wave numbers. It is obvious that for the same value $\langle\varepsilon^2\rangle$ the function $F(0)$ is greater for large-scale than for small-scale non-uniformity, as the integral of the function $F(k)$ is equal to $\langle\varepsilon^2\rangle$. That is why the wave absorption coefficient is greater for large bottom roughness than for a small one. At the same time the value μ is mainly determined by forward scattering for large-scale roughness; but as for small-scale roughness it is determined by scattering forward and backward. The scattering is practically equal in both directions.

As a wave propagates at some angle to the bottom unevenness ($\beta \neq 0$), the absorption coefficient depends on the incidence angle, and at the same time backward scattering disappears at $\beta = 45°$. This can be explained by noting that for the angle of $45°$, the scattered wave propagates perpendicularly to the main one, without affecting its amplitude. It is interesting to note that the scattered wave also propagates forward for $\beta > 45°$. The value μ is unlimitedly increased as $\beta \to 90°$. This result is easily explained by the resonance character of scattering. The effective correlation scale is increased sharply ($L_{\text{eff}} = L/\cos\beta$) for a wave propagating under the small angle $\pi/2 - \beta$, so the unevenness becomes more large-scaled, and the scattering is increased sharply.

Now the two-dimensional isotropic roughness of the bottom relief with the spectrum: $\Phi(\boldsymbol{k}_2) = \Phi(k_2)$ will be considered. In this case at small-scale roughness the following formulas can be written: $\Phi(2k\sin(\theta/2)) \approx \Phi(0) \sim \langle\varepsilon^2\rangle L^2$ and $\mu \sim -\frac{\sqrt{gH_0}k^3L^2}{\sqrt{1+1/3k^2H_0^2}}$. It is seen that the power dependence of the absorption coefficient is changed (k^3 instead of k^2) in the case of small-scale roughness. At large scale roughness the spectrum $\Phi(k_2)$ is different from zero within the small area ($\beta = 0°$). The scattering takes place backwards, and it can be found that $\mu \sim -\frac{\sqrt{gH_0}k^2L^2}{\sqrt{1+1/3k^2H_0^2}}$.

Thus, with this approach, the classic problem of wave scattering by a rough bottom can be solved more precisely taking into account wave dispersion.

6.5 Numerical Model of Wind Wave Transformation in a Coastal Area

Formulation of the problem. Some cases, when the solution of the problem of wave refraction in shallow water and in non-uniform currents can be obtained analytically, are given in the previous sections. It is impossible to obtain such a solution in the general case with the water depth and current velocity being changed ad arbitrium. However, a correct estimation of wind wave transformation in a certain natural basin is of the most practical interest. Estimation difficulties are connected with the problem formulation as it is and its numerical implementation, including the availability of necessary information about wind speed, current fields, morphometry, etc.

There have been numerous attempts to solve this problem. Some of them have already been mentioned in the review of this monograph. As a rule, the problem results in determining the wave elements for a certain coastal area, if the initial wave values are known at open sea in deep water. The traditional method of solving the problem according to GOST (the Russian State Building Standards and Rules, 1983) is based on defining the energy flow along a ray tube originated from the initial boundary. However this can lead to caustics and singularities in the problem solution. The method of developing the characteristics for the spectral wave energy balance equation without resulting in such peculiarities is given in this monograph. This method obtains rather an accurate solution, producing fine details of wave refraction, because it is not connected with numerical diffusion present in most numerical schemes.

The problem is solved using the balance equation of the wave action spectral density, written in the general form (1.84)–(1.89). One can pass from the spectral density of the wave action N to the spectral density of the wave energy $S = N\sigma$, which is a function of the wave number $k = |\boldsymbol{k}|$ and the angle β. As for the source function G, the non-linear wave interaction in the spectrum can be neglected in the case of wave transformation in a limited local coastal area, as shown in Sect. 6.2. Taking into account the known mechanism of wave energy dissipation G_{ds}, connected with bottom friction, the balance equation of the energy spectral density can be presented in the following form:

$$\frac{\mathrm{d}S}{\mathrm{d}t} = \left[\frac{1}{2} \frac{\tilde{f}-1}{H} \left(\frac{\partial H}{\partial \vartheta} \frac{\mathrm{d}\vartheta}{\mathrm{d}t} + \frac{\partial H}{\partial \varphi} \frac{\mathrm{d}\varphi}{\mathrm{d}t} \right) + \frac{\tilde{f}+2}{k} \frac{\mathrm{d}k}{\mathrm{d}t} \right] S - G_{\mathrm{ds}}, \qquad (6.59)$$

where $\tilde{f} = 1 + 2kH/\mathrm{sh}\,(2kH)$.

Algorithm for problem solution. In order to obtain the energy wave spectrum and its mean elements (heights, lengths, periods, etc.), it is necessary to integrate the set of equations (6.59) with the characteristics (1.86)–(1.89). This integration should be fulfilled for a discrete set of wave numbers k_i and directions β_j (see Fig. 6.16). However, the initial ray points $\{\varphi_{ij}^*, \vartheta_{ij}^*, k_i^*, \beta_j^*\}$ of phase space, making a contribution to the estimated spectrum and serving as initial values for the ray integration, are not known beforehand. That is why there are two stages to the numerical implementation of the problem. In the first stage the "reverse problem" will be solved, i.e. proceeding from the estimated point $\{\varphi^0, \vartheta^0, k_i, \beta_j\}$, the rays and the corre-

6.5 Numerical Model of Wind Wave Transformation in a Coastal Area

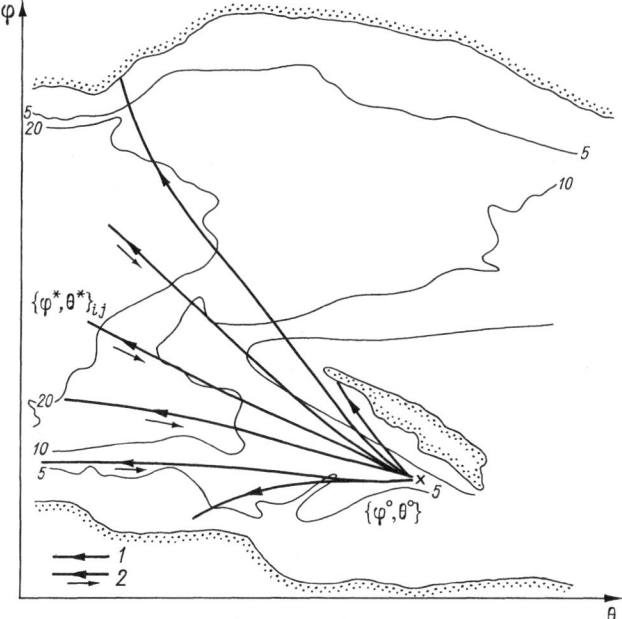

Fig. 6.16. Calculation of wave rays in the eastern part of the Finnish Gulf: 1 – reverse rays; 2 – energy-supplying rays

sponding initial points $\{\varphi_{ij}^*, \vartheta_{ij}^*, k_i^*, \beta_j^*\}$ can be found. At the same time the equation set (1.86)–(1.89) is solved by integrating "backwards", i.e. changing $\Delta t \to -\Delta t$. After obtaining the point values $\{\varphi_{ij}^*, \vartheta_{ij}^*, k_i^*, \beta_j^*\}$, the characteristic equations (1.86)–(1.89) are integrated "forward" along the rays, jointly with the energy equation (6.59). After that the spectral density of energy in calculated point is estimated.

Test calculation results. Taking into account the multi-factor character of the model, i.e. the great number of different mechanisms forming the wave spectrum, the wave transformation within a coastal area without currents will be considered. Waves propagation from deep to shallow water in a model basin with constant bottom slope will be investigated as a test. The depth variation is supposed to be described by the ratio:

$$H(x) = 28.0 - 1.05 \times 10^{-3} x, \tag{6.60}$$

where $x = R(\vartheta - \vartheta_0)\cos(\varphi_0)$; φ_0 and ϑ_0 are the origin of the local coordinate system; and R is the Earth's radius. The dimension of the water basin in the Ox direction is $0 < x < 27 \times 10^3$ m. The choice of cross-dimensions of the water basin (along the Oy axis, $y = R(\varphi_0 - \varphi)$) is determined to be larger enough so that the influence of transversal boundaries can be neglected

$(-130 \times 10^3 \text{ m} < y < 130 \times 10^3 \text{ m})$ for wave transformation at the point under consideration.

The initial wave spectrum in deep water (for $x = 0$) is accepted according to (5.16), with the angular distribution taken in the form of the specified approximation (4.14) and (4.15). The frequency $\sigma_{\max}$ of the maximum spectrum is accepted to be equal to 1.0 rad s^{-1}, and the wave number is presented in discrete form (i.e. the set of 17 values is varied within the range $0.06 < k_i < 1.26 \text{ m}^{-1}$). The set of angle directions β_j consists of 30 values within the range $-0.96\pi/2 < \beta_j < 0.96\pi/2$.

Applying this algorithm, the equation set (1.86)–(1.89) is integrated "forward" and after that "backward" jointly with the wave energy balance equation, using the Runge–Kutta method of fourth-order accuracy with automatic choice of integration step.

As a result of numerical calculations made for a general wave direction coinciding with the Ox axis the following values: spectrum $S(k, \beta)$, frequency angular spectrum $S(\sigma, \beta) = S(k, \beta) \, \mathrm{d}k/\mathrm{d}\sigma$, spectrum of wave numbers $S(k) = \int S(k, \beta) \, \mathrm{d}\beta$ and frequency spectrum $S(\sigma) = \int S(\sigma, \beta) \, \mathrm{d}\beta$ are obtained for different points along the Ox axis (for $y = 0$).

The frequency angular spectrum $S(\sigma, \beta)$ appears to reveal a tendency to increase at low frequencies and decrease somewhat at high frequencies (for $\beta \approx 0$). The spectral density decrease takes place at every frequency for the angles β, being sufficiently different from the general directions. The spectral density is decreased to zero at low frequencies for large angles ($\beta > 30°$). This is explained by the absence of proper spectral components and by narrowing the angular distribution of the wave energy.

The behaviour of the spectrum $S(\sigma, \beta)$ is determined by diminishing the depth with the group velocity being decreased at first and then increased, whereas the wave number is monotonically increased, and the propagation rays of the spectral components are turned perpendicular to the shore line. The frequency spectrum $S(\sigma)$ is not essentially changed (at least up to 5 m depth) and the frequency of the spectral maximum actually remains the same. The spectrum $S(k)$ is shifted to the area of large wave numbers, becoming wider with its maximum being slightly decreased. Such spectrum behaviour is proved by field observations (Krylov et al., 1976).

After obtaining the spectrum value at the point under consideration, it is integrated over angles and wave numbers using the Simpson cubature method. Its statistical moments and the wave element estimations, such as the height h, length λ and period τ, are obtained. Relative values of the mean height $\tilde{h} = \bar{h}/\bar{h}_0$, length $\tilde{\lambda} = \bar{\lambda}/\bar{\lambda}_0$ and period $\tilde{\tau} = \bar{\tau}/\bar{\tau}_0$, presented as functions of the relative depth $H/\bar{\lambda}_0$, practically coincide with accurate spectral solutions (see Fig. 6.10, Sect. 6.3). In particular for a straight isobaths, the relative height $\bar{h}/\bar{h}_0$ is decreased at first to the value ≈ 0.9, and then it is increased. The mean wave period is weakly decreased (up to 3 per cent), then slowly increased. The wavelengths are monotonically decreased.

6.5 Numerical Model of Wind Wave Transformation in a Coastal Area

As shown by the calculation results, the mean wave heights and periods for the parallel and straight isobaths are not changed, at least, up to depths where the non-linear wave transformation takes place. A comparison of the calculated values and field measurements fulfilled in a water basin with constant slope (Krylov et al., 1976) reveals their good conformity.

Estimation of wave transformation in a coastal area. Now the wave transformation in a gulf with isobaths being different from straight lines (see Fig. 6.16), will be considered. A two-dimensional grid (100×38 points) is chosen for covering a natural water basin. The distance between the grid points along the Ox axis is equal to 372 m, and along the Oy axis it is 743 m. At the grid points, the water depth H_{ij} is accepted according to the regional bathimetric map.

As in the previous case, the spectrum is given in the form (5.16) at the initial boundary (for $x = 0$). The calculation is made for the point with coordinates: $x = 20.09 \times 10^3$; $y = 6.32 \times 10^3$. It should be noted that this point is situated in a hollow created by the bottom relief (see Fig. 6.16).

The characteristic equations (1.86)–(1.89) are integrated numerically. At the same time the depth H and its gradients along the rays are determined using interpolation.

As shown by the numerical results the spatial spectrum $S(k,\beta)$ is considerably different from the spectrum calculated in the case of a bottom of constant slope. The spectrum becomes narrow-directed for most wave numbers, excluding the direction coinciding with the isobath directions at the calculated point. The spectral density is sharply decreased, especially for small wave numbers. As a result the spectral density $S(k)$ is essentially decreased and the wave number of the maximum is significantly increased. The relative mean values of the wave elements are as follows: $\tilde{h} = 0.31$; $\tilde{\tau} = 0.85$; $\tilde{\lambda} = 0.66$. As can be seen, the spectrum and mean wave elements calculated for the chosen bottom relief are principally different from the corresponding values of the test case with the constant slope basin (6.60).

The reason of this difference can be explained as follows. Firstly, it is the influence of the geometrical divergence of the wave energy connected with side boundaries of the water area and limited by the area dimensions, where the initial spectrum is given. Secondly, this difference can be caused by the depth variations along the energy-supply ray tube of wave propagation.

Although these two factors are interconnected, an attempt is undertaken to single out their relative effect. In order to do so some additional calculations are made, in which the former side boundaries of the water area and the depth are taken in accordance with the formula for depth variations with a constant slope (6.60). The depth H is accepted to be equal to the mean depth of the natural basin at the initial point ($x = 0$) according to (6.60). It also coincides with the depth at the point under consideration with $x = 20.8 \times 10^3$ m. In other words, in this calculation the real depth is exchanged for the depth

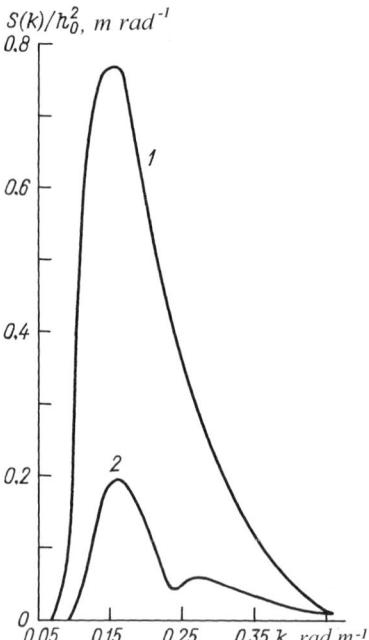

Fig. 6.17. Wave number spectra at a coastal area for modelled (1) and natural (2) depths

with constant slope. The following wave elements are obtained: $\tilde{h} = 0.65$; $\tilde{\lambda} = 0.91$; $\tilde{\tau} = 0.98$.

The spectra $S_1(k)$ and $S_2(k)$ normalized by the square mean height of the initial wave and obtained for natural and modeled depths, correspondingly, are shown in Fig. 6.17. The wave spectrum S_1 in the vicinity of its maximum is approximately 4 times less than the appropriate value of the spectral density S_2. There are similar values of both spectra in the range of large wave numbers. Also, the second maximum is observed in the spectrum S_1, while there is no second maximum in the spectrum S_2 and in the initial spectrum. It indicates that wave refraction at wave numbers less than the wave number of the second maximum results in decreasing wave energy. It does not take place at wave numbers greater than the mentioned value. The spectral density is not actually changed there.

Isolines of the spectral density $S_1(k,\beta)$ and $S_2(k,\beta)$ are shown in Fig. 6.18. Due to the limited linear dimensions of the initial boundary, where the non-zero initial spectral density is given, the narrowing of the angular spectral density distribution takes place. As a result (see Fig. 6.18) the angular range of non-zero energy density is narrowed (from $-90°-+90°$ for the initial spectrum, to $-43.5°-0°$ for the final spectrum). At the same time, this range is narrowed to a greater degree due to the real depth influence, especially for small wave numbers.

6.5 Numerical Model of Wind Wave Transformation in a Coastal Area   299

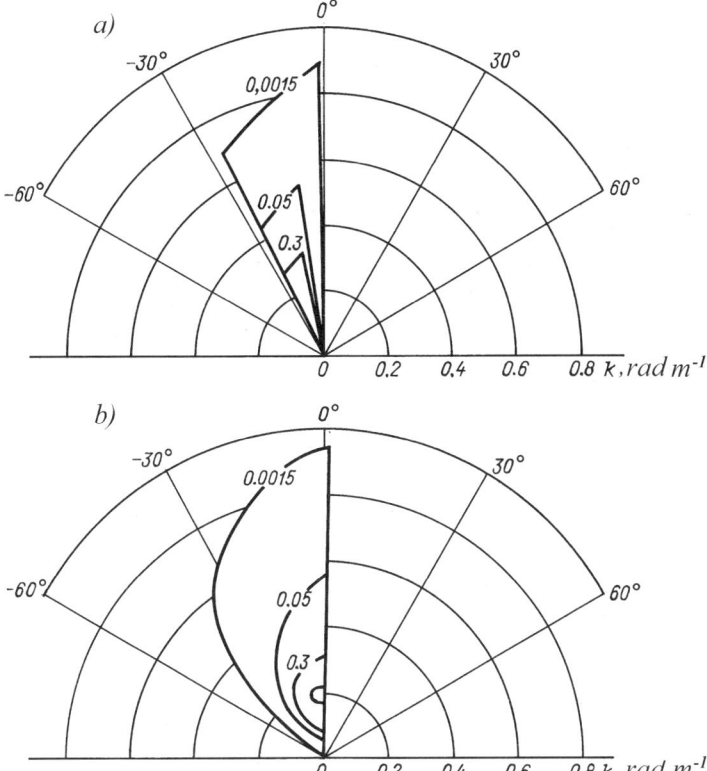

Fig. 6.18. Isolines of spectral density values due to wave refraction for natural (**a**) and modelled (**b**) depths

Consequently, as the waves propagate from deep to shallow water, a filtration of spectral components takes place, caused by the influence of depth. Long-wave components are deviated to the small depth values, whereas the short-wave components penetrate to the calculated point area without any essential changes.

As shown by this example the transformation of the spectrum and mean wave elements can be principally different in field conditions from the wave evolution case with straight isobaths, which is accepted to be a typical one.

Wave estimation in a coastal area with current. Now another example of wave wind transformation in a coastal area with current velocities greater than $1.0\,\mathrm{m\,s^{-1}}$ will be considered. This situation takes place at the shelf of Sakhalin Island.

The rectangular coordinate system Oxy is chosen for numerical simulation of the wave transformation (see Fig. 6.19). Its vertical Oy axis is directed along the meridian, and the horizontal Ox axis is chosen along the parallel

300   6 Wave Transformation in Shallow Water

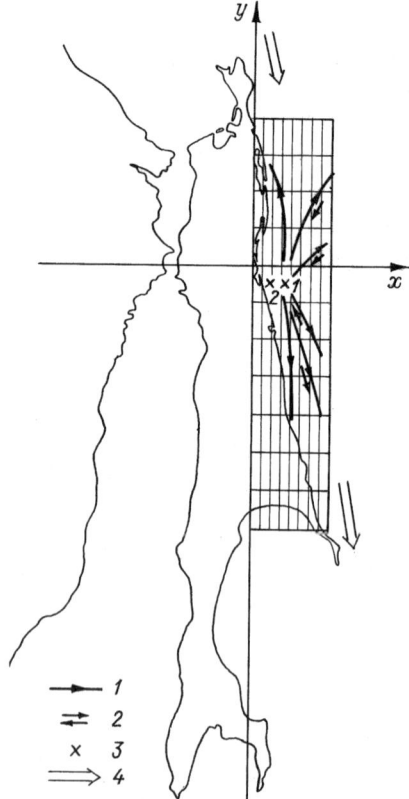

Fig. 6.19. Numerical simulation of wind wave transformation in a coastal area with current: 1 – wave rays; 2 – energy-supply rays; 3 – considered points; 4 – a current direction

(with the spatial step $\Delta x = 4.93$ km, $\Delta y = 24.67$ km). The depth values $H(x,y)$ are set in grid points in accordance with the regional bathemeric map of that area (see Fig. 6.19).

In order to produce numerical simulations the current is accepted to be directed along the coastal line, i.e. the angle between the current velocity direction and the Ox axis is equal to 280°. It is also accepted that the current value is not changed along this direction. The current velocity profile is changed in the Oy direction (see Fig. 6.20). The numerical simulation of wave spectrum is made for the point with the coordinates $x_0 = 32.5 \times 10^3$ m, $y_0 = -54.3 \times 10^3$ m (see Fig. 6.19), using the above described algorithm. The approximation (5.16) is taken as the initial spectrum with the following parameters: $n = 5$; $\omega_{\max} = 0.39$, corresponding to the wave mean period of 11.5 s. The initial general direction of wave propagation is taken to be equal to $\beta = 150°$, $180°$, $210°$ correspondingly (i.e. which coincides with the 120°, 90° and 60° directions in the geographical coordinate system).

6.5 Numerical Model of Wind Wave Transformation in a Coastal Area

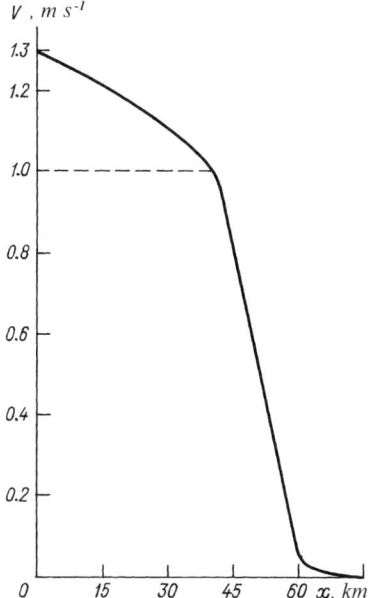

Fig. 6.20. Current velocity variation (transverse profile) away from shore

As shown by the calculations the main influence of wave transformation is due to the current and shallow water effects, whereas the wind influence and bottom friction create less effect due to the limited wave transformation area dimensions. At the same time the current effects are mainly on the short spectrum range. On the contrary, shallow water effects operate on its long-wave range.

Wave heights, periods and lengths are estimated by numerical integration of the calculated two-dimensional spectrum. The error of numerical integration is estimated to be in no more than 2–3 per cent. The numerical simulations are made for two points with 42 m and 12 m depths, correspondingly. The calculations are also made for the case without any current ($V - 0$) and with the current ($V \neq 0$), separately. As shown by the obtained results there is rather a weak effect of the current and depth on waves at the first calculated point. The mean wavelength is subjected to transformation mostly, decreasing by 10 per cent (with the initial wave direction of 60°) without any current, and decreasing by 19 per cent with the current. The mean wave period is practically not changed. As for the mean height, it can be increased by 7 per cent with the initial wave direction being 120°.

In this case the suggestion is justified that the extreme wind wave elements are not effected by the current (Davidan et al., 1985). However, as the waves approach the shore, the depth is decreased and the current effect on waves becomes greater. The wavelength is decreased, and the height is increased

302   6 Wave Transformation in Shallow Water

with the initial general direction of wave propagation being equal to 120°. At the second point under consideration the relative increase of wave height reaches 20 per cent. The greatest height is observed without any current, when the initial wave direction is equal to 90°, and the height increase reaches 8 per cent. The wavelength is decreased by 20–30 per cent at this point.

In conclusion it should be noted that the problem of wave spectrum transformation due to wave refraction in shallow water and in a non-uniform current without wind is studied in this chapter. As shown, the joint effect of a non-uniform current and basin bottom unevenness can lead to such wave field variations, which cannot be presented by simple superposition of the effects caused by current and depth separately.

6.6 Wind Wave Spectrum Transformation According to the Black Sea International Experiment

Description of the experiment. The problems of wave transformation due to the influence of bottom and current variations have been considered in the monograph till now. At the same time wave refraction played a decisive role, whereas the wind influence was neglected. Such an attitude was justified in case of small wind speed and the non-linear effects could be neglected. In reality a strong wind may have a considerable impact on wave evolution in a coastal area. In this connection it is interesting to consider the results of wind wave spectrum transformation, obtained with the Black Sea International Experiment (BSIE), in which the author of the present monograph took part under the supervision of Prof. I.N. Davidan. The BSIE, devoted to investigation of wave transformation in coastal shallow water at the area south of Varna (Bulgaria), was held in November–December, 1999. Scientific workers of the Oceanlogy Institute of the Bulgarian Academy of Sciences (under the supervision of Prof. Z. Belberov), the Saint-Petersburg Branch of the State Oceanography Institute, the Water Problem Institute of the Polish Academy of Sciences and the Woods Hall Oceanography Institute (USA), participated in the experiment.

The coastal area was suitable to produce reliable measurements of waves, propagating from eastern and north-eastern directions without any coastal or shallow disturbances on their way.

The bottom area made up a flat slope surface with an averaged slope about 0.025 in the depth zone 0–10 m, and 0.006 in the depth zone 10–20 m. Submerged shoals were practically absent, so wave reflection could be neglected. The bottom profile is shown in Fig. 6.21.

The location of measuring devices is shown in Fig. 6.22. One-string wave measuring equipment was installed on the platform along 250 m at eight points, where the depth was 1.3, 1.6, 2.6, 2.8, 3.2, 3.4, 3.8, 4.2 m, correspondingly. A five-string wave measuring device was installed at the end of the

6.6 Black Sea International Experiment   303

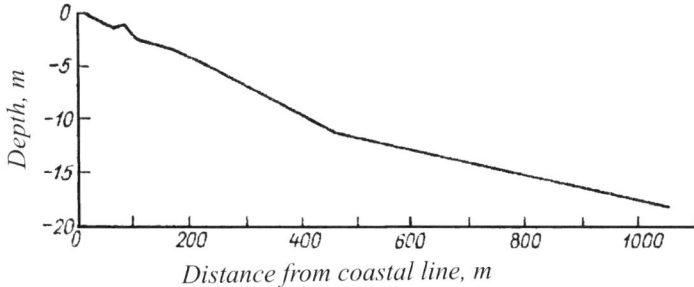

Fig. 6.21. Bottom profile in coastal area

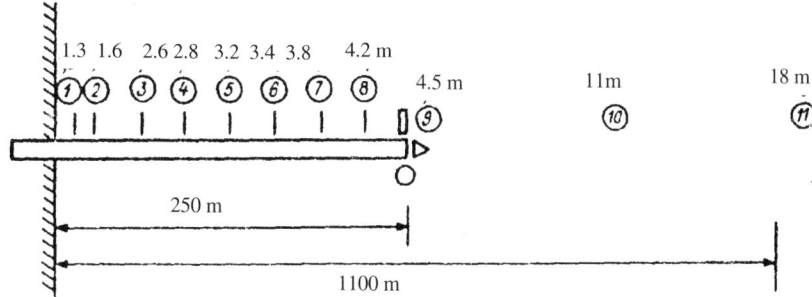

Fig. 6.22. Scheme of experimental wave measuring device locations

platform, where the depth was 4.5 m. Another five-string wave measuring devices and the equipment, measuring wave pressure and current speed in two planes perpendicular to each other, were installed at separate points with depth 11 and 18 m. The frequency and frequency-angular wave spectra were estimated using the obtained measurements.

Observation results. The measurements, carried out at the beginning of December, were the most interesting ones. In that time the waves propagated in the direction 60–70°, in which the measuring devices were installed (see Fig. 6.22), nearly perpendicular to the isobath.

The time variations of wave heights (of 3 per cent probability) and the wave period of spectrum maximum registered every hour from 11 am on December 2 to 04 am on December 3 are presented in Figs. 6.23 and 6.24. Unfortunately, due to a storm the string measuring devices were distorted after 4 am and the measurements were carried out after that up to December 12, only at the point with depth 11 m.

The data, shown in Fig. 6.24, indicate that the frequency of the spectrum maximum was not much different in wave propagation from deep to shallow water. As for the wave energy there were three zones, different in transfor-

304   6 Wave Transformation in Shallow Water

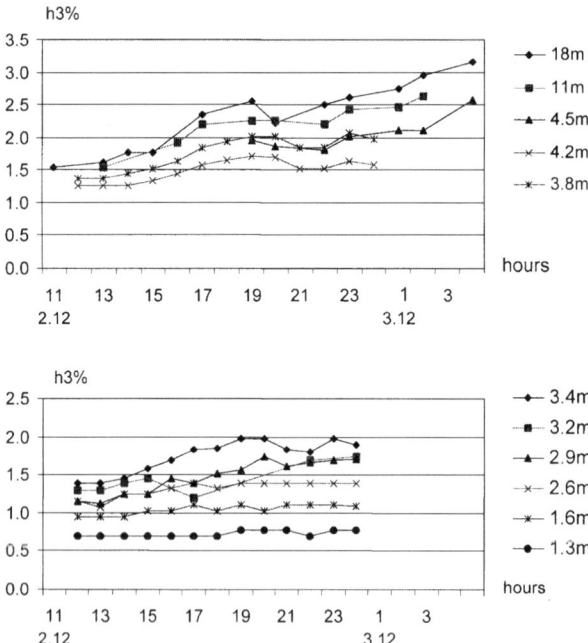

Fig. 6.23. Wave height (3 per cent probability) time variation at points with different depth

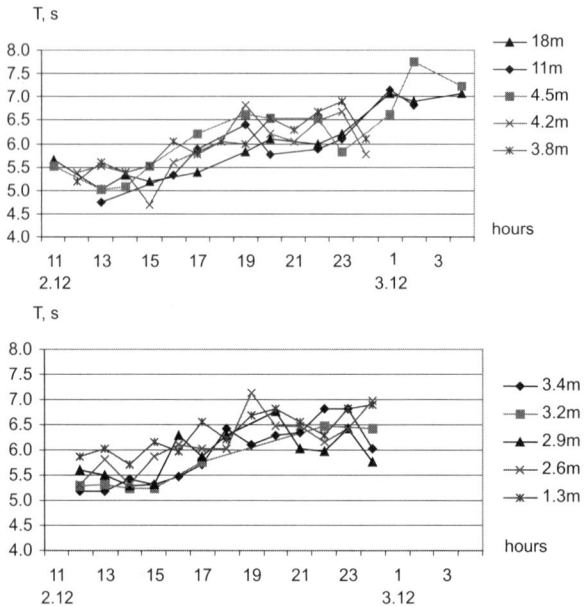

Fig. 6.24. Wave period time variations

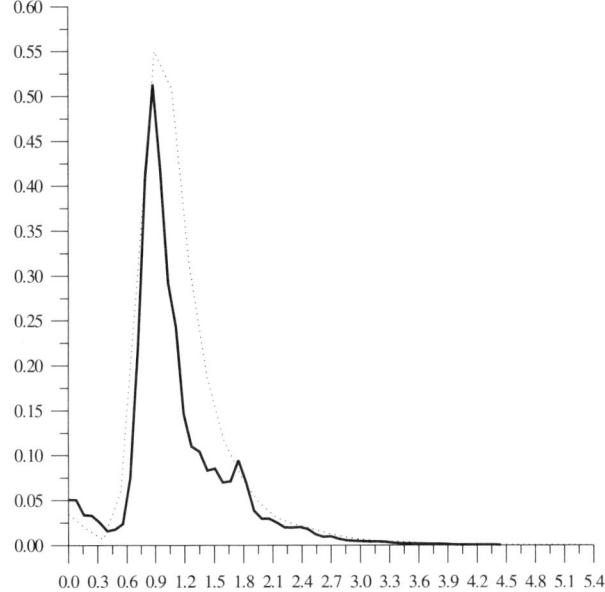

Fig. 6.25. Frequency wave spectra recorded at 04 am on December 3, at a point with depth 4.5 m (*thick line*) and 18 m (*dashed line*)

mation of the wave frequency spectra. These zones coincided with details of the bottom slope variations in the measured area. There was a reduction of the wave energy (or spectrum zero moment) in the area with the depth variation from 18 to 11 m, where the slope made up mainly 0.006. To a lesser degree this was observed at the point with depth 4.5 m. The value of the spectrum zero moment was sharply increased between two points with depth 3.8 and 3.4 m. It reached its initial value, which was the same as at the point with the depth 18 m. This indicated a considerable increase of wave steepness as soon as the wavelength became shorter. A significant reduction of the spectrum zero moment started from the point with depth 3.2 m. Finally the value of the zero momentum reached its limited value at depth 1.3 m and it was not changed during the whole experiment. The dependence of the non-dimensional spectrum zero moment was changed in all enumerated transformation zones depending on the non-dimensional wave number of the spectrum maximum.

An example of the frequency spectrum variation with different depths is shown in Fig. 6.25.

Interpretation of observational data. In order to interpret the data, obtained in the shallow water case, the following standard normalization of the spectrum zero moment $\tilde{m}_0$, the frequency of the spectrum maximum ω_H and the wave number of the spectrum maximum $\tilde{k}_m$ are used:

$$\tilde{m}_0 = \frac{g^2 m_0}{u_{10}^4} \qquad (6.61)$$

$$\omega_H = \omega \left(\frac{H}{g}\right)^{1/2} \qquad (6.62)$$

$$\tilde{k}_m = \frac{k_m u_{10}^2}{g}, \qquad (6.63)$$

where m_0 is the spectrum zero moment, ω is the frequency, k_m is the wave number of the spectrum maximum, and H is the water depth.

As shown in Problems of Research and Mathematical Modeling of Sea Wind (1995), a normalized value of the spectrum zero moment at the point with depth 11 and 18 m is correlated with the non-dimensional value of the spectrum maximum as follows:

$$\tilde{m}_0 = \tilde{a} k_m^{-1.47} \qquad (6.64)$$

where $\tilde{a} = \alpha \times 10^{-3}$, $\alpha = 2.0$.

In order to check this correlation all results, obtained in December 2–4, are used. The estimation results are shown in Table 6.1.

It is seen (Table 6.1) that the coefficient value α is practically independent of the non-dimensional frequency at least at ω_H 0.6–1.6 with depth 11–18 m, where the bottom slope makes up 0.006. The mean value α is equal to 2.04, corresponding to the value obtained earlier. The mean square deviation is 0.22. This is less in comparison with the observational analysis made in other conditions (shown below).

The mean value α is increased to 2.57 at the depth 4.5 m, where the bottom slope is different and makes up 0.025. As seen (see Figs. 6.23 and 6.24), the wave height is not only reduced, but it is considerably increased at the depth 3.8 and 3.4 m. The mean values are increased to 3.85, after which wave breaking starts and the wave height is again reduced. The mean value α is equal to 3.08 at the depth 2.9 m.

In order to analyse the obtained results some other experimental data will be noted. The main results of the ARSLOE experiment (Bouws et al., 1985, 1987), devoted to the investigation of wave transformation in a coastal zone lead to the conclusion that it is possible to use the correlation obtained for the Phillips spectrum (Kitaigorodsii et al., 1975) for approximation of the one-dimensional spectrum using the wave number spectrum. Having such results, it is possible to use the following transfer function passing from the frequency spectrum in deep water to that in shallow water:

$$\Phi(\omega_H) = \frac{(k(\omega, H))^{-3} \partial k(\omega, H)/\partial \omega}{(k(\omega, \infty))^{-3} \partial k(\omega, \infty)/\partial \omega}, \qquad (6.65)$$

where ∞ denotes deep water, and H is the shallow water depth.

Table 6.1. Coefficient value α at points with different depth.

	Depth 11 and 18 m					
ω_H	0.63	0.63	0.68	0.68	0.77	0.84
α	2.38	2.12	1.98	2.16	1.78	1.97
ω_H	1.11	1.22	1.22	1.27	1.36	1.42
α	2.25	2.09	2.12	1.95	1.98	2.35
	Depth 11 and 18 m					
ω_H	0.97	0.99	0.99	1.06	1.06	
α	2.21	1.85	1.82	1.88	2.00	
ω_H	1.49	1.62				
α	1.90	1.90				
	Depth 4.5 m					
ω_H	0.60	0.60	0.64	0.68	0.71	
α	2.9	2.99	2.37	2.43	2.85	
	Depth 3.8 m					
ω_H	0.57	0.65	0.62			
α	3.7	4.7	4.1			
	Depth 3.4 m					
ω_H	0.65	0.62	0.60			
α	3.6	4.9	4.4			
	Depth 2.9 m					
ω_H	0.60	0.57	0.54			
α	2.3	3.3	3.6			

As a result of using the JONSWAP spectrum $S_{\text{Jonswap}}(\omega)$ in deep water (4.13), the following spectrum in shallow water due to wave transformation is obtained:

$$S_{\text{TMA}}(\omega, H) = S_{\text{Jonsawp}}(\omega)\Phi(\omega_H). \tag{6.66}$$

According to the measurement result, the correlation, connecting parameters of the TMA spectrum approximation with non-dimensional wave numbers of the spectrum maximum, is found to be:

$$y = a\tilde{k}_m^b, \tag{6.67}$$

where $\tilde{k}_m = k_m u^2/g$ and k_m is the wave number of the spectrum maximum.

The coefficient values are determined for different values of the non-dimensional frequency and on average for all obtained spectra. In particular, for the non-dimensional zero spectrum moment the coefficients a and b in the correlation (6.67) turn out to be equal on average to 1.98×10^{-3} and -1.34, correspondingly (Bouws et al., 1987).

308   6 Wave Transformation in Shallow Water

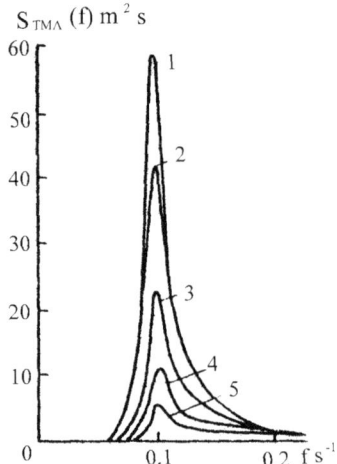

Fig. 6.26. TMA spectrum for different depths. 1 – deep water, 2 – $H = 40$ m, $\omega_H = 1.27$, 3 – $H = 20$ m, $\omega_H = 0.90$, 4 – $H = 10$ m, $\omega_H = 0.63$, 5 – $H = 5$ m, $\omega_H = 0.45$

Miller & Vincent (1991), obtained a correlation between spectrum parameters and the wave number of the spectrum maximum, using a correlation of the spectral density in the equilibrium range as a function of the frequency to the power of minus four. Thus, coefficients in the correlation (6.67) for zero momentum of the spectrum are varied for different depth, and their averaged values are accepted to be equal to 1.91×10^{-3} and -1.24, correspondingly.

The value of the TMA spectrum for different depths (from deep water to shallow water) with relative value $\omega_H = 0.45$ is estimated and presented in Fig. 6.26. The initial spectrum in deep water is described by the JONSAWP spectrum with $\omega_{\max}/2\pi = 0.1$, $\alpha = 0.01$, $\gamma = 3.3$. As seen, the spectrum maximum value is decreased due to wave propagation in shallow water. The spectrum equilibrium range is shifted to high frequency. These trends hold true up to the value $\omega_H = 0.45$.

Using the results, obtained in this monograph for the wave spectrum transformation, and the measurements in the North Sea and Atlantic coast (USA) (Bouw et al., 1985, 1987; Miller & Vincent, 1991), the following conclusion can be made:

1. The results, obtained in the BSIE, are rather close to those obtained earlier in the area, where the bottom slope was similar to the one in the aforementioned experiments, i.e. at depth 11 and 18 m. Thus, for example, at $\tilde{k}_m = 0.4$ the wave height according to (6.67) is different by ± 7 per cent from those obtained by Bouws et al., (1987); Miller & Vincent (1991).

2. There are principal differences between the BSIE results from those obtained in the aforementioned papers for the same non-dimensional values of the spectral maximum frequency.

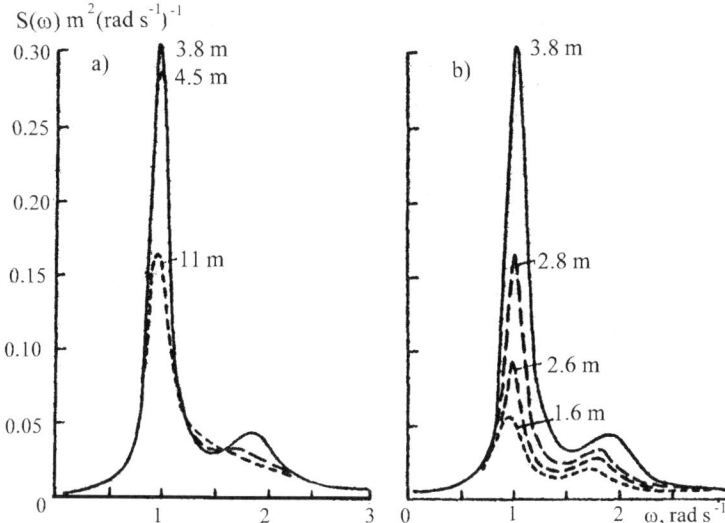

Fig. 6.27. Measured frequency spectra obtained for wave growth (**a**) and breaking (**b**) stages

3. The same conclusions are valid for the frequency spectrum approximation, concerning possible use of the TMA approximation, remaining valid for a coastal region with small bottom slope.

It should be noted that in our experiment (BSIE) the frequency spectrum transformation originates in a more complex way. Frequency spectra obtained at different points with depths form 1.6 to 11 m and wind speed 15 m s^{-1} at February 2 are presented in Fig. 6.27. The non-dimensional value ω_H is changed in the range $0.36 \leq \omega_H \leq 1.1$. As seen, the spectrum is increased in wave propagation between two points with depths 4.5 and 4.2 m. After that there is the opposite case, i.e. starting from the depth $H < 4.2$ m the spectrum maximum is decreased due to wave breaking. A second spectrum maximum is observed in the high-frequency range at double the value of the main peak spectrum frequency. This maximum is stable and its contribution to the total wave energy is increased with decrease in depth. This is evidence of the principal difference between spectra obtained in the BSIE and TMA spectrum.

As for the frequency spectrum obtained at the points with depth 11–18 m, where the bottom slope is 0.006 (see Fig. 6.28), it is rather close to the TMA spectrum. Thus, the spectrum differences, estimated according to the measurement data using the TMA spectrum, become significant at the depth where the bottom slope is 0.025.

Not only the frequency spectra, but also the frequency angular spectra, are estimated at the depth 4.5, 11 and 18 m, correspondingly (see Fig. 6.29).

It is seen that the frequency angular spectrum is not symmetric with respect to the general direction at the depth 18 m. This asymmetry can be

310   6 Wave Transformation in Shallow Water

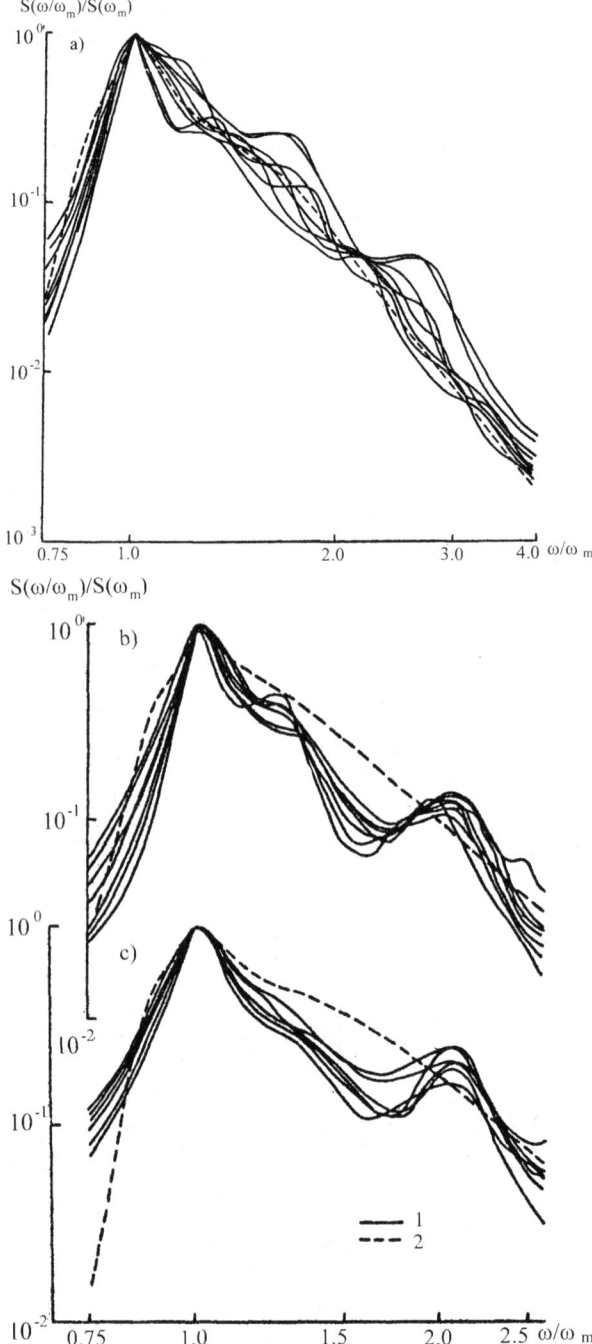

Fig. 6.28. Normalized empirical (*solid line*) and TMA (*dotted line*) spectra at points with 18 and 11 m (**a**); 3.4–4.5 m (**b**) and less than 2.9 m (**c**)

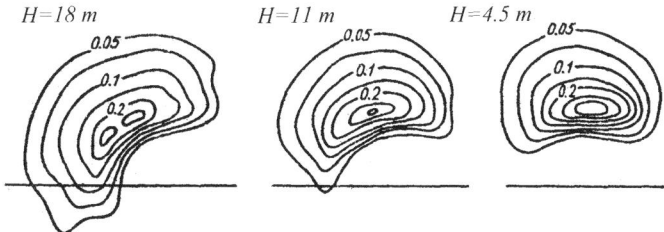

Fig. 6.29. Frequency angular spectra measured at points with depth 18, 11 and 4.5 m. Figures on isolines are values of $2\pi S(\omega, \theta)$; the horizontal straight line is the direction parallel to the coast line

explained by the different direction propagation of spectral components. The spectrum becomes more symmetric at the intermediate depth 11 m. At the depth 4.5 m the spectrum is practically symmetric due to a sharper turn of the wave spectral components. At the same time the general direction of wave propagation becomes perpendicular to the coast line.

As a result of the frequency angular spectrum analysis the estimations of angular energy distributions for the different non-dimensional frequencies ω_H are obtained. A normalized function of the angular energy distribution is approximated with the help of the following correlation:

$$q(\omega, \theta) = \left\{ \cos \left[\frac{(\theta - \theta_m(\omega))}{2} \right] \right\}^{2\beta}, \qquad (6.68)$$

where $\beta = \beta(\omega)$ is the power degree, depending on the frequency ω, and θ_m is the direction relative to the maximum frequency angular spectrum value at the frequency ω.

Mean values of the power degree β in the approximation (6.68) for different non-dimensional frequency values are given in Table 6.2. It is seen from Table 6.2 that the angular energy distribution in shallow water is dependent on the frequency and relative depth. The frequency dependence is similar to the obtained regularities for deep water (Davidan et al., 1985). The largest value of the parameter β corresponds to the frequency of the maximum spec-

Table 6.2. Mean values of power degree β in (6.68)

Relative depth	ω/ω_m					
	0.7	0.8	1.0	1.2	1.5	2.0
Deep water	5	7	10	8	6	3
$1.3 < \omega_H < 1.6$	6	8	16	10	6	3
$1.0 \leq \omega_H < 1.3$	8	10	24	14	8	3
$0.7 \leq \omega_H < 1.0$	10	12	30	18	10	3
$0.5 \leq \omega_H < 0.7$	12	16	34	20	12	3

trum. It is reduced with deviation from this frequency both to the low and high frequency ranges.

The dependence on the relative depth is as follows:
– values of the power degree β, corresponding to the frequency of the maximum spectrum, and to those higher and lower than the frequency of the maximum spectrum, exceed similar values for the deep water case;
– values of the power degree β are increased at all frequencies with depth decrease.

These regularities of the angular distribution can be approximated as follows:

$$\beta \equiv \beta(\omega) = A\hat{\omega}^{-2.5}, \quad A = 45 - 18\omega_H, \quad 0.5 < \omega_H < 1.6; \quad (6.69)$$

$$\hat{\omega} = \begin{cases} \frac{\omega}{\omega_m} & \text{for } \omega \leq \omega_{\max}; \\ \frac{\omega_m}{\omega} & \text{for } \omega > \omega_{\max}. \end{cases} \quad (6.70)$$

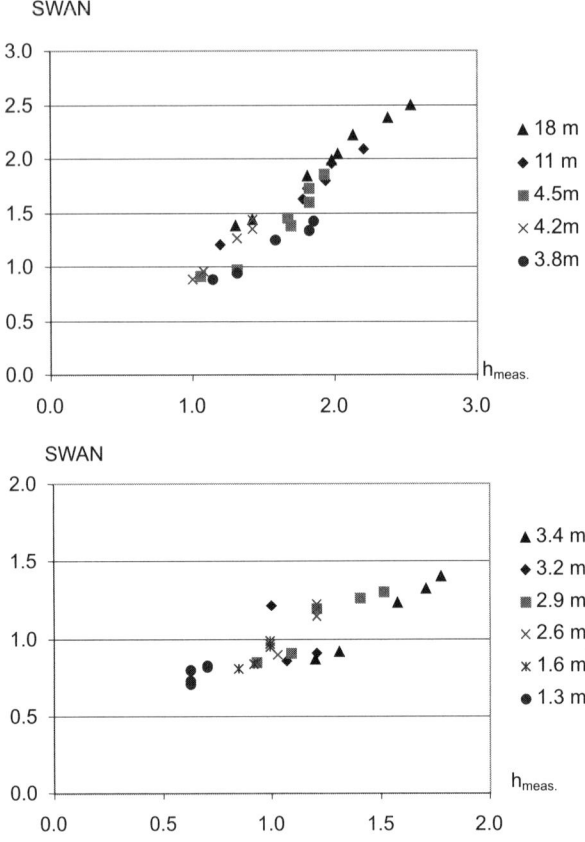

Fig. 6.30. Comparison of wave height estimated using the SWAN model with results obtained in BSIE for different depths

Comparison of wave measurements with SWAN model numerical simulations. The results obtained in our experiments can be compared with the SWAN model (Ris et al., 1999) numerical simulations. Thus, the estimations were fulfilled using a numerical grid with the 400 m step and covering a coastal area with 10 km width and 20 km length.

Results of numerical simulations are shown in Figs. 6.30 and 6.31. The data analysis allows drawing the following conclusions.

The wave height discrepancies between measurements and numerical simulations make up on average 4 per cent (from 1 per cent to 8 per cent) in the areas, where the bottom slope is 0.06.

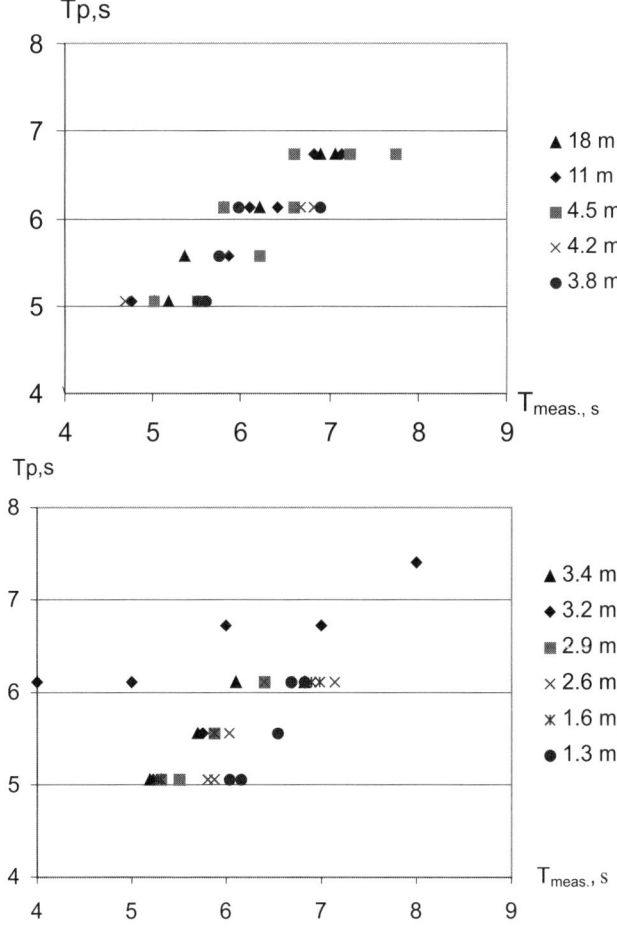

Fig. 6.31. Comparison of wave period estimated using the SWAN model with results obtained in BSIE for different depths

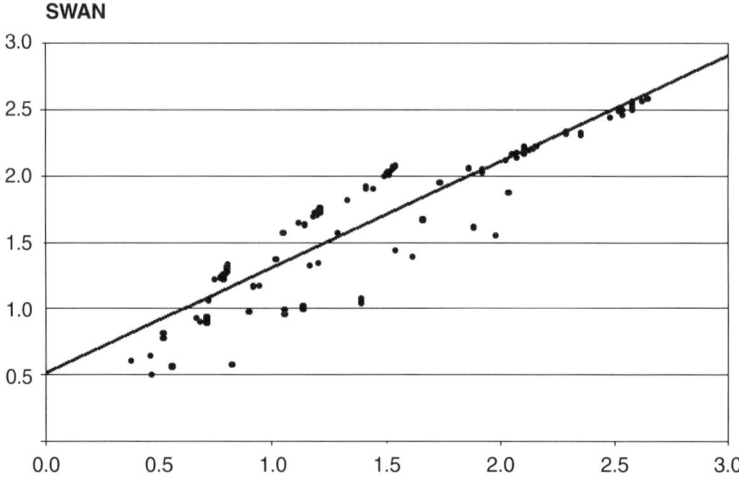

Fig. 6.32. Comparison of wave height $\bar{h}_{1/3}$ obtained by numerical simulations and correlation (6.64)

The results of numerical simulation are considerably underestimated (up to 20–30 per cent) in coastal areas with the bottom slope being 0.025 and when the wave height starts increasing (at the depth 3.8 and 3.4 m).

Discrepancies between numerical simulations and experimental heights are again reduced and they do not overestimate in general by 5–10 per cent in the area of wave height decrease after breaking (at a depth less than 2.9 m).

Wave period differences are not large (generally less than 5 per cent) in all observational areas, where the depth is more than 2.6 m; the period differences are increased up to 10–15 per cent at smaller depth.

It is interesting to make a comparison using the correlation (6.64), obtained for definite conditions of the measured area at depth 11 m with the SWAN results for varied wind speeds and depth in the BSIE. The calculation results are given in Fig. 6.32.

As seen (Fig. 6.32), there is good correspondence between the SWAN wave height estimations and the correlation (6.64). But, there are some discrepancies for the coastal area in some parts. These discrepancies are considerably reduced for waves with height more than 2 m.

7 Wave Transformation in Ice-Covered Water

7.1 Wave Problem Formulation in Water with Ice Cakes

Boundary conditions in a water surface. In the Arctic sea when the waves propagate from the open water surface into the region covered with ice, they meet ice of different density, reducing their intensity considerable. Now the cases of wave interaction with ice cases will be considered.

Let the ice cakes be accepted as floating masses without any interaction between them (Kheysin, 1967). This idealization can be valid when omitting the interaction forces between separate ice floes. At the same time their sizes are supposed to be small enough compared with the wavelength, so the ice floe is not bent. In order to use a comparatively simple mathematical approximation of the wave pattern, a supposition about the continuity of the ice cake area will be applied. The mean ice thickness L, distributed evenly over the local area, is usually introduced. Actually, the problem is reduced to the study of two-layer medium motion with the upper layer describing a continuous system of floats, simulating the ice dynamics.

The equations of motion make up the initial point of the problem formulation, taking into account the presence of ice in the water surface. The following suppositions are used:

- water is covered evenly with ice cakes;
- ice dimensions are much less than the wavelength;
- wave frequencies are essentially less than the frequency of the ice floe eigen-oscillation in water. This means the constancy of ice floe deepening.

The equations of motion are written with the help of the potential approximation, for which the Laplace equation (1.10) is used. In a water surface covered with ice cakes the boundary conditions can be presented in the following form (Bukatov & Bukatova, 1993):

$$\Lambda \frac{\partial}{\partial z}\left[\frac{\partial \phi}{\partial t} + \frac{1}{2}\left(\frac{\partial \phi}{\partial x_i}\right)^2\right] + g\eta + \frac{\partial \phi}{\partial t} + \frac{1}{2}\left(\frac{\partial \phi}{\partial x_i}\right)^2 + \frac{1}{2}\left(\frac{\partial \phi}{\partial z}\right)^2 \Big|_{z=\eta} = 0 \,; \tag{7.1a}$$

$$\frac{\partial \eta}{\partial t} + \left(\frac{\partial \eta}{\partial x_i}\frac{\partial \phi}{\partial x_i}\right)\Big|_{z=\eta} = \frac{\partial \phi}{\partial z}\,, \tag{7.1b}$$

where the following designations are accepted: $\Lambda = L\rho_3/\rho_2$ is the parameter of ice floe deepening, L is the ice thickness, ρ_3 is the ice density; ρ_2 is the water density; $\phi(x, z, t)$ is the liquid velocity potential and $\eta(x, t)$ is the elevation of the water–ice boundary interface; and the index i indicates summation. The traditional condition of non-leakage is assumed at the bottom:

$$\left.\frac{\partial \phi}{\partial z}\right|_{z=-H} = 0, \qquad (7.2)$$

where H is the water depth.

Dispersion wave ratio in water with ice cakes. Using the aforesaid problem formulation with linear boundary conditions, the dispersion ratio connecting the wave number k and the frequency ω can be obtained for surface gravity waves with ice cakes as follows:

$$gk \operatorname{th}[k(H-\Lambda)] - \frac{\rho_2 \omega^2}{\rho_2 g - \rho_3 L \omega_b^2}. \qquad (7.3)$$

In most cases, it can be assumed that $H \gg \Lambda$.

Introducing the value $\omega_b^2 = \rho_2 g/\rho_3 L = g/\Lambda$, the formula (7.3) can be presented as:

$$gk \operatorname{th}(kH) \approx \frac{\omega^2}{1 - \omega^2/\omega_b^2}. \qquad (7.4)$$

It should be noted that the value ω_b in (7.4) is an oscillation frequency of a float in the water surface.

Proceeding from (7.4), the group velocity C_g can be presented in the following form:

$$C_g = C_g^0 \left(1 - \frac{\omega^2}{\omega_b^2}\right)^{\frac{3}{2}}, \qquad (7.5)$$

where C_g^0 is the wave group velocity in water without ice.

In order to obtain the wave number with the help of the ratio (7.4), a polynomial expansion (Jolm, 1979) can be applied. In the shallow water case, the expression for the wave number is simplified and can be presented in the analytical form:

$$k = \sqrt{\frac{\omega^2/gH}{1 - \omega^2/\omega_b^2}}. \qquad (7.6a)$$

In the deep-water case a similar expression can be written as:

$$k = \frac{\omega^2/g}{1 - \omega^2/\omega_b^2}. \qquad (7.6b)$$

As shown in (7.5) and (7.6a), the presence of ice results in diminishing the wavelength and propagating velocity compared with the case without ice. It should be noted that the wave number becomes infinitely large in the case of the wave frequency approaching that of the floe oscillation. This shows the inapplicability of the traditional approach and the necessity to use a more precise approximation.

7.2 Wave Spectrum Evolution in Water with Ice Cakes

Spectral formulation of the problem. Now the problem of wave spectrum transformation in water with ice cakes will be considered. Supposing that the mean ice thickness $L(x, y)$ varies smoothly enough over all space (i.e. the spatial scale of typical variations of the mean ice thickness is much larger than the wavelength), then the geometrical optics approximation can be used for the description of wave packet propagation.

The spectral density balance equation of wave energy evolution is used in its traditional form:

$$\frac{\partial S}{\partial t} + \frac{\partial S}{\partial \boldsymbol{k}} \frac{\mathrm{d}\boldsymbol{k}}{\mathrm{d}t} + \frac{\partial S}{\partial \boldsymbol{r}} \frac{\mathrm{d}\boldsymbol{r}}{\mathrm{d}t} + \frac{\partial S}{\partial \omega} \frac{\mathrm{d}\omega}{\mathrm{d}t} = G, \qquad (7.7)$$

where G is the source function describing physical mechanisms, which form the wave spectrum in water with ice cakes. The characteristics of (7.7) are the solutions to the Hamiltonian equation:

$$\frac{\mathrm{d}\boldsymbol{r}}{\mathrm{d}t} = \frac{\partial \omega}{\partial \boldsymbol{k}}; \quad \frac{\mathrm{d}\boldsymbol{k}}{\mathrm{d}t} = -\frac{\partial \omega}{\partial \boldsymbol{r}}; \quad \frac{\mathrm{d}\omega}{\mathrm{d}t} = \frac{\partial \omega}{\partial t}, \qquad (7.8)$$

where the frequency ω is written in accordance with the ratio (7.3) as follows:

$$\omega^2 = \frac{\sigma^2}{1 + (\sigma/\omega_b)^2}, \qquad (7.9)$$

where $\sigma^2 = gk \, \mathrm{th}(kH)$.

In this section the effects of non-uniform currents and depth variations are neglected.

As for the source function G, the dissipation connected with the friction between ice floes is the most typical physical mechanism forming the wave spectrum in the case of their propagation in water with ice cakes. Wave generation by wind is less effective in this case, as long as the transfer of momentum and energy from the atmospheric boundary layer into the ocean is prevented by ice. The mechanism of on-linear wave energy transfer in the spectrum is of great interest, so it is thoroughly considered below.

Wave energy dissipation in water with ice cakes. At the first stage a wave energy dissipation mechanism, connected with the presence of ice cakes, will be considered. There is no general theory concerning this mechanism and it is researched mainly experimentally (Wadhams et al. 1986). For the first time the simplest approach towards the solution of the problem was suggested by Krylov (1948). He considers the case of a viscous layer simulating ice cakes in an ideal liquid surface. The following expression connecting the frequency and the wave number is obtained as:

$$\frac{i\omega\nu L}{\rho_2 g}k^2 + 1 - \frac{\omega^2}{\omega_b^2} - \frac{\omega^2}{gk\,\text{th}[k(H-\Lambda)]} = 0, \qquad (7.10)$$

where ν is the viscosity coefficient.

It is seen that (7.10) is a transcedental equation with an imaginary value. Contrary to the non-damping oscillations considered above (7.4), this equation provides a finite solution of the wave number k, even for the case $\omega = \omega_b$. Its solution consists of imaginary and real parts. The imaginary part of the wave number k results in a damping coefficient. The value of this coefficient is determined simply enough for every separate case of deep and shallow water. So, for shallow water, a biquadratic equation can be obtained to determine the wave number value. For the deep-water case, a cubic equation can be obtained as well.

Supposing that the damping coefficient is small ($\mu \ll 1.0$) for the low frequencies ($\omega \ll \omega_b$), the wave number can be written in the form $k = k_0(1 + i\mu)$, where the value k_0 is determined by (7.4) or (7.5). Neglecting small values of second order, an approximate expression for the damping coefficient is obtained. It can be written for deep water as follows:

$$\mu = \frac{\nu L}{\rho_2 g^3} \frac{\omega^5}{(1 - \omega^2/\omega_b^2)^3}. \qquad (7.11)$$

Similarly, for shallow water it can be presented as:

$$\mu = \frac{\nu L}{\rho_2 H g^2} \frac{\omega^5}{(1 - \omega^2/\omega_b^2)^2}. \qquad (7.12)$$

It follows from these formulas that the wave damping becomes larger with increase of ice cake thickness or wave frequency. Thus, if a wave perturbation exists in a water area covered with ice cakes, the short-wave components, corresponding to high frequencies, are observed only in the vicinity of the perturbation area and they are damped outside it. These components are also quickly damped in time after the perturbation is over. The undamped long-wave components, moving with large velocity, propagate far away from the perturbation area. The wave energy dissipation is infinitely increased, as the wave frequency ω approaches the ice oscillation frequency ω_b. This reveals some solution trends and shows the limitation of the applicability of the approximation.

When applying the geometrical optics approximation to solve the problem of spectrum evolution in a water area with spatially non-uniform ice cake cover, the Hamiltonian equations (7.8), describing wave packet propagation, are used. It should be noted that these equations can be applied for the description of non-dissipative systems (Landau & Lifshits, 1973). This means the absence of imaginary values of the wave number k and the frequency ω. In order to use the above mentioned description of wave propagation in water with ice cakes, the imaginary values should be excluded from (7.8). At the same time the dissipation may be included into the source function of the kinetic equation (7.7).

The wave energy dissipation in the right part of the kinetic equation can be written in the following form:

$$G_{ds} = -2\mu k C_g S \,. \tag{7.13}$$

Thus, the problem is reduced to solving (7.7), substituting in it the source function (7.13) and using proper initial or boundary conditions.

Solution of the spectral problem of wave transformation in a non-uniform ice cake field. Substituting the dissipative function (7.13) into (7.7), the latter can be written as:

$$\frac{dS}{dt} \cdot \frac{1}{S} = A(\omega, k) \,, \tag{7.14}$$

where $A = -2\mu k C_g$.

The equation (7.14) is assumed to divide the variables. Solving this equation jointly with the characteristic equation set (7.8), the solution can be presented in the form:

$$\tilde{Z} = \tilde{Z}_0 + \int A\left(\boldsymbol{k}(t), \boldsymbol{r}(t), t\right) dt \,, \tag{7.15}$$

where the function $\tilde{Z}$ is determined as $\tilde{Z} = \ln(S)$.

The integral (7.15) is calculated using the characteristics arriving at the point under consideration from the boundary, where the proper boundary conditions are formulated. Thus, the solution to the spectral problem can be written as follows:

$$S(k, \beta, \boldsymbol{r}, t) = \frac{k}{k_0} S_0(k_0, \beta_0, \boldsymbol{r}_0, t_0) \exp\left\{\int A(k, \beta, \boldsymbol{r}, t) \, dt\right\} \,, \tag{7.16}$$

where S_0 is the spectrum value at the initial boundary, and the values k_0, β_0 and $\boldsymbol{r}_0, t_0$ are calculated using (7.8). They are functions of the values $k, \beta, \boldsymbol{r}, t$ defined at the calculated point. It is necessary to use numerical methods for solving this problem in the general case.

A special solution, which can be presented in explicit form, will be considered. The ice thickness is assumed to vary smoothly enough along the Ox axis: $L = L(x)$. This situation can be seen, for example, in the near-edge area of ice fields.

7 Wave Transformation in Ice-Covered Water

The movement equations for wave packets can be written in the following form:

$$\frac{dx}{dt} = C_{gx} = C_g \cos(\beta) \; ; \quad \frac{dy}{dt} = C_{gy} = C_g \sin(\beta) \; ; \qquad (7.17)$$

$$\frac{dk_x}{dt} = \frac{1}{2}\omega \left(\frac{\omega}{\omega_b}\right)^2 \frac{\partial L}{\partial x} \; ; \quad \frac{dk_y}{dt} = 0 \; . \qquad (7.18)$$

An attempt will be undertaken to find movement integrals for the equation set (7.17), (7.18). It can be seen that the coordinate y is cyclic and the component k_y of the wave vector $\mathbf{k}$ remains constant along the trajectory of wave packet propagation. So it can be written as:

$$k_y = k\sin(\beta) = k_0 \sin(\beta_0) \; . \qquad (7.19)$$

Preservation of the frequency ω is the second integral of movement. It can be presented as follows:

$$\frac{\sigma_0^2 \omega_{b_0}^2}{\omega_{b_0}^2 + \sigma_0^2} = \frac{\sigma^2 \omega_b^2(x)}{\omega_b^2(x) + \sigma^2} \; . \qquad (7.20)$$

The ratios (7.19) and (7.20) turn out to be sufficient to determine the wave number k and the angle β along the trajectory depending on variations of the ice thickness $L(x)$. If the ice thickness is monotonically increased along the positive direction of the Ox axis from zero to some definite values, the wave number is increased. According to (7.19) the wave angle propagation β is decreased, i.e. it is turned to the positive direction of the Ox axis. The waves are directed along the Ox axis accurately at the point where the wave frequency coincides with the float oscillation frequency.

Similar wave behaviour is observed in shallow water, with waves propagating to the shore. However, there are some differences. In shallow water, the shore line (i.e. depth equal to zero) is a special point for every wave, where the group velocity is equal to zero. As waves propagate in this situation, the eigenfrequency of ice oscillation is varied along the Ox axis, i.e., the position of the special point is different for different waves.

Using (7.14), the solution of frequency angular spectrum evolution along the wave propagation trajectory can be written as:

$$S(\omega, \beta, x) = \frac{kC_g^0}{k_0 C_g} S_0(\omega, \beta_0, x_0) \exp\left[E(\omega, \beta_0, x)\right]$$

$$= \left(1 - \frac{\omega^2}{\omega_b^2}\right)^{-\frac{5}{2}} S_0(\omega, \beta_0, x_0) \exp\left[E(\omega, \beta_0, x)\right] , \qquad (7.21)$$

where $E(\omega, \beta_0, x)$ is the integral value of the dissipative function along the characteristic in (7.16).

7.2 Wave Spectrum Evolution in Water with Ice Cakes

Due to the expression in round brackets (7.21), the frequency angular spectrum $S(\omega, \beta, x)$ might seem to become infinitely large with the wave packet arriving at the point, where $\omega^2 = \omega_b^2$. However, as shown below, this does not happen due to energy dissipation.

In order to obtain the complete spectrum (7.21) at some point $\{x_0, y_0\}$, it is necessary to bring together all the rays there. To do this, the set of equations (7.17), (7.18) must be solved. However, there are other ways to do this using the following kinematic notions. It should be noted that a filtration of wave components takes place due to dissipation as waves propagate within the ice cake area, whereas its thickness is increased in the wave propagation direction. The wave, which value $\gamma(x) = 1 - \omega^2 \rho_3 L(x)/g\rho_2$ becomes less than or equal to zero, and ceases to exist in the area under consideration. When the complete set of equations for the characteristics (7.17), (7.18) cannot be solved, it is necessary to impose an additional kinematic condition onto the solution (7.21), describing the absence of proper frequency components in the spectrum at the considered point. This is attained by multiplying the spectrum (7.21) by the Heaviside function $\Theta(\omega)$:

$$S(\omega, \beta, x) = \left(1 - \frac{\omega^2}{\omega_b^2}\right)^{-\frac{5}{2}} S_0(\omega, \beta_0) \exp\left[E(\omega, \beta, x)\right] \Theta\left[1 - \frac{\omega^2}{\omega_b^2}\right]. \quad (7.22)$$

The initial angle β_0 is determined using the ratio (7.19) and can be written as:

$$\beta_0 = \arcsin\left(\frac{\sin(\beta)}{1 - \omega^2/\omega_b^2}\right). \quad (7.23)$$

The arcsine argument value in (7.23) must be no larger than 1.0, so there is an additional restriction on the spectrum arguments:

$$\left|\frac{\sin(\beta)}{1 - \omega^2/\omega_b^2}\right| \leq 1. \quad (7.24)$$

A breach of this condition means that the corresponding spectral component is absent at this point. In this case it should be taken as:

$$S(\omega, \beta, x) \equiv 0. \quad (7.25)$$

If the initial spectrum is accepted to be:

$$S_0(\omega, \beta_0) = \tilde{S}_0(\omega) \cos^n(\beta_0), \quad (7.26)$$

it can be seen that the value $1 - \omega^2/\omega_b^2$ decreases with increasing ice thickness L and the angular distribution of wave energy is narrowed. If the angle range is originally limited as $|\beta| \leq \pi/2$, then the range is reduced in the presence of ice as $|\beta| \leq \arcsin\left(1 - \omega^2/\omega_b^2\right)$, and the following can be written:

$$\cos^n(\beta_0) = \left[1 - \left(\frac{\sin\beta}{1-\omega^2/\omega_b^2}\right)^2\right]^{\frac{n}{2}} \approx (\cos(\beta))^{n(1-\omega^2/\omega_b^2)^{-2}}. \qquad (7.27)$$

Thus, with increasing value L, the power of the cosine function is increased and the energy angular distribution becomes narrower.

It follows from the ratio (7.22), when the dissipation is neglected, that the spectrum $S(\omega,\beta,x)$ is not dependent on the ice thickness gradient, but is determined only by its value. In this case the spectral density ratio $S(\omega,\beta,x)/S_0(\omega,\beta_0)$ is increased with increasing L and at the special point where $L(x) = g\rho_2/\omega^2\rho_3$, it becomes infinitely large. Actually, this never happens due to wave energy dissipation.

Now this problem will be considered in detail. Returning to the solution of the balance wave energy equation (7.7) and using (7.6a) and (7.11), the argument of the exponential function in (7.16) can be presented as follows:

$$E(x) = -2\int_0^x \mu k(\omega)/\cos(\beta)\,dx = -\frac{2\nu\omega^7}{\rho_2 g^4}\int_0^x \frac{L(x)\,dx}{(1-\omega^2/\omega_b^2)^4 \cos(\beta)}. \qquad (7.28)$$

In this case it should be taken into account that the angle β is a function of x: $\beta = \beta(x)$, as variations of the angle β, depending on the ice thickness $L(x)$, take place along the trajectory of wave packet propagation.

The ratio (7.28) can be integrated analytically in the case of linear dependence of the ice thickness linear the coordinate: $L = \alpha x$. Omitting intermediate calculations, the final result can be written as:

$$E(\omega,\beta,x) = -\frac{\nu\omega^3\rho_2}{g^2\rho_3^2\alpha}\left\{\frac{\sqrt{1-\gamma^2 b^2}}{\gamma^2}\left(\frac{2+4\gamma^2 b^2}{3\gamma}-1\right)\right.$$
$$\left. - \sqrt{1-b^2}\left(\frac{4b^2-1}{3}\right) - b^2 \ln\left|\frac{1+\sqrt{1-\gamma^2 b^2}}{\gamma(1+\sqrt{1-b^2})}\right|\right\}, \qquad (7.29)$$

where $b = \sin(\beta_0) = \sin(\beta)/\gamma = \text{const}$; $\gamma(x) = 1 - \alpha\rho_3\omega^2 x/g\rho_2 = 1 - \omega^2/\omega_b^2$ and $\gamma > 0$.

It is easy to see that the function $|E(x)|$ is monotonically increased. In the case of $x \to g\rho_2/(\alpha\omega^2\rho_3)$, the parameter $\gamma \to 0$ and $|\cos^n(\beta_0)| \approx 1$, the function $|E(x)|$ takes infinitely large values: $E \approx -A(1/3 + (2/3\gamma - 1)/\gamma^2) \approx -2A/(3\gamma^3) \to -\infty$, where A is the multiplier before the bracket in (7.29). The value of the expression (7.22) becomes zero in spite of the fact that the multiplier connected with the group velocity variations placed before the spectral density in (7.22) takes an infinitely large value.

In order to get real value estimations it is necessary to define the viscous coefficient ν. Its value is dependent on a number of determining physical parameters such as ice and surrounding temperatures, ice salinity and age, etc. For ocean conditions the value of the viscous kinematic coefficient ν/ρ_3 can

7.2 Wave Spectrum Evolution in Water with Ice Cakes

change by three orders of magnitude within the range from 1 to 10^3 cm^2 s^{-1} (Kheysin, 1967).

The transfer function $\Phi(\omega, \beta, x) = S(\omega, \beta, x)/S_0(\omega, \beta_0)$ at $\beta = 0$ for different values of the kinematic viscosity coefficient ν (for $\partial L/\partial x = 0.25 \times 10^{-4}$, $L = 1$ m) is shown in Fig. 7.1. It is seen that at first the transfer function is equal to 1.0 at $\omega_0 = 0$, and later on with an increase of the frequency ω, it is slightly increased, reaching its maximum value. After that it is decreased and turns to zero, i.e. the spectral density is equal to zero. The maximum of the function Φ is increased and displaced to the large frequency area with the viscous coefficient ν and the local ice thickness L are decreased. As noted, the initial increase of the function Φ is connected with decreasing the group velocity of wave propagation due to the influence of ice. This results in increasing the energy spectral density. The further decrease of the transfer function is caused by dissipation. The question arises whether the increase of the energy spectral density really takes place in spite of the influence of the dissipation mechanism.

The ratio of the frequency spectrum to its initial value for small values of the angle β can be written as follows:

$$S(\omega, x)/S_0(\omega) \approx \gamma^{-\frac{5}{2}} \exp\left\{-A \left[1/3 + (2/3\gamma - 1)/\gamma^2\right]\right\}. \qquad (7.30)$$

The value (7.30) increases with waves propagating in the positive direction of the Ox axis. It reaches its maximum at $\gamma \approx [4/(5A)]^{1/3}$.

The energy spectral density $S(\omega, x)$ for different values of the viscosity coefficient (with mean period of initial waves being equal to 5 s) is shown in Fig. 7.2. The spectral density is decreased with wave propagation in ice cover (see Fig. 7.2, Curve 1). However, there is an opposite situation (see Fig. 7.2 Curves 2 and 3), i.e. the increase of wave energy density can be observed at

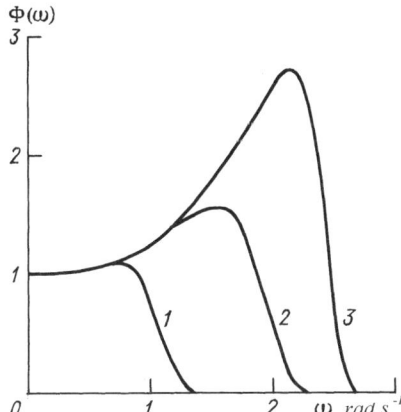

Fig. 7.1. Transfer function for different values of the viscosity kinematic coefficient: $1 - \nu/\rho_3 = 1.0$ cm^2 s^{-1}; $2 - \nu/\rho_3 = 100$ cm^2 s^{-1}; $3 - \nu/\rho_3 = 1000$ cm^2 s^{-1}

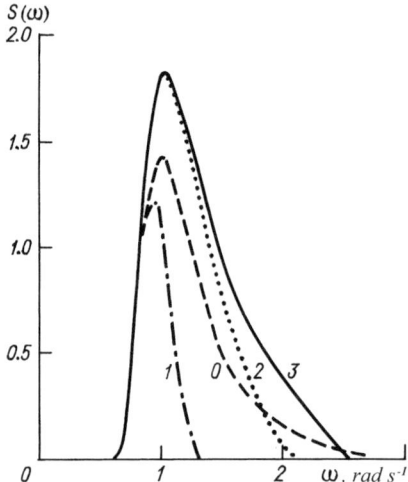

Fig. 7.2. Frequency spectra in water with ice cakes for different values of viscosity coefficient: 0 – initial spectrum; $1 - \nu/\rho_3 = 1.0 \text{ cm}^2 \text{s}^{-1}$; $2 - \nu/\rho_3 = 100 \text{ cm}^2 \text{s}^{-1}$; $3 - \nu/\rho_3 = 1000 \text{ cm}^2 \text{s}^{-1}$

the initial stage. This takes place with relatively small values of the mean ice thickness and small values of the viscosity coefficient ν. It is seen (see Fig. 7.2) that the energy increase takes place at low frequencies and it is decreased at high frequencies.

7.3 Kinetic Equation for Non-linear Waves in Sea with Ice Cakes

Introduction. As shown by numerous studies of wave processes, non-linear interactions are key mechanisms of wave evolution in water. The results of such studies obtained with both physical and statistical descriptions of random wave fields are well known. In spite of the obtained results concerning investigations in a sea covered with ice (Bukatov & Bukatova, 1993; Bukatov & Cherkesov, 1971; Kheysin, 1967; Liu et al., 1991; Marchenko, 1988; Masson & Le Blond, 1989; Meylan et al., 1994; Shuchman & Rufenach, 1994; Wadhams et al., 1986; Lavrenov & Novakov, 2000), the problems of non-linear wave evolution have not been studied properly. In particular, there is no theoretical derivation of the non-linear wave–wave interaction coefficients of the kinetic equation for waves in a water surface covered with ice cakes. This does not allow creating a well-grounded numerical model of wave spectrum evolution in a water surface covered with ice. The study of this problem is of real importance due to practical human activity in the polar seas.

The main purpose of this section is to deduce the explicit form of the interaction coefficients in the kinetic equation describing the spectrum evolution

7.3 Kinetic Equation for Non-linear Waves in Sea with Ice Cakes

of non-linear waves in water with ice cakes. Methods of the Hamiltonian formalism, developed in the paper by Zakharov (1968) for waves in non-linear media with dispersion will be applied. This technique allows passing from the initial dynamic equations expressed by physical variables to the simpler Hamiltonian equations expressed by special canonical variables. Such equations written in terms of Fourier variables give an exhaustive description of non-linear waves both in the dynamic and statistical sense for small wave non-linearity parameters: $\varepsilon \sim ak$, where a is the amplitude, and k is the wave number. After deducing these equations, the transition to the kinetic equation is made on the grounds of general theory results (Zakharov, 1968; Krasitskii & Kalmykov, 1993).

General theses of the Hamiltonian formalism. In order to apply the Hamiltonian formalism it is necessary for the system to preserve its energy. The total system energy function in physical variables can be presented as:

$$\mathcal{H} = K + P, \tag{7.31}$$

where K is the sum of wave and ice kinetic energy:

$$K = \frac{1}{2} \int \int_{-H}^{\eta} \left[\left(\frac{\partial \Phi}{\partial x_i} \right)^2 + \left(\frac{\partial \Phi}{\partial z} \right)^2 \right] \mathrm{d}\boldsymbol{x}\,\mathrm{d}z + \Lambda \frac{1}{2} \int \mathrm{d}\boldsymbol{x} \left(\frac{\partial \Phi}{\partial z} \right)^2 \bigg|_{z=\eta} \mathrm{d}\boldsymbol{x},$$

and P is the potential energy:

$$P = \frac{g}{2} \int \mathrm{d}\boldsymbol{x}\, \eta^2.$$

Since $\mathcal{H}$ is the Hamiltonian function, it may be expressed by two canonically conjugated variables. It is known (Zakharov, 1968) that the functions $\eta(\boldsymbol{x},t)$ and $\tilde{\phi}(\boldsymbol{x},t) = \phi[\boldsymbol{x}, \eta(\boldsymbol{x},t), t]$ (i.e. displacement of the water surface and the potential value at this boundary) are such variables in the problem of waves in a water surface without ice.

In the problem under consideration this is not so, due to the first item in (7.1a), describing ice cakes. There are reasons to suppose that the previous function $\eta(\boldsymbol{x},t)$ and a new function defined in the surface $z = \eta(\boldsymbol{x},t)$, namely:

$$\chi(\boldsymbol{x},t) = \tilde{\phi}(\boldsymbol{x},t) + \Lambda \frac{\partial \phi(\boldsymbol{x},z,t)}{\partial z}\bigg|_{z=\eta(\boldsymbol{x},t)}, \tag{7.32}$$

may be a canonically conjugated couple.

If the adduced statement is valid, the movement equations can be presented in the form:

$$\frac{\partial \eta}{\partial t} = \frac{\delta \mathcal{H}}{\delta \chi}, \quad \frac{\partial \chi}{\partial t} = -\frac{\delta \mathcal{H}}{\delta \eta}, \tag{7.33}$$

where the symbol δ means the functional derivative and the Hamiltonian $\mathcal{H}$ can be written as a function of the variables: $\eta(\boldsymbol{x}, t)$ and $\chi(\boldsymbol{x}, t)$.

This statement can be checked using the Fourier space with the help of the expansion of the unknown function into a series by the power of the non-linearity parameter ε.

The Fourier presentations can be introduced in the following form:

$$\eta(\boldsymbol{x}) = \frac{1}{2\pi} \int \eta(\boldsymbol{k}) \exp(\mathrm{i}\boldsymbol{k}\boldsymbol{x}) \, \mathrm{d}\boldsymbol{k} \; ; \qquad \eta(\boldsymbol{k}) = \eta^*(-\boldsymbol{k}); \qquad (7.34\mathrm{a})$$

$$\tilde{\phi}(\boldsymbol{x}) = \frac{1}{2\pi} \int \phi(\boldsymbol{k}) \exp(\mathrm{i}\boldsymbol{k}\boldsymbol{x}) \, \mathrm{d}\boldsymbol{k} \; ; \qquad \phi(\boldsymbol{k}) = \phi^*(-\boldsymbol{k}); \qquad (7.34\mathrm{b})$$

$$\chi(\boldsymbol{x}) = \frac{1}{2\pi} \int \chi(\boldsymbol{k}) \exp(\mathrm{i}\boldsymbol{k}\boldsymbol{x}) \, \mathrm{d}\boldsymbol{k} \; ; \qquad \chi(\boldsymbol{k}) = \chi^*(-\boldsymbol{k}), \qquad (7.34\mathrm{c})$$

where $\boldsymbol{k}$ is the horizontal wave vector, the integration is made over all the horizontal plane, the star means a complex-conjugated value, and the explicit function dependencies and their Fourier transformations over time are omitted to simplify the designations. The functions themselves and their Fourier transformations are designated using the same signs, differing only by their arguments. The Fourier transformation is canonical, so the equations (7.33) are reduced to similar ones in the form of:

$$\frac{\partial \eta(\boldsymbol{k})}{\partial t} = \frac{\delta \mathcal{H}}{\delta \chi^*(\boldsymbol{k})}, \qquad \frac{\partial \chi(\boldsymbol{k})}{\partial t} = -\frac{\delta \mathcal{H}}{\delta \eta^*(\boldsymbol{k})} \qquad (7.35)$$

with the canonically conjugated variables $\eta(\boldsymbol{k})$, $\chi^*(\boldsymbol{k})$ and $\chi(\boldsymbol{k})$, $\eta^*(\boldsymbol{k})$.

Additional canonical transformations for a new couple of the so-called canonically conjugated normal variables $a(\boldsymbol{k})$ and $a^*(\boldsymbol{k})$ can be made as follows:

$$\eta(\boldsymbol{k}) = [\omega(\boldsymbol{k})/2g]^{1/2} [a(\boldsymbol{k}) + a^*(-\boldsymbol{k})],$$

$$\chi(\boldsymbol{k}) = -\mathrm{i} \, [g/2\omega(\boldsymbol{k})]^{1/2} [a(\boldsymbol{k}) - a^*(-\boldsymbol{k})], \qquad (7.36)$$

where $\omega(k)$ is the linear wave dispersion ratio (7.9) used in the form $\omega^2(k) = \frac{gq(k)}{1+Aq(k)}$, where $q = gk \, \mathrm{th}(kh)$. The dispersion ratio reduces (7.35) to:

$$\mathrm{i}\frac{\partial a(\boldsymbol{k})}{\partial t} = \frac{\delta \mathcal{H}}{\delta a^*(\boldsymbol{k})}, \qquad (7.37)$$

where $\mathcal{H}$ is already a function of $a(\boldsymbol{k})$ and $a^*(\boldsymbol{k})$. The equation (7.37) and its complex conjugate form are the final couple of Hamiltonian canonical equations, being really a single equation determining the non-linear wave evolution.

Waves of small, but finite amplitudes (i.e. non-linear waves) are investigated below. If the wave steepness is supposed to be small, the Hamiltonian

7.3 Kinetic Equation for Non-linear Waves in Sea with Ice Cakes

$\mathcal{H} = \mathcal{H}(a, a^*)$ can be formally decomposed into integral-power series by the degrees $a(\boldsymbol{k})$ and $a^*(\boldsymbol{k})$. In this case the expansion is considered with accuracy up to fourth-order terms. The standard presentation of the expansion of the Hamiltonian $\mathcal{H} = \mathcal{H}(a, a^*)$ is as follows:

$$\mathcal{H} = \int \omega_0 a_0^* a_0 \, d\boldsymbol{k}_0 + \int U_{0,1,2}^{(1)} (a_0^* a_1 a_2 + c.c.) \, \delta_{0-1-2} \, d\boldsymbol{k}_{0,1,2}$$

$$+ \frac{1}{3} \int U_{0,1,2}^{(3)} (a_0 a_1 a_2 + c.c) \, \delta_{0+1+2} \, d\boldsymbol{k}_{0,1,2}$$

$$+ \int V_{0,1,2,3}^{(1)} (a_0^* a_1 a_2 a_3 + c.c) \, \delta_{0-1-2-3} \, d\boldsymbol{k}_{0,1,2,3}$$

$$+ \frac{1}{2} \int V_{0,1,2,3}^{(2)} (a_0^* a_1^* a_2 a_3 + c.c) \, \delta_{0+1-2-3} \, d\boldsymbol{k}_{0,1,2,3}$$

$$+ \frac{1}{4} \int V_{0,1,2,3}^{(4)} (a_0 a_1 a_2 a_3 + c.c.) \, \delta_{0+1+2+3} \, d\boldsymbol{k}_{0,1,2,3} + \ldots, \qquad (7.38)$$

where $\delta_{0-1-2} = \delta(\boldsymbol{k} - \boldsymbol{k}_1 - \boldsymbol{k}_2)$ is the Dirac delta function; $c.c$ is the complex-conjugate value; and $U_{0,1,2}^{(n)}$, $V_{0,1,2,3}^{(n)}$ are the Hamiltonian expansion coefficients describing the non-linear interaction intensity. These coefficients satisfy certain conditions of symmetry in rearranging lower indexes, expressing the reality of the Hamiltonian value.

The abbreviated designations are used for the arguments $\boldsymbol{k}_j$ of the Hamiltonian expansion coefficients $U_{0,1,2}^{(n)}$, $V_{0,1,2,3}^{(n)}$, being replaced by the indexes j, whereas the zero index belongs to the vector $\boldsymbol{k}$. Thus, the following designations are applied:

$$U_{0,1,2}^{(n)} = U^{(n)}(\boldsymbol{k}, \boldsymbol{k}_1, \boldsymbol{k}_2),$$
$$\omega_j = \omega(\boldsymbol{k}_j), \qquad (7.39)$$
$$a_j = a(\boldsymbol{k}_j, t),$$
$$\delta_{0-1-2} = \delta(\boldsymbol{k} - \boldsymbol{k}_1 - \boldsymbol{k}_2).$$

As for the differentials, the following designations are used: $d\boldsymbol{k}_0 = d\boldsymbol{k}$, $d\boldsymbol{k}_{01} = d\boldsymbol{k} \, d\boldsymbol{k}_1$, etc. The integrals are estimated within the range: $-\infty$ to $+\infty$.

It follows from (7.37) that the movement equation can be presented in the following form, using the Hamiltonian (7.38):

$$i \frac{\partial a_0}{\partial t} = \omega_0 a_0 + \int U_{0,1,2}^{(1)} a_1 a_2 \delta_{0-1-2} \, d\boldsymbol{k}_{1,2} + \ldots, \qquad (7.40)$$

where two second-order terms and four third-order terms in the amplitude a are designated by the dots. It should be noted that (7.40) can be deduced directly with the help of the dynamic equations (7.1) without the Hamiltonian calculation (7.38) by passing to the Fourier representation for the variables

$\eta(\mathbf{k})$ and $\chi(\mathbf{k})$. Evidence of canonical conjugation for the couple of variables $\eta(\mathbf{x},t)$, $\chi(\mathbf{x},t)$ may be considered to obtain equations coinciding with each other with an accuracy up to complex conjugation after passing to the normal variables $a(\mathbf{k})$, $a^*(\mathbf{k})$ and proper symmetrization of the interaction coefficients. The same expansion coefficients occur in the final equation as in (7.40).

Expansion coefficients of the Hamiltonian. The technique described by Krasitskii & Kalmykov (1993) for water without ice is used below.

The general solution of the Laplace equation, satisfying the bottom boundary condition (7.2), can be expressed in the form of the Fourier integral:

$$\phi(\mathbf{x}, z) = \frac{1}{2\pi} \int \phi(\mathbf{k}) \frac{\text{ch}\,[k(z+H)]}{\text{sh}\,(kH)} \exp(\mathrm{i}\mathbf{k}\mathbf{x})\, \mathrm{d}\mathbf{k}, \tag{7.41}$$

Substitution of (7.34) and (7.41) into (7.31), using the expansion of exponents by the small parameter $\varepsilon \equiv k\eta \ll 1$, results in the Hamiltonian representation as terms of the amplitude degrees of the Fourier components $\eta(\mathbf{k})$ and $\phi(\mathbf{k})$. With fourth-order accuracy by non-linearity, the Hamiltonian can be written as:

$$\mathcal{H} = K^{(2)} + K^{(3)} + K^{(4)} + P^{(2)}, \tag{7.42}$$

where

$$P^{(2)} = \frac{1}{2} \int g\eta_1 \eta_2 \delta_{1+2}\, \mathrm{d}\mathbf{k}_{1,2}; \tag{7.43}$$

$$K^{(2)} = \int B^{(2)}_{1,2} \phi_1 \phi_2 \delta_{1+2}\, \mathrm{d}\mathbf{k}_{1,2}; \tag{7.44}$$

$$K^{(3)} = \int B^{(3)}_{1,2} \phi_1 \phi_2 \eta_3 \delta_{1+2+3}\, \mathrm{d}\mathbf{k}_{1,2,3}; \tag{7.45}$$

$$K^{(4)} = \int B^{(4)}_{1,2} \phi_1 \phi_2 \eta_3 \eta_4 \delta_{1+2+3+4}\, \mathrm{d}\mathbf{k}_{1,2,3,4}. \tag{7.46}$$

In this case it is important to consider the general form of integrands as the function of the Fourier variables $\eta(\mathbf{k})$ and $\phi(\mathbf{k})$ without detailing the kernel B. The upper indexes in brackets mean the order of summation of non-linearity.

Now, it is necessary to pass to the variable $\chi(\mathbf{k})$ in (7.44)–(7.46) using the following formula:

$$\chi(\mathbf{k}) = \frac{1}{2\pi} \int \chi(\mathbf{x}) \exp(-\mathrm{i}\mathbf{k}\mathbf{x})\, \mathrm{d}\mathbf{x}. \tag{7.47}$$

The ratio (7.32) and the determination (7.34c) should be used to express $\phi(\mathbf{k})$ in terms of $\chi(\mathbf{k})$. Fulfilling the transformations in (7.47) taking consideration of the small steepness expansion, the expression $\chi(\mathbf{k})$ can be obtained

7.3 Kinetic Equation for Non-linear Waves in Sea with Ice Cakes

by a series of the variables $\eta(\mathbf{k})$ and $\phi(\mathbf{k})$. Expanding this ratio relative to $\phi(\mathbf{k})$ with an accuracy up to third-order terms the following formula can be obtained:

$$\phi(\mathbf{k}) = \left(\frac{\text{th}(kH)}{1 + \Lambda q}\right) \left[\chi(\mathbf{k}) + \int D_1^{(2)} \chi_1 \eta_2 \delta_{0-1-2} \, d\mathbf{k}_{1,2} \right.$$

$$\left. + \int D_{0,1,2,3}^{(3)} \chi_1 \eta_2 \eta_3 \delta_{0-1-2-3} \, d\mathbf{k}_{1,2,3} \right]. \tag{7.48}$$

The following designations are used:

$$D_0^{(2)} \equiv D^{(2)}(k) = -\frac{1}{2\pi}\left[q\left(\frac{1 + \Lambda k \coth(kH)}{1 + \Lambda q}\right)\right] \equiv -cQ_0 \tag{7.49}$$

$$D_{0,1,2,3}^{(3)} = -\frac{1}{2\pi}\left(\frac{1}{2\pi}\right)^2 \left[k_1^2 - Q_1 Q_{0-3} - Q_1 Q_{0-2}\right]. \tag{7.50}$$

Substituting (7.48)–(7.50) into (7.44)–(7.46), the following is obtained:

$$K_2 = \left(\int E_0^{(2)\text{w}} \chi_0 \chi_0^* d\mathbf{k} + \int E_0^{(2)\text{a}} \chi_0 \chi_0^* d\mathbf{k}\right)$$

$$= \frac{1}{2}\left(\int \frac{q \chi_0 \chi_0^*}{(1 + \Lambda q)^2} d\mathbf{k} + \Lambda \int \frac{q \chi_0 \chi_0^*}{(1 + \Lambda q)^2} d\mathbf{k}\right) = \frac{1}{2}\int \frac{q \chi_0 \chi_0^*}{(1 + \Lambda q)} d\mathbf{k}, \tag{7.51}$$

where

$$E_0^{(2)\text{w}} \equiv E^{(2)\text{w}}(\mathbf{k}) = \frac{1}{2}\frac{q}{(1 + \Lambda q)^2}; \quad E_0^{(2)\text{a}} \equiv E^{(2)\text{a}}(\mathbf{k}) = \frac{1}{2}\frac{\Lambda q^2}{(1 + \Lambda q)^2};$$

$$K_3 = \int \left[\left(A_{0,1}^{(3)\text{w}} + E_{0,1}^{(3)\text{w}}\right) + \left(A_{0,1}^{(3)\text{a}} + E_{0,1}^{(3)\text{a}}\right)\right] \chi_0 \chi_1 \eta_2 \delta_{0+1+2} \, d\mathbf{k}_{0,1,2} \tag{7.52}$$

$$K_4 = \int \left[\left(B_{0,1,2,3}^{(4)\text{w}} + C_{0,1,2,3}^{(4)\text{w}} + E_{0,1}^{(4)\text{w}}\right) \right.$$

$$\left. + \left(B_{0,1,2,3}^{(4)\text{a}} + C_{0,1,2,3}^{(4)\text{a}} + E_{0,1}^{(4)\text{a}}\right)\right] \chi_0 \chi_1 \eta_2 \eta_3 \delta_{0+1+2+3} \, d\mathbf{k}_{0,1,2,3} \tag{7.53}$$

Designations similar to those used by Krasitskii (1974) are used here. They are added by modifications taking ice into account (the upper indexes w and a belong to water and ice, correspondingly).

It is important to note that the integrand in the right-hand side of (7.51) provides the exact linear dispersion ratio $\omega(k)$ of the form (7.3) in the first term of the final Hamiltonian representation (7.38).

The core expressions of the third and fourth order items are written in a rather complicated form. That is why they are introduced as formulas of complicated functions convenient for numerical programming. The kernel expressions of the cubic energy term are as follows.

Firstly:

$$A_{0,1}^{(3)\mathrm{w}} = 2E_0^{(2)\mathrm{w}} D_1^{(2)} \equiv E_0^{(2)\mathrm{w}} D_1^{(2)} + E_1^{(2)\mathrm{w}} D_0^{(2)}, \qquad (7.54)$$

where $E_0^{(2)\mathrm{w}}$ is defined by the expression (7.51) and $D_1^{(2)}$ by the expression (7.49). It should be noted that expressions of the functions $D^{(n)}$ of all non-linear orders n are similar for ice and waves, as they are determined only by the expansion (7.48).

Secondly:

$$E_{0,1}^{(3)\mathrm{w}} = -\frac{1}{4\pi} \frac{[\boldsymbol{k}\boldsymbol{k}_1 - q(k)q(k_1)]}{(1 + \Lambda q(k))(1 + \Lambda q(k_1))}. \qquad (7.55)$$

Similar ratios for ice are:

$$A_{0,1}^{(3)\mathrm{a}} = 2E_0^{(2)\mathrm{a}} D_1^{(2)} \equiv E_0^{(2)\mathrm{a}} D_1^{(2)} + E_1^{(2)\mathrm{a}} D_0^{(2)}, \qquad (7.56)$$

where $E_0^{(2)\mathrm{a}}$ is approximated by the expression (7.51) and

$$E_{0,1}^{(3)\mathrm{w}} = \frac{\Lambda}{4\pi} \frac{q(k)k_1^2 + q(k_1)k^2}{(1 + \Lambda q(k))(1 + \Lambda q(k_1))}. \qquad (7.57)$$

Now, it should be noted that the identities in (7.54), (7.56) mean the symmetrization of the kernels A by indexes 0 and 1, necessary for the Hamiltonian representation in standard form. However, this symmetrization is of no great importance for numerical simulations of the non-linear mechanism. In order to simplify numerical codes, the unsymmetrized kernels can be used.

The kernel expressions of the fourth-order non-linear terms are the following. Firstly:

$$\begin{aligned} B_{0,1,2,3}^{(4)\mathrm{w}} &= \frac{1}{4} D_0^{(2)} D_1^{(2)} \left(E_{0+2}^{(2)\mathrm{w}} + E_{0+3}^{(2)\mathrm{w}} + E_{1+2}^{(2)\mathrm{w}} + E_{1+3}^{(2)\mathrm{w}} \right) \\ &\quad + E_0^{(2)\mathrm{w}} D_{-0,1,2,3}^{(3)} + E_1^{(2)\mathrm{w}} D_{-1,0,2,3}^{(3)}, \end{aligned} \qquad (7.58)$$

where all proper factors are determined using the formulas (7.49), (7.50) and (7.51). Secondly:

$$\begin{aligned} C_{0,1,2,3}^{(4)\mathrm{w}} &= 2D_1^{(2)} E_{0,1+2}^{(3)\mathrm{w}} \\ &\equiv \frac{1}{2} \left[D_0^{(2)} E_{1,0+2}^{(3)\mathrm{w}} + D_1^{(2)} E_{0,1+2}^{(3)\mathrm{w}} + D_0^{(2)} E_{1,0+3}^{(3)\mathrm{w}} + D_1^{(2)} E_{0,1+3}^{(3)\mathrm{w}} \right], \end{aligned} \qquad (7.59)$$

where the symmetrized kernel is given as an example. Thus, the following is obtained:

7.3 Kinetic Equation for Non-linear Waves in Sea with Ice Cakes

$$E_{0,1}^{(4)\text{w}} = -\frac{1}{4}\left(\frac{1}{2\pi}\right)^2 \frac{(\boldsymbol{k}\boldsymbol{k}_1 - k_1^2)q(k) + (\boldsymbol{k}\boldsymbol{k}_1 - k^2)q(k_1)}{(1 + \Lambda q(k))(1 + \Lambda q(k_1))}. \tag{7.60}$$

There are similar expressions for the kernels B and C as for ice and for water. The only difference is that the proper kernels E are taken for ice. So, it is enough for ice to get only the last fourth-order term, namely:

$$E_{0,1}^{(4)\text{a}} = \Lambda\frac{1}{4}\left(\frac{1}{2\pi}\right)^2 \frac{[2k^2 k_1^2 + (k^2 + k_1^2)q(k)q(k_1)]}{(1 + \Lambda q(k))(1 + \Lambda q(k_1))}. \tag{7.61}$$

Thus, the Hamiltonian is completely written in the Fourier representation of canonical variables. Now it is possible to write it in the form of normal variables.

Interaction coefficients and the kinetic equation. According to standard methods (Zakharov, 1968), the following expressions for the Fourier components of canonical variables can be written with the help of normal variables using the formulas (7.36). Substituting (7.36) in the Hamiltonian terms into (7.43) and (7.51)–(7.53), under the condition that the quadratic terms are presented in their standard form (7.38), the final expressions for the interaction coefficients U and V can be obtained. Using the results of Krasitskii & Kalmykov (1993), the coefficients of three-wave interactions can be presented as:

$$\begin{aligned} U_{0,1,2}^{\text{w,a}} &= -L_0 L_1 M_2 \left[A_{0,1}^{(3)\text{w,a}} + E_{0,1}^{(3)\text{w,a}}\right] \\ &= -(g\omega_2/8\omega_0\omega_1)^{1/2}\left[A_{0,1}^{(3)\text{w,a}} + E_{0,1}^{(3)\text{w,a}}\right] \end{aligned} \tag{7.62}$$

and

$$U_{0,1,2}^{(1)} = \sum_{\text{w,a}} \left[-U_{-0,1,2}^{\text{w,a}} - U_{-0,2,1}^{\text{w,a}} + U_{1,2,-0}^{\text{w,a}}\right], \tag{7.63}$$

$$U_{0,1,2}^{(3)} = \sum_{\text{w,a}} \left[U_{0,1,2}^{\text{w,a}} + U_{0,2,1}^{\text{w,a}} + U_{1,2,0}^{\text{w,a}}\right]. \tag{7.64}$$

It should be noted that the summation in (7.63), (7.64) is taken over the indexes w, a. The proper coefficients of four-wave interactions are as follows:

$$V_{0,1,2,3}^{\text{w,a}} = -2L_0 L_1 M_2 M_3 \left[B_{0,1,2,3}^{(4)\text{w,a}} + C_{0,1,2,3}^{(4)\text{w,a}} + E_{0,1}^{(4)\text{w,a}}\right], \tag{7.65}$$

and they can be written for unsymmetrized coefficients in the final Hamiltonian as:

$$V^{(1)}_{0,1,2,3} = \sum_{w,a} \left[V^{w,a}_{1,2,-0,3} - V^{w,a}_{-0,1,2,3} \right], \tag{7.66}$$

$$V^{(2)}_{0,1,2,3} = \sum_{w,a} \left[V^{w,a}_{-0,-1,2,3} + V^{w,a}_{2,3,-0,-1} - V^{w,a}_{-0,2,-1,3} \right.$$

$$\left. - V^{w,a}_{-1,2,-0,3} - V^{w,a}_{-0,3,-1,2} - V^{w,a}_{-1,3,-0,2} \right], \tag{7.67}$$

$$V^{(4)}_{0,1,2,3} = 2 \sum_{w,a} V^{w,a}_{0,1,2,3}. \tag{7.68}$$

At this stage the derivation of the interaction coefficients for the standard Hamiltonian of the wave–ice system is complete. For further calculations of the kinetic integral one should use standard formulas (Zakharov, 1968). The kinetic integral can be presented as:

$$\frac{\partial N_0}{\partial t} = 4\pi \int T^2_{0,1,2,3} \, \delta(\boldsymbol{k} + \boldsymbol{k}_1 - \boldsymbol{k}_2 - \boldsymbol{k}_3) \, \delta(\omega_0 + \omega_1 - \omega_2 - \omega_3)$$

$$\times N_0 N_1 N_2 N_3 \left[\frac{1}{N_0} + \frac{1}{N_1} - \frac{1}{N_2} - \frac{1}{N_3} \right] d\boldsymbol{k}_{1,2,3} \tag{7.69}$$

where the determination of the wave action spectrum is as follows:

$$N(\boldsymbol{k}) \, \delta(\boldsymbol{k} - \boldsymbol{k}') \; = \; < a(\boldsymbol{k}) \, a^*(\boldsymbol{k}') >, \tag{7.70}$$

and the matrix element $T(k, \boldsymbol{k}_1, \boldsymbol{k}_2, \boldsymbol{k}_3)$ is determined only by the interaction coefficients $U^{(1)}_{0,1,2}$, $U^{(3)}_{0,1,2}$ and $V^{(2)}_{0,1,2,3}$, as in the paper by Krasitskii & Kalmykov (1993).

In conclusion, the Hamiltonian approach to the description of non-linear wave evolution in water with ice cakes was presented. The Hamiltonian H is applied for waves in water with ice, the canonical variables are found and the Hamiltonian H is written in the Fourier representation form for canonical variables. After that the transition to the normal variables $a(\boldsymbol{k})$, $a^*(\boldsymbol{k})$ in Fourier space is fulfilled, and the Hamiltonian is obtained in the form (7.38). This describes the intensity of non-linear wave–wave interactions up to fourth order, allowing us, finally, to write the kinetic integral in the explicit form (7.69).

7.4 Estimation of Non-Linear Energy Transfer in the Wave Spectrum in Water with Ice Cakes

Algorithm for calculation. In the previous section the matrix elements $T_{1,2,3,4} \equiv T(\boldsymbol{k}_1, \boldsymbol{k}_2, \boldsymbol{k}_3, \boldsymbol{k}_4)$ were obtained for the kinetic equation of non-linear evolution of the wave action spectrum $N(\boldsymbol{k})$ in water with ice cakes (7.69).

7.4 Non-Linear Transfer in the Spectrum for Waves with Ice Cakes

The algorithm for calculating the integral (7.69) described in Chap. 4 for the case of deep water without ice. The situation is more complicated for finite depth water with ice. Due to the difficulty in using the accurate dispersion ratio determined by the formula (7.9) for waves in water with ice, its approximate expression can be used as follows:

$$g^2 k^2 = \sigma^4 + \sigma^2 \omega_0^2 , \qquad (7.71)$$

where $\omega_0^2 = g/H$ is the typical frequency defined by water depth, and the frequency σ is found using the formula:

$$\sigma^2 = \omega^2/(1 - \omega^2/\omega_b^2) .$$

It should be noted that (7.71) is valid for every frequency and wave number with an error of no more than 10 per cent.

There are two parameters: ω_0 and ω_b in this problem. In other respects the algorithm for calculating the integral (7.69) coincides completely with that used in Chap. 4.

Estimation results of non-linear energy transfer in the wave spectrum. According to the above mentioned algorithm the calculations of non-linear energy transfer in the wave spectrum are fulfilled, taking into account water depth and the presence of ice cakes. The wave spectrum is used in the form of the JONSWAP approximation, and the angle distribution function is accepted as the cosine. The calculations are made for different values of the peakedness parameter equal to $\gamma = 1.0$ and 3.3, and the cosine power in the angle distribution function is varied from 2 to 8. The water depth is accepted to be equal to 10 m and the ice cover thickness is 0.3, 0.5 and 0.8 m, correspondingly.

An example of such a calculation for the peakedness parameter equal to 1.0 is shown in Fig. 7.3. The value of the non-linear transfer is given in the normalized form:

$$\tilde{G}_{nl}(\omega) = G_{nl}(\omega)/\left((\pi/16) S_{max}^3 \omega^{11} g^{-4}\right) .$$

For comparison, the results are shown in the case of infinitely deep water without ice cover. The results show that the character of the non-linear energy transfer is qualitatively the same as in the traditional case of deep water without ice. The energy flow is directed from the central spectrum area to the high and low frequency ranges. The depth decrease results in increasing the non-linear energy transfer, otherwise, the increase of ice cover thickness diminishes the intensity of the non-linear energy transfer.

It can be concluded from the aforesaid that the presence of ice causes narrowing of the frequency range of the wave spectrum. The positive low-frequency and negative areas are shifted to low frequencies in comparison with the traditional case without ice. This shows the intensive spectrum shift

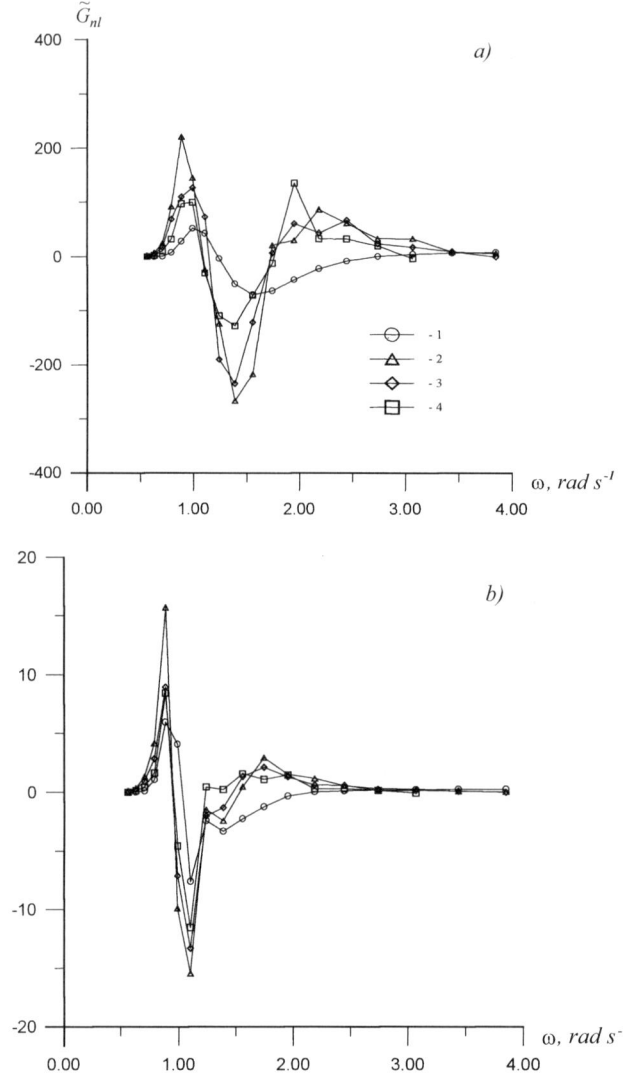

Fig. 7.3. Non-linear energy transfer in water with ice cakes for different values of the spectrum peakedeness $\gamma = 1.0$ (**a**) and $\gamma = 3.3$ (**b**): 1 – deep water without ice; 2 – basin depth is 10 m, ice thickness is 0.3 m; 3 – water depth is 10 m; ice thickness is 0.5 m, 4 – water depth is 10 m; ice thickness is 0.8 m

to the low-frequency range and narrowing of the wave frequency spectrum due to ice. Calculations reveal that the intensity of energy flow is increased to the high frequency range in the presence of ice, particularly to the frequency equal to the double frequency of the spectrum maximum. This creates tendencies for wave monochromatization.

It should be noted in conclusion that an attempt is undertaken in this chapter to describe the principal physical mechanisms forming the sea wave spectrum in the presence of ice cakes. They are: wave refraction, wave energy dissipation connected with friction between ice floes and non-linear energy redistribution in the wave spectrum. The obtained results can be a basis for developing a mathematical model describing the spectral evolution of waves in sea with ice. This model can be used for wave calculations and forecasts in the Arctic seas and, particularly, for numerical simulation of phenomena such as "ice storms", caused by storm wind waves propagating to marginal ice zones. These are very dangerous for oil platforms and hydraulic structures.

7.5 Numerical Simulation of Surface Gravity Wave Interaction with Elastic Ice Floes

Surface waves can penetrate from the open sea into regions with marginal and continuous ice cover. Numerous experimental data of the wave regime in the presence of ice were obtained during the MIZEX experiment (Wadhams et al., 1986). Observations of waves penetrating under fast ice were described by Squire (1984). In particular, it is noted that, under ice cover, long waves can penetrate large distances, reach considerable amplitudes, and even break up solid ice fields (Liu & Mollo-Christensen, 1988; Smirnov, 1996).

The methods of simulating the interaction between gravity waves and an ice field or a semi-infinite ice plate are fairly well worked out (Bukatov et al., 1984; Fox & Squire, 1991; Lavrenov and Novakov, 2000; Sturova 1998, 2000). In order to estimate the evolution of the wave spectrum in the presence of ice, an attempt to simulate surface waves in water covered with floating ice floes of finite length was developed by Masson & LeBlond (1989). The sea state is described by a two-dimensional discrete spectrum. Time-limited wave growth is obtained by numerical integration of the wave balance equation using the exact non-linear transfer integral without taking into account the presence of ice. Wave scattering by a single floe is presented in terms of far-field expressions of diffracted and forced potentials obtained by the Green function.

The solution of the problem of gravity waves interacting with ice of finite length encounters certain difficulties. The problem is reduced to simultaneous solution of the Laplace equation for the fluid and the bending equation for a thin elastic plate, which includes fourth-order spatial derivatives.

In this monograph a numerical solution of the problem based on the combined method of finite and boundary elements is developed. A problem formulation similar to Meylan & Squire (1994) is used. This makes it possible to compare calculation results with the analytical solution for some special cases.

7 Wave Transformation in Ice-Covered Water

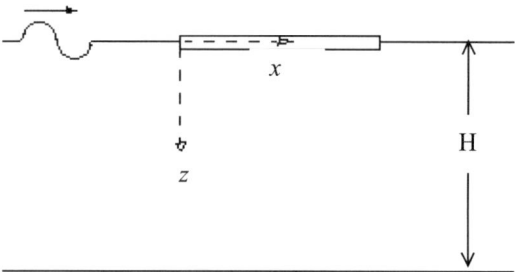

Fig. 7.4. Problem formulation scheme

Problem formulation. Let a homogeneous isotropic ice plate of length L float on the surface of an ideal fluid (see Fig. 7.4). The coordinate origin is fixed in the free surface of the fluid and the Z axis is directed vertically downward. The linearized boundary problem for the velocity potential $\phi(x,z,t)$ is presented as follows:

$$\nabla^2 \phi = 0; \quad 0 < z < H;$$

$$\frac{\partial \phi}{\partial z} = 0; \quad z = H;$$

$$\frac{\partial \phi}{\partial z} = -\frac{\partial \zeta}{\partial t}; \quad z = 0;$$

$$-\rho \left(g\frac{\partial \phi}{\partial z} - \frac{\partial^2 \phi}{\partial t^2} \right) = -\frac{\partial p}{\partial t}; \quad z = 0 \tag{7.72}$$

where g is the gravitational acceleration, p is the pressure on the fluid surface, ρ is the water density, ζ is the deviation of the free surface, and H is the fluid depth.

As for the homogeneous and isotropic plate, its displacement from equilibrium satisfies the biharmonic oscillation equation:

$$\frac{\partial^2 \zeta}{\partial t^2} + \mu^2 \frac{\partial^4 \zeta}{\partial x^4} = \frac{p}{\rho' h}; \quad z = 0; \quad 0 < x < L.$$

Thus, the boundary condition beneath the ice plate can be written in the form:

$$-\rho \left(g\frac{\partial \phi}{\partial z} - \frac{\partial^2 \phi}{\partial t^2} \right) = \rho' h \left(\frac{\partial^3 \phi}{\partial^2 t \partial z} + \mu^2 \frac{\partial^5 \phi}{\partial^4 x \partial z} \right);$$

$$z = 0; \quad 0 < x < L. \tag{7.73}$$

As for the open water area, the assumption $\partial p / \partial t = 0$ is used and the boundary condition can be written as follows:

7.5 Surface Gravity Wave Interaction with Elastic Ice Floes

$$g\frac{\partial \phi}{\partial z} - \frac{\partial^2 \phi}{\partial t^2} = 0\,; \quad z = 0\,; \quad -\infty < x < 0\,; \quad L < x < \infty\,. \tag{7.74}$$

The free condition should be fulfilled at the ice edge, i.e. the bending moment and the shearing force are equal to zero (Krasil'nikov, 1960):

$$\frac{\partial^2 \varsigma}{\partial x^2} = \frac{\partial^3 \varsigma}{\partial x^3} = 0\,; \quad x = 0\,; \quad x = L\,; \quad z = 0 \tag{7.75}$$

In (7.72)–(7.75) the following notation is used: ρ' is the ice density, $\mu^2 = Eh^2/12\rho'(1-\nu^2)$, E is the Young modulus, h is the ice-plate thickness, and ν is the Poisson ratio.

Now a solution is sought as a function periodic in time: $\phi = \phi' e^{i\omega t}$. Thus, the problem can be rewritten as follows (the primes are subsequently omitted):

$$\nabla^2 \phi = 0\,; \quad \infty < z < 0\,;$$

$$\frac{\partial \phi}{\partial z} = 0\,; \quad z \to \infty\,;$$

$$\frac{\partial \phi}{\partial z} + \frac{\omega^2}{g}\phi = 0\,; \quad z = 0\,, \quad -\infty < x < 0\,, \quad L < x < \infty\,;$$

$$\frac{\partial^5 \phi}{\partial x^4 \partial z} - \alpha^4 \frac{\partial \phi}{\partial z} + \beta \phi = 0\,; \quad z = 0\,, \quad 0 < x < L\,;$$

$$\frac{\partial^3 \phi}{\partial x^2 \partial z} = \frac{\partial^4 \phi}{\partial x^3 \partial z} = 0\,; \quad x = 0\,, \quad x = L\,, \quad z = 0\,;$$

$$\alpha^4 = -\frac{\omega^2}{\mu^2} + \frac{\rho g}{\rho' h \mu^2}\,; \quad \beta = -\frac{\omega^2 \rho}{\rho' h \mu^2}\,.$$

(7.76)

Assuming that the potential of the arriving wave is of unit amplitude, the boundary condition as $x \to \pm\infty$ can be written in the following form:

$$\phi = T \exp(-ikx - kz)\,; \qquad\qquad x \to \infty\,;$$
$$\phi = \exp(-ikx - kz) + R \exp(ikx - kz)\,; \qquad x \to -\infty \tag{7.77}$$

where R and T are the reflection and transmission coefficients, respectively.

An analytical solution of the problem (7.76), (7.77) is given by Meylan & Squire (1994) in the form of the Fredholm integral equation:

$$R = ik \int_0^L e^{-ik\xi} \left[\phi(\xi,0) + \beta \int_0^L \delta(\xi,\iota) \phi(\iota,0) \, d\iota \right] d\xi$$

$$T = 1 + ik \int_0^L e^{ik\xi} \left[\phi(\xi,0) + \beta \int_0^L \delta(\xi,\iota) \phi(\iota,0) \, d\iota \right] d\xi \tag{7.78}$$

$$\phi(x,0) = e^{-ikx} + k \int_0^L [G(\xi,0\,;\varsigma,0) + i\cos(kr)]$$

$$\times \left[\phi(\xi,0) + \beta \int_0^L \delta(\xi,\iota) \phi(\iota,0) \, d\iota \right] d\xi$$

where G is the Green function for the Laplace equation, and δ is the Green function for the elastic plate equation (7.73).

Numerical solution. In order to implement the problem numerically, a combination of the boundary element methods (for approximation of the Laplace equation) and finite elements (for approximation of the elastic plate equation) is used.

Taking into account the following property of the Green function:

$$\frac{\partial \phi(x,0)}{\partial z} = -\beta \int_0^L g(x,\xi) \phi(\xi,0) \, d\xi$$

the equation for the velocity potential (7.78) is rewritten in the form:

$$\phi(x,0) = e^{-ikx} + \int_0^L [G(\xi,0\,;\varsigma,0) + i\cos(kr)] \left[k\phi(\xi,0) + \frac{\partial \phi(\xi,0)}{\partial z} \right] d\xi \,. \tag{7.79}$$

This equation is treated as the boundary-element formulation of the problem (Brebbia et al., 1984).

The Green function G in this equation is assumed to satisfy the boundary conditions in the free surface and in the lower boundary. Meylan & Squire (1994) propose the following relation for G:

$$G(\xi,\varsigma\,;x,z) = \frac{1}{4\pi} \ln\left[(\xi-x)^2 + (\varsigma-z)^2\right] - \frac{1}{4\pi} \ln\left[(\xi-x)^2 + (\varsigma+z)^2\right]$$

$$- \frac{1}{2\pi} \int_{-\infty}^{\infty} \frac{1}{|\omega|-k} e^{-|\omega|(\varsigma+z)} e^{i\omega(\xi-x)} \, d\omega \,.$$

7.5 Surface Gravity Wave Interaction with Elastic Ice Floes

On the surface ($z = \zeta = 0$) this relation can be transformed as follows:

$$G(\xi, 0 ; x, 0) = -\frac{1}{\pi}\left(\text{Ci}(kr)\cos(kr) + \sin(kr)\left(\text{Si}(kr) + \frac{\pi}{2}\right)\right).$$

After spatial discretization, (7.78) can be written in the matrix form:

$$M_{ij}\phi_i + N_{ij}\frac{\partial \phi_i}{\partial z} = B_i . \tag{7.80}$$

The form of the matrices M and N in this equation depends on the numerical integration formulas chosen.

The finite-element formulation for the ice-plate equation (7.73) can be written as follows:

$$\int_{S(x)}\left[W_i\frac{\partial^5 \phi'}{\partial x^4 \partial z} + \alpha W_i\frac{\partial \phi'}{\partial z}\right] dS(x) - \beta \int_{S(x)} W_i \phi' \, dS = 0 ,$$

$$\alpha = \frac{\omega^2}{\mu^2} - \frac{\rho g}{\rho' h \mu^2}$$

$$\phi'(x) = \varphi_1(x)\phi_1 + \varphi_2(x)\phi_2 + \ldots$$

where $\varphi_i(x)$ are the basic functions; and W_i are the weighting functions, taken equal to the basic functions in accordance with the Bubnov–Galerkin method.

Evaluating the double integration of the first integral by parts, the following relation is obtained:

$$\int_{S(x)}\left[\frac{\partial^2 W_i}{\partial x^2}\frac{\partial^2 \varphi_j}{\partial x^2} + \alpha W_i \varphi_j\right]\frac{\partial \phi_i}{\partial z} dS(x) - \beta \int_{S(x)} W_i \varphi_j \phi_i \, dS(x)$$

$$+ \left[W_i\frac{\partial^4 \phi'}{\partial x^3 \partial z} - \frac{\partial W_i}{\partial x}\frac{\partial^3 \phi'}{\partial x^2 \partial z}\right]_{x=0,L} = 0 . \tag{7.81}$$

Taking into account the boundary conditions at the ends of the plate (7.75), the last term on the right side of (7.81) is eliminated and then (7.81) is rewritten in the form:

$$M'_{ij}\phi_i + N'_{ij}\frac{\partial \phi_i}{\partial z} = 0$$

$$M'_{ij} = \int_{S(x)}\left[\frac{\partial^2 W_i}{\partial x^2}\frac{\partial^2 \varphi_j}{\partial x^2} + \alpha W_i \varphi_j\right] dS(x) , \quad N'_{ij} = \beta \int_{S(x)} W_i \varphi_j \, dS(x) . \tag{7.82}$$

Using the finite element method it is necessary to ensure the continuity not only of the potential, but also of its first derivative with respect to x at the boundaries (Ciarlet, 1978).

Thus, in order to solve the boundary problem for the Laplace equation (7.72)–(7.76), simultaneous solution of the system (7.80), (7.82) is required. To test the numerical algorithm, the calculation results are compared with the analytical solution. The deviation of the numerical results from the analytical ones obtained by solving the system (7.78) does not exceed 10^{-6}.

Numerical simulation results of the interaction between waves and ice floe. In calculations the following values of the physical parameters are used: the Young modulus $E = 6 \times 10^9$ Pa, the Poisson ratio for the ice $n = 0.3$, sea-water and ice densities of 1025 and 922.5 kg m^{-3}, respectively.

The dependence of the absolute value of the reflection coefficient $|R|$ on the ice-plate length L for different wavelengths (10, 50 and 200 m) is shown in Fig. 7.5. Each curve has an approximately identical structure formed by a succession of upturned segments with $|R| = 0$ at both ends. This means that for these system parameters, the wave propagates under the plate without

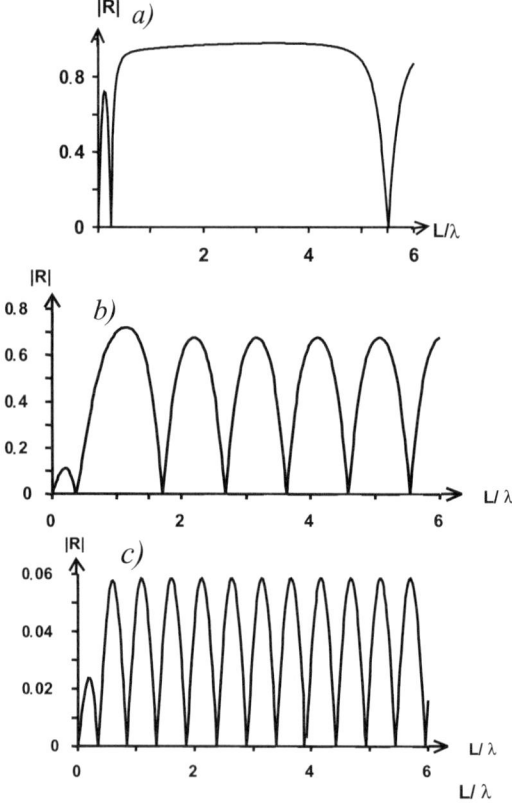

Fig. 7.5. Reflection coefficient $|R|$ vs ice-field length L for ice thickness $h = 1$ m and wavelengths $\lambda = 10$ m (**a**), 50 m (**b**), 200 m (**c**)

any reflection. According to Meylan & Squire (1994), these zeros are called resonance points.

If $L/\lambda \gg 1$, the distance between the resonance points tends to half a wavelength beneath the ice. In the case of small plate lengths the structure of the curve is disturbed. This may be attributed to fringe effects on the penetration of a wave under ice cover. As shown by calculation, these effects become noticeable at plate lengths less than the value $c_1 \lambda'$ (where λ' is the wavelength beneath the ice, $c_1 = 0.80$–0.85 is an empirical coefficient). The absolute value of the reflection coefficient is calculated to reach a maximum for the following relationship between the wave and plate lengths: $L \approx c_2 \lambda_2$ ($c_2 = 0.55$–0.60).

The velocity potential beneath the ice plate as a function of the dimensionless coordinate x/L (at $z = 0$) is shown in Fig. 7.6 for a plate length of about 366 m, which is one of the resonance lengths. It should be noted that the amplitude of the velocity potential at the edge region exceeds that in the middle of the plate by 10–12 per cent. This value is varied with the wave-plate length ratio and reaches 20 per cent in the non-resonance case. At the edge of the ice plate, the real part of the potential reaches a maximum, while the imaginary part vanishes. This indicates that at the boundary (at $z = 0$) the horizontal fluid velocity vanishes and the wave phase changes sign.

The wavelength beneath the ice plate $\lambda = 2\pi/k$ can be estimated using the following dispersion relation obtained by substituting the solution in the form $\phi(x,z) = \exp(-ikx - kz)$ in the initial equations (7.76):

$$\omega^2 \left(1 + \frac{\rho' h k}{\rho}\right) = gk + \frac{Eh^3 k^5}{12\rho(1-\nu^2)}. \tag{7.83}$$

The wavelength beneath the ice cover λ' versus the ice thickness, calculated using (7.83) for wavelengths $\lambda = 50$, 100 and 200 m in open water, is shown in Fig. 7.7. Thus, the waves penetrating under the ice cover are longer than those in the open water; the shorter waves are transformed more strongly.

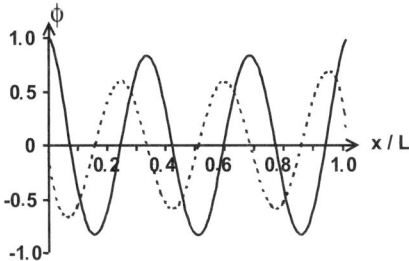

Fig. 7.6. Velocity potential ϕ beneath the ice plate vs the dimensionless coordinate x/L ($L = 366$ m). Wavelength $\lambda = 100$ m, ice thickness $h = 1$ m. The *Continuous* and *dotted curves* represent the real and imaginary potential parts, respectively

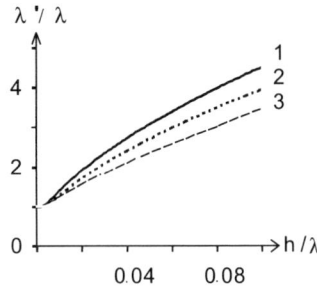

Fig. 7.7. Wavelength λ'/λ beneath ice cover vs ice thickness h/λ for open-water wavelength $\lambda = 50, 100, 200$ m (*curves* 1–3)

As the ice thickness is increased, the reflection coefficient grows. The presence of the ice thickness as a parameter in the dispersion equation (7.83) indicates that the wavelength becomes greater beneath thicker ice. The distance between resonance points is also increased.

In order to solve the problem of wave–ice interaction with the internal stress in the ice cover taken into account, the ice-plate equation should be modified as follows:

$$\rho' h \mu^2 \frac{\partial^5 \phi}{\partial x^4 \partial z} + hP \frac{\partial^3 \phi}{\partial x^2 \partial z} + \left(\rho' h \omega^2 - \rho g\right) \frac{\partial \phi}{\partial z} - \omega^2 \rho \phi = 0 \qquad (7.84)$$

where P is the internal stress in the ice cover.

Internal compression in the ice cover leads to reduction in gravity wave reflection by the plate and, consequently, internal tension leads to an increase in absolute value of the reflection coefficient (see Fig. 7.8).

The wavelength beneath the ice is also dependent on the internal ice stress. The wavelengths λ' calculated for the ice tension $P = -10^6$ N m^{-2} and the ice compression $P = 10^6$ N m^{-2}, are equal to 138.2 m and 114.2, respectively. Thus, the distance between the resonance points is increased with internal tension and decreased with compression.

The absolute value of the reflection coefficient is dependent on the wavelength not only qualitatively, but also quantitatively. This is most pronounced in the case of transformation of the frequency spectrum under the influence of a floating ice floe. Whereas for a wavelength of 10 m, the reflection coefficient reaches almost unity, it hardly exceeds 0.06 for a wavelength of 200 m (i.e. reflection by the ice plate is negligible at large wavelengths). It may be concluded from this fact that the high-frequency component of the wave spectrum is more strongly reflected by the ice floe, but the presence of resonance points can lead to part of the spectrum penetrating under the ice plate almost unimpeded.

A change in the Pierson–Moskowitz spectrum of waves, penetrating under an ice plate 110 m long, is shown in Fig. 7.9a. It is seen that the spectrum acquires a second maximum at the frequency 0.23 Hz.

7.5 Surface Gravity Wave Interaction with Elastic Ice Floes

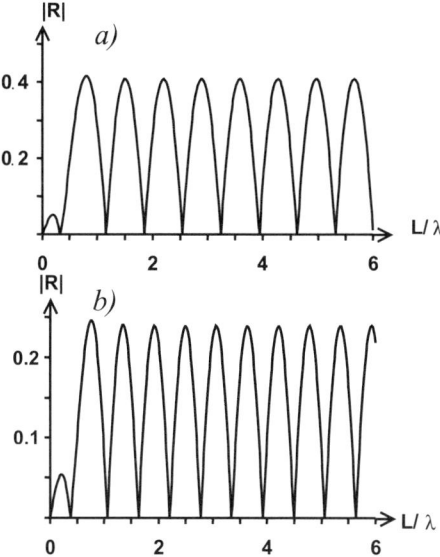

Fig. 7.8. Reflection coefficient $|R|$ vs ice-field length L for wavelength $\lambda = 10$ m and ice thickness $h = 1$ m: in the presence of ice tension with internal stress $P = -10^6 \,\mathrm{N\,m^{-2}}$ (**a**); in the presence of ice compression with internal stress $P = 10^6 \,\mathrm{N\,m^{-2}}$ (**b**)

A more detailed picture of the high-frequency part of the spectrum (see Fig. 7.9b) also makes it possible to isolate a third maximum at the frequency 0.45 Hz. This frequency corresponds to a wavelength slightly greater than 7 m, i.e., in some cases even such short waves can penetrate under the ice almost non-reflected.

Numerical simulation results of wave interaction with several ice floes. The aforementioned method of solving the problem of gravity wave interaction with an ice floe can be generalized for the case of several plates with edge coordinates x_j, L_j (j is the number of the plate):

$$\phi(x,0) = \mathrm{e}^{-ikx} + \sum_j \int_{x_j}^{L_j} [G(\xi,0\,;\varsigma,0) + \mathrm{i}\cos(kr)] \left[k\phi(\xi,0) + \frac{\partial \phi(\xi,0)}{\partial z} \right] \mathrm{d}\xi \,. \tag{7.85}$$

The calculation results are shown for two, three and four ice plates in Fig. 7.10. It is interesting to note that a system of several ice floes of finite length may display zero reflection even in the case of each plate being not absolutely transparent to waves. This situation, however, occurs only when all the plates are of the same length. For arbitrary plate lengths, the reflection

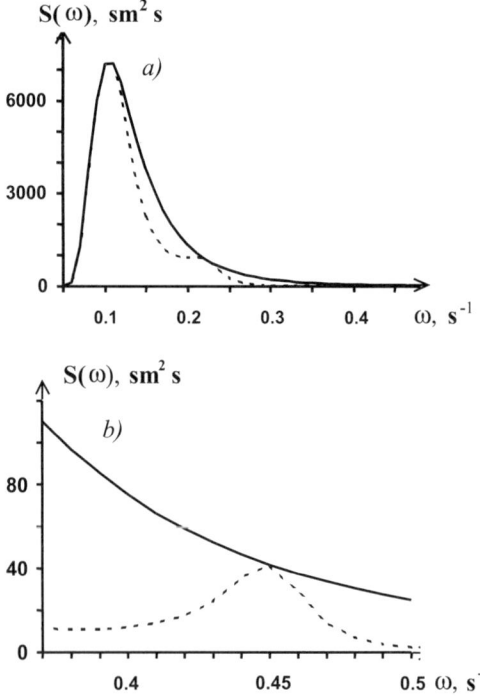

Fig. 7.9. Pierson–Moskowitz spectrum of waves before and after (*dotted line*) penetrating beneath an ice plate with $L = 110$ m and $h = 1$ m (**a**). A more detailed representation of the high-frequency part of the spectrum is shown in (**b**)

coefficient of the system does not reach zero, as illustrated in Fig. 7.10a, where the dependence of the reflection coefficient on the inter-plate distance is shown for two-plate systems. It is seen that this dependence can degenerate into an almost straight line (for example, for plates 100 and 175 m long).

The general structure of the curves is also represented as a segmented structure. The distance between the resonance points corresponds to half a wavelength in open water. This value is the same for three or four plates. Additional maxima appear between the main maximum: one for three plates (see Fig. 7.10b) and two for four plates (see Fig. 7.10c). The maximum value of the reflection coefficient grows with increasing the number of plates.

Summary. When a gravity wave penetrates under ice plates, it is possible to have a combination of system parameters for which there is no reflection at all. The reflection coefficient is strongly dependent on the wavelength. This may vary from practically total reflection to a few percent at wavelengths of 200 m and more. The maximum reflection coefficient corresponds to the ice plate length $L \approx c_2 \lambda'$ ($c_2 = 0.55$–0.60). As the ice thickness grows, the

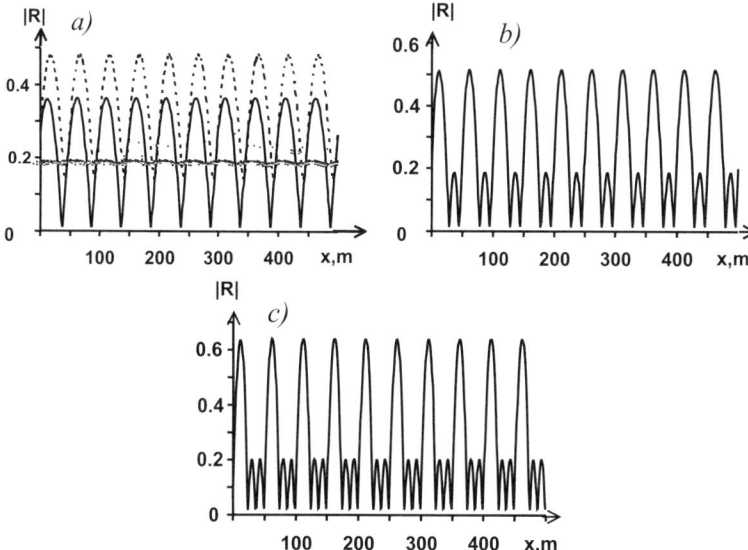

Fig. 7.10. Reflection coefficient $|R|$ vs distance between two (**a**), three (**b**) and four (**c**) ice plates for wavelength $\lambda = 100$ m and ice thickness $h = 1$ m. a: *continuous curve* – two plates 100 m in length, *broken curve* – plates 100 and 150 m in length, *dotted line* – plates 100 and 175 m in length

maximum reflection coefficient and the distance between resonance points are increased. The maximum reflection coefficient and the distance between resonance points are decreased in the presence of internal compression in ice cover and, consequently, increased in the case of the development of internal tension.

The reflectance of a system of several ice plates may be less than that of each plate separately. It may reach zero for a system of plates with identical length. The dependence of the absolute value of the reflection coefficient on the inter-plate distance has a period equal to a half-length of the wave in open water.

There is a filtering effect of the ice floes on the surface wave spectrum: high-frequency components are reflected by the ice floe much more strongly than low-frequency components. But due to the presence of resonance points, local maxima may appear in the spectrum, including its high-frequency range.

8 Conclusion

Wind wave investigations in the World's oceans are carried out by scientists in many countries. Now it is a subject of international research. This is manifested by a great number of papers and monographs (Komen et al., 1994; Problems of Research and Mathematical Modeling of Wind Sea. 1995; Massel, 1996; Young, 1999), published in the last few years and devoted to the study of wind waves and numerical simulation. The present monograph is a logical continuation of those publications. The main attention is paid to investigation of the effect of spatial inhomogeneous media on wind waves, including the effect of the Earth's surface sphericity, the presence of non-uniform currents, uneven bottom, ice cover, etc, which affect wind wave spectral structure formation in the World's oceans.

The principal conclusions are as follows.

1. In order to investigate the influence of the inhomogeneity of the World's oceans on wind waves the general problem formulation is stated based on principal physical notions. The problem is reduced to solving the kinetic equation on a spherical surface, considering the main wave-forming factors: wave energy input, non-linear energy transfer in the wave spectrum, energy dissipation due to wave breaking and bottom friction, wave transformation in non-uniform currents, shallow water and ice cover.

Taking into account the solution difficulties, an approach based on the analysis of spatial and temporal variability scales of the natural phenomenon, is suggested. The scale estimation of different physical mechanisms forming the wind wave spectral structure simplifies the solution and reveals the most effective mechanisms for every separate case. It should be noted that at the same time the geometric aspect of the solution is analysed (this means a description of wave packet propagation in phase space) and described not only by the right, but also by the left-hand side of the kinetic balance equation.

2. Great attention is paid in the monograph to the description of wave evolution in a non-uniform current. Both wind wave generation and wave transformation are affected by currents. The latter not only cause essential spectrum and wave parameter variations (the wave height can sometimes be changed in a countercurrent), non-uniform currents can result in generating freak waves in the ocean, whose catastrophic effects on oceanic vessels are notoriously known.

Using the geometrical optics approximation method, combined with the wave action density balance equation, the most adequate description of wave behaviour is obtained. This approach excludes solution singularities, appearing around caustics. It gives the possibility of obtaining detailed results of wave parameter transformation in a non-uniform current with different ways of determining the problem parameters. The combined effect of non-uniform current and bottom unevenness is shown to cause such wave field variations, which are impossible to describe by simple superposition of current and depth effects taken separately.

A drawback of the method is that the traditional spectral approach does not allow the description of the wave shape, because the wave phase is not taken into consideration. But the phase data are very important, for example, for the explanation of a wave form in the ocean. Also, some singularities of the solution can arise with the frequency angular spectrum becoming very narrow, i.e. for monochromatic waves. That is why the asymptotic estimations obtained in the diffraction approximation are given simultaneously with the solution of the spectral problem.

An attempt is undertaken in this monograph to describe the principal physical mechanisms forming the wave spectrum in sea with marginal ice. There are the following mechanisms: wave energy scattering, wave refraction, wave energy dissipation due to friction between ice floes and weak non-linear transfer in the wave spectrum. These results are supposed to be the basis for elaborating a mathematical model of wave spectrum evolution in seas covered with marginal ice. In particular, it can be used for numerical modelling of natural phenomena such as "ice storms", which can be generated by storm sea waves in the marginal ice zone. This is an extremely dangerous phenomenon for oil platforms and other hydraulic engineering structures in Arctic seas.

3. With the exception of the aforesaid problems, some physical mechanisms forming the wave spectrum in deep water are investigated. Thus, one of the most effective algorithms for calculating the collision integral describing non-linear interaction in the wind wave spectrum is proposed. The reliable calculation results are obtained with the help of numerical integration methods of the highest accuracy. In particular, some fine effects, such as non-linear energy transfer to spectral components propagating against the wind are described, explaining the physics of the generation of these waves. The non-linear evolution of the wave spectrum is estimated for large time periods and a self-similar solution is found.

At the same time some important questions still remain unsolved, in particular, the investigation of wind wave energy dissipation. The absence of a well-grounded theory of wave dissipation allows various groundless hypotheses concerning this mechanism. The investigation shows a principally different character of dissipation mechanisms within different frequency spectral ranges. Thus, as for the low-frequency ranges, the dissipation is caused by interaction of waves and turbulence. As for the high-frequency ranges, it is connected with wave breaking, being essentially a non-linear mechanism.

8 Conclusion   349

There is another unsolved problem connected with the investigation of meso-scale effects in forming the wind wave spectral structure. In particular, it is revealed that the frequency spectrum quasi-oscillations can essentially increase the effectiveness of non-linear energy transfer in the wave spectrum. This calls into question some principal conclusions concerning the balance and quantitative values of other physical mechanisms forming the wind wave spectrum.

When solving practical problems of wave calculations in a sea, a spatial step in tens or hundreds of miles is used and the surface wind speed in a 3 and more hour temporal step is introduced. But this traditional approach does not take into account the so-called "sub-grid effects", i.e. describing small-scale details of wind wave field evolution, which can be principally important. The question of obtaining a detailed spatial and temporal description of the wind field is connected with the general problem of meteorology, depending on the state of this science on the whole, and the quality of the initial synoptic data. As for the problem of calculating the wind waves, there is a series of specific problems, for example, how to take into account meso-scale velocity variations, such as wind flows and gusts; how the stratification of boundary layer is affected; and how to parameterize the meso-scale effects in wind wave models.

The investigation of these and other problems is continuing.

Bibliography

Abreu M., A. Larraza, E. Thornton, 1992: Nonlinear transformation of directional wave spectra in shallow water, J. Geophys. Res., **97**, C**10**, 15579–15589

Abuzyarov Z.K., 1981: *Wind Sea and its Prediction* (Gidrometeoizdat, Leningrad), in Russian

Action of Large-Scale Internal Waves On The Sea Surface, 1982, in: *Papers*, Ye.N. Pelinovskiy (Ed.) (IPF Acad. Sc. USSR, Gorkiy) pp. 251, in Russian

Aleshkov Yu.Z., 1981: *Wave Theory for Heavy Liquid Surface* (Leningrad State University, Leningrad), in Russian

Aleshkov Yu.Z., 1990: *Theory of Wave and Obstacle Interaction* (Leningrad State University, Leningrad), in Russian

Andreev B.M., 1988: *Variability of Spectral Characteristics of Wave Developing. Theoretical Foundations and Calculation Methods of Wind Waves* (Gidrometeoizdat, Leningrad), 75–81, in Russian

Arnold V.I., 1989: *Mathematical Methods of Classical Mechanics* (Nauka, Moscow), in Russian

Arushanyan O.B., S.F. Zaletkin, 1990: *Numerical Solution of Ordinary Differential Equations in FORTRAN* (Moscow State University, Moscow), in Russian

Babanin A.V., Yu.P. Solovyev, 1998: Field investigation of transformation of the wind wave frequency spectrum with fetch and stage of development, J. of Phys. Ocean., **28**, **3**, 563–576

Backus G. F., 1962: The effect of the earth's rotation on the propagation of ocean waves over long distances, Deep-Sea Res. **9**, 25–32

Banner M.I., 1990: Equilibrium spectra of wind waves, J. Phys. Ocean. **20**, 966–984

Banner et al. (2002) to appear.

Barber N.F., 1961: Discussion 2. Ocean wave spectra. National Academy of Sciences, Proc. of Easton conference, May 1–4 (Maryland) pp. 186–189

Barenblatt G.I., 1984. Self-similarity and intermediate asymptotics, Hydrometeoizdat, pp. 256

Barenblatt G.I., I.A. Leykin, A.S. Kazmin. et al., 1985: Rips in the White Sea, Rep. Acad. Sc. USSR, 281, **6**, 1435–1439, in Russian

Barnett T.P., 1968: On the generation, dissipation and prediction of ocean wind waves, J. Geophys. Res., **73**, **2**, 513–529

Barnett T.P., C.H. Holland, Jr. Yager, 1969: A general technique for wind wave prediction, with application to the South China Sea, Final Report, U.S. Naval Oceanographic Office

Basovich A.Ya., V.V. Bakhanov, 1979: On transformation of surface waves in nonuniform current, Proc. Acad. Sc. USSR, FAO, **15**, **6**, 660–665, in Russian

Basovich A.Ya., V.I. Talanov, 1977: On transformation of surface waves in nonuniform current, Proc. Acad. Sc. USSR, FAO, **13**, **7**, 660–665, in Russian

Basovich A.Ya., V.V. Bakhanov, V.I. Talanov, 1982: Influence of intensive internal waves on wind sea, in *Action of Large-Scale Internal Waves on Sea Surface* (IPF, Acad. Sc. USSR, Gorkiy), pp. 8–30, in Russian

Bauer E. et al., 1992: Validation and assimilation of SEASAT altimeter wave heights using the WAM wave model, J. Geophys. Res., **97**, 12671–12682

Beji S., J.A. Battjes, 1993: Experimental investigation of wave propagation over a bar, Coastal Engineering, **19**, 151–162

Belberov Z.K. et al., 1990: Principles of making the wind-wave atlas for the Bulgarian region of the Black Sea, Theses of sc. techn. conference Problems of Coastal Protection of the Black Sea Near Bulgaria, Varna, in Russian

Belevich M.Yu., I.A. Neelov, 1998: Estimation of the wind-wave interaction parameter, Proc. Acad. Sc., FAO, **34**, **3**, 435–440, in Russian

Belevich M.Yu., I.A. Neelov, 2000: Evalution of the mutual influence of the wind-wave components on the energy interchange with the wave boundary layer, Meteorology and Hydrology, **1**, 70–78, in Russian

Benoit M., F. Marcos, F. Becq, 1996: Development of a third-generation shallow water wave model with unstructured spatial meshing, Proc. 25th Int. Conf. Coastal Engineering, ASCE (Orlando)

Berkhoff J.C.W., 1972: Computation of combined refraction–diffraction, Proc. 13th Inst. Conf. on Coastal Eng. (Vancouver) pp. 471–490

Biesel F., 1950: Etude theorigue de la houle en eau courante, Houille Blanche, **5A**, 279–285

Bitner-Gregersen E., S. Gran, 1983: Local properties of sea waves derived from a wave record. Appl. Ocea Res., **5**, **4**, 210–214

Bolshakov A.N., V.M. Burdyukov, S.A. Grodskiy, V.N. Kudryavtsev, 1988: Determining the spectrum of energy surface waves using a pattern of sunlight patch, The Earth's Examination from the Cosmos, **5**, 11–18, in Russian

Booij N.et al., 1984: A numerical model for wave boundary conditions in port design, in Inst. Conf. on Numerical and Hydraulic Modeling of Port and Harbors, BHRA/IAHR, 23–25 April (Birmingham) pp. 263–268

Booij N., L.H. Holthuijsen, 1987: Propagation of ocean waves in discrete spectral wave model, J. of Comput. Phys, **68**, 307–326

Booij N., R.C. Ris, L.H. Holthuijsen, 1999: A third-generation wave model for coastal regions, J. Geophys. Res., **87**, C8, 7667–7681

Bouws E. et al. 1985: Similarity of the wind wave spectrum in finite depth water Part 1: spectral form. J. Geophys. Res. **90**, (C1), 975–986

Bouws E. et al., 1987; Similarity of the wind wave spectrum in finite depth water Part 2: Statistical relationships between shape and growth stage parameters. Deutsch. Hydrog. Z., **40**, 1–24

Braslavskiy A.P., 1952: Wind wave calculation, in: Trans. State Hydrological Inst., **35**, pp. 94–158 (in Russian)

Brebbia C.A., J.C.F. Telles, L.C. Wrobel, 1984: *Boundary Element Techniques* (Springer, Berlin)

Bretherton F.P., 1971: The general linearized theory of wave propagation, Mathematical Problems in the Geophys. Sci., **1**, 61–102

Bretherton F.P., C.J.R. Garret, 1968: Wave trains in inhomogeneous moving media, Proc. Roy. Soc., Ser. A, **302**, 529–554

Bretschneider C.L., 1958: The generation and decay of wind waves in deep water, Trans. Amer. Geophys. Union, **33**, **3**, 381–389

Bretschneider C.L., 1973: Designing Hurricane Waves for the Island of Oahu, Hawaii, with special application to the Sand Island ocean out fall system, Vol. **3**, **2**, 112–118

Brink-Kjer O., 1984: Depth current refractions of wave spectra, Sympos. on Description and Modeling of Directional Sea. 18–20 June, C.**7**

Brink-Kjer O., I.G. Jonsson, 1975: Wave height and set-down of water on a shear current over a weakly varying bed, Inst. Hydrodyn. and Hydraulic Eng. Techn. Univ. Denmark, Prog. Rep. 37, pp. 17–24

Brovikov I.S., 1954: Statistical characteristics of wind wave parameters, in Trans. State Oceanographic Inst., **26**, 147–194, in Russian

Building Standards and Rules. Loads and Forcing Onto Hydrotechnical Constructions, 1983 (Stroyizdat, Moscow) in Russian

Bukatov A.Ye., O.M. Bukatova, 1993: Surface waves of finite amplitude in a basin with broken ice, Proc. Russian Acad. Sc., FAO, **29**, **3**, 421–425, in Russian

Bukatov A.Ye., L.V. Cherkesov, 1971: Influence of ice on wave movements, in *Marine Hydrophysical Research* (Marine Hydrophysical Institute, Sevastopol) pp. 113–114 (in Russian)

Bukatov A.E., L.V. Cherkesov, A.A. Yaroshenko, 1984: Flexural-gravity waves due to moving perturbations. Prikl. Mat. Tekhn. Fiz., **2**, 151–157

Bunting D.C., 1970: Evaluating forecasts of ocean-wave spectra, J. Geophys. Res., **21**, 4131–4143

Burgers G., 1990: A guide to the Nedwam wave model, Sci. Rep. KNMI, No. WR-90-04, De Bilt

Burgers G., V.K. Makin, 1992: Boundary layer model results for wind-sea growth, J. Phys. Ocean, **23**, 372–385

Burgers G. et al., 1992: Wave data assimilation for operational wave forecasting at the North Sea, Third International Workshop on Wave Hindcasting and Forecasting (Montreal) pp. 202–209

Catalogue of Applied Computer Programs for Shore Protection, 1989 (VNIITS, Moscow) pp. 14–16 (in Russian)

Cavaleri L., 1999: *The Ozeanographic Tower Acqua Alta: More than a Quarter of a Century of Activity*, Cartiere del Garda S.p.A., Riva del Garda (TN)

Cavaleri L., P.M. Rizzoli, 1981: Wind wave prediction in shallow water: theory and applications, J. Geophys. Res., **86**, 10961–10973

Chalikov D.V., 1986: Spectrum of energy flow to waves, Oceanology, **26**, **2**, 199–203, in Russian

Chalikov D.V., 1993: The parameterization of the wave boundary layer, J. Phys. Oceanogr., **25**, **6**, Part 1, 1333–1349

Chalikov D.V., M.Yu. Belevich, 1995: *Theory of Wave Fields. Problems of Investigation and Numerical Simulations of Wind Waves* (Gidrometeoizdat, Leningrad), in Russian

Charnock H., 1955: Wind stress on a water surface, Quart. J. Roy. Meteorol. Soc., **81**, 639–640

Christoffersen J.B., I.G. Jonsson, 1985: Bed friction in a combined current and wave motion, Ocean Engineering, **12**, 387–423

Ciarlet Ph., 1978: *The Finite Element Method for Elliptic Problems* (North-Holland, Amsterdam)

Collins I.J., 1972: Prediction of shallow-water spectra, J. Geophys. Res., **77**, **15**, 2693–2707

Crombie D.D., 1972: Resonant backscatter from the sea and its application to physical oceanography, Proc. IEEE Conf. Engineering in Ocean Environment

Crombie D.D., K. Hasselmann, W. Sell, 1978: High-frequency radar observations of sea waves travelling in opposition to the wind, Boundary-Layer Meteorol., **13**, p. 45

Darbyshire M., J. Simpson, 1967: Numerical prediction of wave spectra in the North Atlantic, Deutsch. Hydrog. Z., H.**1**, 18–22

Davidan I.N., 1969: Frequent spectrum of wind sea, in Trans. State Oceanographic Inst., **96**, 185–210, in Russian

Davidan I.N. et al., 1975: Calculation of wind sea spectral characteristics using numerical and analytical solutions for the wave energy balance equation in spectral form, Marine Hydrophysical Researches, **4**, 73–82, in Russian

Davidan I.N. et al., 1977: *Methods of Wave Spectrum Calculation. Oceanology* (VNIIGMI-MTsD, Obninsk), in Russian

Davidan I.N., I.V. Lavrenov, 1991: On the problem of energy "disbalance" in low frequency area of developed wind sea, Proc. Acad. Sc. USSR, FAO, **27**, **8**, 853–861, in Russian

Davidan I.N., L.I. Lopatukhin, 1982: *To Meet with Storms* (Gidrometeoizdat, Leningrad), in Russian

Davidan I.N., I.V. Lavrenov, T.A. Pasechnik et al., 1988b: Numerical method of calculation of wind waves at the Baltic sea, Proc. of the XYI Conf. of Baltic Oceanogr. 2–5 Sept. (Kiel)

Davidan I.N., I.V. Lavrenov, T.A. Pasechnik. et al., 1988c: Spectral parameter model of wind sea, Marine Hydrophysical Journal, **5**, 21–27, in Russian

Davidan I.N., L.I. Lopatukhin, V.A. Rozhkov, 1978: *Wind Sea as a Probabilistic Hydrodynamic Process* (Gidrometeoizdat, Leningrad), in Russian

Davidan I.N., L.I. Lopatukhin, V.A. Rozhkov, 1985: *Wind Sea in the World Ocean*(Gidrometeoizdat, Leningrad), in Russian

Dobrokhotov S.Yu., P.N. Zhevandrov, 1988a: *Asymptotic Decompositions and Maslov's Canonical Operator in the Linear Theory of Surface Gravitational Waves: Principal Equations and Constructions*, N 328 (Inst. Mechanics Problems, Acad. Sc. USSR, Moscow), in Russian

Dobrokhotov S.Yu., P.N. Zhevandrov, 1988b: *Cauchy–Poisson's Problem and Seized Waves*, N329, (Inst. Mechanics Problems, Acad. Sc. USSR, Moscow), in Russian

Dobson F.W., 1971: Measurements of atmospheric pressure on wind-generated sea waves, J. Fluid Mech., **48**, 91–127

Donelan M.S., J. Hamilton, W.H. Hui, 1985: Directional spectra of wind-generated waves, Phil. Tras. R. Soc., A **315**, 509–562

Dreyzis Yu.I., I.G. Kantargi, Ye.N. Pelinovskiy, 1986: Wave filtration by displacement current in shallow water, Oceanology, **XXVI**, 6, 907–913, in Russian

Dubrovin B.A., S.N. Novikov, A.T. Fomenko, 1986: *Modern Geometry. Methods and Applications* (Nauka, Moscow), in Russian

Dulov V., V.N. Kudryavtsev, 1989: Current non-uniformity reflection in oceanic state, Marine Hydrophysical Journal, **4**, 3–13, in Russian

Dulov V.A., V.N. Kudryavtsev, 1992: Wave breaking and non-uniformities of atmosphere and ocean. *Nonlinear Water Waves*. Ed. D.H. Peregrine. Bristol (UK) Univ. Bristol, p. 75–82.

Dulov V.A., S.A. Grodsky, V.N. Kudruavtsev, 1996: Effect of ocean non-uniformities on wave breaking. *The Air–Sea Interface, Radio and Acoustc Sensing, Turbulence and Wave Dynamics.* Eds. M.A. Donelan, W.H. Hui, W.J. Plant. Miami (US); RSMAS, Univ. Miami, p. 283–288

Dulov V.A., V.N. Kudruavtsev, O.G. Sherbak, S.A. Grodsky 1998: Observation of wave breaking in the Gulf Stream frontal zone, Global Atmosphere and Ocean System, **6**, 209–242

Dungey J.C., W.H. Hui, 1985: Nonlinear energy transfer in a narrow gravity-wave spectrum, Proc. R. Soc., A **368** (London) pp. 239–265

Dzhenyuk S.L., 1976: Numerical calculation of waves, using wind fields, Trans. Hydrometorol. Centre USSR, **164**, 3–10, in Russian

Eldeberky Y., J.A. Battjes, 1995: Parametrization of trial interaction in wave energy models, Proc. Coastal Dynamics Conf. (Gdansk) pp.140–148

Ewing J.A., 1971: A numerical wave prediction method for the North Atlantic Ocean, Deutsch. Hydrogr. Zeitschrift, **24**, 241–261

Fons C., 1966: Prevision de la houle par methode des densities spectroanqularies No. 5, Cahier Oceanogr., **18**, **1**, 15–33

Fox M.J.H., 1976: On the nonlinear transfer of energy in the peak of the gravity-wave spectrum, Proc. Roy. Soc., **A348**, 467–483

Fox M.J.H., V.A. Squire, 1991: Strain in shore fast ice due to incoming ocean waves and swell. J. Geophys. Res., **96**, **C3**, 4531–4547

Francis J.R.D., C.R. Dudgeon, 1967: An experimental study of wind-generated waves on a shear current, Quart. J. Roy. Meteorol. Soc., **93**, 247–253

Gelci R., E. Devillaz, 1975: Le calcul numerique de l'etat de la mer, Meteorologie, **2**, 157–180

Gelci R., E. Devillaz, 1969: Le calcul numerique de l'etat de la mer, Notes de l'etablissment d'etat de recheres meteorologiques, **268**, 1–74

Gelci R., P. Chavy, E. Devillaz, 1963: Traitment numerique de l'etat de la mer, Cahier Oceanogr., **15**, **3**, 158–160

Gent P.R., P.A. Taylor, 1976: A numerical model of the air flow above water waves, J. Fluid Mech., **77**, 105–128

Glukhovskiy B.Kh., 1966: *Research on Wind Sea* (Gidrometeoizdat, Leningrad), in Russian

Golding B.W., 1983: A wave prediction system for real time sea state forecasting, Quart. J. Roy. Meteor. Soc., **109**, 393–416

Goncharov V.V., I.A. Leykin, 1983: Waves on current with velocity shift, Oceanology, **23**, **2**, 210–216, in Russian

Graber H.C., O.S. Madsen, 1988: A finite depth wind wave model, J. Physical Oceanogr., **18**, **11**, 1465–1483

Grodskiy S.A., V.A. Dulov, V.N. Kudryavtsev, O.V. Shulgina, 1989: Experimental research of wind wave evolution on non-uniform currents, Marine Hydrophysical Journal, **5**, 36–47, in Russian

Grundligh M., M.Rossouw, 1995: Wave attenuation in the Agulhas Current, Research Letters of South African Journal of Science, **91**, 357–359

Grushin V.A., Yu.L. Il'in, A.L. Lazarev., et. al., 1986: Synchronous optical and contact research of spatial-spectral characteristics of wind sea, Research of the Earth from the Cosmos, **2**, 57–67, in Russian

Guide to Wave Analysis and Forecasting, 1988, **702** (WMO)

Gunther H., W. Rosenthal, 1983: Self similarity of surface wave spectra in water of finite depth, in *Proc. Sixth Australian Conf. on Coast. and Ocean Eng.*, pp. 264–272

Gunter H., K. Hasselmann, J.A. Ewing, 1979: A hybrid parametrical wave prediction model, J. Geophys. Res., **89**, **9**, 5727–5738

Gutshabash Ye.Sh., I.V. Lavrenov, 1986: Swell transformation on current at the Igolny Cape, Proc. Acad. Sc. USSR, FAO, **22**, **6**, 526–531, in Russian

Gutshabash Ye.Sh., I.V. Lavrenov, 1987: Spectral model of wind sea transformation on horizontal non-uniform current with changing basin depth, in *Theses of papers of the III meeting of Soviet oceanologists. Section of physics and ecology of ocean.* (Gidrometeoizdat, Leningrad) pp. 79–80, in Russian

Gutshabash Ye.Sh., I.V. Lavrenov, 1988: Spectral model of wind sea transformation in coastal zone with horizontal non-uniform currents, Marine Hydrophysical Journal, **5**, 15–21, in Russian

Hasselmann D.E., M. Dunckel, J.A. Ewing, 1980: Directional wave spectra observed during JONSWAP 1973, J. Phys. Oceanogr.,**10**, 1264–1280

Hasselmann K., 1960: Grundgleichungen der Seegangsvoraussage, Schiffstechnik **7**, **39**, 191–195

Hasselmann K., 1962: On the nonlinear energy transfer in a gravity wave spectrum. Part 1, J. Fluid Mech., **12**, 481–500

Hasselmann K., 1963: On the nonlinear energy transfer in a gravity wave spectrum. Part 2, J. Fluid Mech., **15**, 273–281

Hasselmann K., 1965: On the nonlinear energy transfer in a gravity-wave spectrum. Conservation theorem, wave particle correspondence, irreversibility, J. Fluid Mech., **15**, 273–281

Hasselmann K., 1966: Feymann diagrams and interaction rules for wave–wave scattering, Rev. Geophys., **4**, 1–32

Hasselmann K., 1974: On the spectral dissipation of ocean waves due to white capping, Bondary Layer Met., **6**, **1–2**, 107–127

Hasselmann K., 1979: Description of non-linear interaction by methods of theoretical physics (with application to wave generation by wind), in: *Non-linear Theory of Wave Propagation* (Nauka, Moscow) pp. 106–136, in Russian

Hasselmann K. et al., 1973: *Measurements of Wind-Wave Growth and Swell Decay During the Joint North Sea Wave Project (JONSWAP)*(Deutsch.Hydrogr. Inst., Hamburg)

Hasselmann K. et al. 1976: A parametric wave prediction model, J. Phys. Oceanogr., **6(2)**, pp. 200–228

Hasselmann K., J.I. Collins, 1968: Spectral dissipation of finite-depth gravity waves due to turbulent bottom friction, J. Mar. Res., **26**, 1–12

Hasselmann K., S. Hasselmann, L.R. Jung, 1987: Computation of the response of a wind spectrum to a sudden change in wind direction, J. Phys. Oceanogr., **17**, 1317–1338

Hasselmann S., K. Hasselmann, 1981: A symmetrical method of computing the nonlinear transfer in a gravity wave spectrum, in *Hamburger Geophys. Einzelschriften* (Hamburg) pp. 52–172

Hasselmann S., K. Hasselmann, T.P. Barnett, 1985: Computation and parameterization of the nonlinear energy transfer in gravity wave spectrum, J. Phys. Oceanogr., **15**, pp. 1378–1391

Hasselman S., K. Hasselmann, J.H. Allender, T.P.Barnett, 1985: Computation and parameterization of the nonlinear energy transfer in a gravity wave spectrum, J. Phys. Oceanogr., **15**, 1378–1391

Hayes J.G., 1980a: Ocean current interaction study, J. Geophys. Res., **85**, C9, 152–157

Hayes J.G., 1980b: Ocean current interaction, Study. J. Physical Oceanography. **97**, 215–224

Herterich K., K. Hasselmann, 1980: A similarity relation for the nonlinear energy transfer in a finite-depth gravity wave spectrum, J. Fluid Mech., **97**, 215–224

Hidy G.M., E.J. Plate, 1966: Wind action on water standing in a laboratory channel, J. Fluid Mech., **26**, 4, 651–687

Holthuijsen L.H., N. Booij, T.H.C. Herbers, 1989: A prediction model for stationary short-crested waves in shallow water with ambient currents, Coastal Engineering, **13**, 23–54

Hsiao S.V., O.H. Shemdin, 1980: Interaction of ocean waves with a soft bottom, J. Phys. Oceanogr., **10**, 605–610

Huang N.E. et al., 1986: A study of the relationship among wind speed, sea state and drag coefficient for a developing wave field, J. Geophys. Res., **91**, 7735–7742

Huang N.E., P.T. Chen, C.C. Tung, J.R. Smith, 1972: Interaction between nonuniform current and gravity waves with application for current measurements, J. Phys. Oceanogr., **2**, 420–431

Hurghes B., H.L. Grant, 1978: The effect of internal waves on surface wind waves. 1. Experimental measurements and 2. Theoretical analysis, J. Geophys. Res., **83**, C1, 443–465

Hydrometeorological Conditions for the Shelf Zone of the Seas in the USSR. The Barents Sea, 1985, **6**. (Gidrometeoizdat, Leningrad), in Russian

Inoue T., 1966: On the growth of spectrum of wind generated sea according to a modified Miles–Phillips mechanism and its application to wave forecasting, Geophys. Sci. Lab. Rep., TRG7-5

Isozaki I., T. Uji, 1973: Numerical prediction of ocean wind waves, Jap. Met. Geophys., **24**, 2, 207–231

Isozaki I., T. Uji, 1974: Numerical model of marine surface wind and its application to the prediction of ocean wind waves, Pap. Met. Geophys., **25**, 3, 197–239

Ivanenkov G.V., G.V. Matushevskiy, G.V. Rzheplinskiy, 1977: Effect of wave generation by a moving atmospheric cold front, Proc. Acad. Sc. USSR, FAO, **13**, 1, 80-87, in Russian

Janssen P.A.E.M., G.J. Komen, W.J.P de Voogt, 1984: An operational coupled hybrid wave prediction model, J. Geophys. Res., **89**, C3, 3635–3654

Jeffreys H., 1971: On the formation of water by wind, Proc. Roy. Soc., **9**, 1, 1–11

Jolm N., 1979: Hunt for direct solution of wave dispersion equation, J. of Waterway Art Coast and Ocean Division, **105**, **WW4**, 112–116

Kajiura K., 1968: A model of bottom boundary layer in water waves, Bull. Earthquake Res. Inst., **46**, 75–123

Kamenkovich V.M., A.S. Monin, 1978: Small oscillations in oceans, in *Physics of the Ocean. Vol. 2 Hydrodynamics of Oceans* (Nauka, Leningrad) pp. 5–48, in Russian

Kantargi I.G, 1995: Effect of depth current profile on wave parameters, Coastal Engineering **26**, 195–206

Kantargi I.G., I.L Makarova, Ye.N. Pelinovskiy, 1989: Wave transformation by current with linear shift of velocity in depth, Oceanology, **29**, 2, 198–204, in Russian

Kantargi I.G., Ye.I. Mass, K.I. Shevchenko, 1990: Experimental research of conditions for sediment drift in mixed flows, Water Resources, **3**, 54–62, in Russian

Kantargi I.G., N.Sh. Tsivtsivadze, Kh.L. Akmuratov, 1984: *Hydraulics of Wind Waves in Channels* (Tbilisi University, Tbilisi), in Russian

Kato H., S. Sato, 1978: Experimental study of wind waves generated on currents, Proc. 16th Coast. Eng. Conf. Hamburg, **1**, **4**, 742–755

Kato H., H. Tsuruya, 1978: Experimental study of wind waves generated on currents, Proc. 16th Coast. Eng. Conf. ASCE, **1**, 742–755

Kato H., H. Tsuruya, H. Terakawa, 1981: Experimental study of wind waves generated on water currents. Wave forecasting methods and its experimental confirmation, Rep. Port. Harbor Res. Inst., **20**, **3**, 94–129

Kato H., H. Tsuruya, T. Doi, Y. Mijarari, 1976: Experimental study of wind waves generated on water currents, Rep. Port. Harbor Res. Inst, **15**, **4**, 3–46

Kats A.V., I.S. Spevak, 1980: Reconstruction of wind sea spectra using moving meter's measurements, Proc. Acad. Sc. USSR, FAO, **16**, **3**, 294–304, in Russian

Kelvin (W. Thomson), 1871: On stationary waves in flowing water, Phil. Mag., Ser. 4., Vol. **42**, 362

Kenyon K., 1981: Wave refraction in ocean currents, Deep Sea Res., **18**, 1023–1034

Khandekar M.L., 1989: *Operational Anlysis and Prediction of Ocean Wind Waves. Coastal and Estuarine Studies* (Springer-Verlag)

Kheysin D.Ye., 1967: *Ice Cover Dynamics* (Gidrometeoizdat, Leningrad), in Russian

Kirby T.J., T. Chen, 1989: Surface waves on vertically sheared flows: approximate dispersion relations, J. Geophys. Res., **94** (C1), 1013–1027

Kitaigorodskii S.A., V.P. Krasitskii, M.M. Zaslavskii, 1975: On the Phillips theory of equilibrium range in the spectra of wind-generated gravity waves, J. Phys. Oceanogr., **5**, 410–420

Komatsu K., A. Masuda, 1996: A new scheme of nonlinear energy transfer among wind waves: RIAM method – algorithm and perfomance. J. Oceanology., **52**, 509–537

Komen G.J., S. Hasselmann, K. Hasselmann, 1984: On the existence of a fully developed wind-sea spectrum, J. Phys. Oceanogr.,14, 1271–1285

Komen G.J., L. Cavaleri, M. Donelan, K. Hasselmann, S. Hasselmann., P.A.E.M. Janssen, 1994: *Dynamics and Modelling of Ocean Waves* (University Press, Cambridge)

Kononkova G.Ye., K.V. Pokazeyev, 1982: *Spectral Characteristics of Wind Waves for Small Fetches and Currents*. Proceedings of Conferences and Meetings in Hydrotechnics. Methods of Research and Calculations of Wave Actions on Hydrotechnical Constructions and Shore (VNIIG, Leningrad), in Russian

Kononkova G.Ye., K.V. Pokazeyev, 1985: *Sea Wave Dynamics* (Moscow State University, Moscow), in Russian

Kononkova G.Ye., L.V. Poborchaya, K.V. Pokazeyev, 1977: Laboratory examination of wind wave generation on a wake, Proc. Acad. Sc. USSR, FAO, **13**, **9**, 991–993, in Russian

Korobov V.B., I.V. Lavrenov, 1989: Estimations of tide current effluence on the regime-climatic functions in the range of small value probabilities, Metereology and Hydrology, **10**, 73–78

Kostichkova D.P., V.P. Krasitskii, Zh.I. Cherneva, 1990: Methods of recalculating wind wave spectra in coastal zone into deep-water ones, Oceanology, **2**, 211–216 in Russian

Krasil'nikov V.N., 1960. The effect of a thin elastic layer on sound propagation in a fluid half-space. Akust. Zhurn., **6**, No.2, 220–228

Krasitskii V.P., 1974: On theory of wave spectrum transformation with wind wave refraction, Proc. Acad. Sc. USSR, FAO, **1**, 72–82, in Russian

Krasitskii V.P., V.A. Kalmykov, 1993: On four-wave reduced equations for surface gravity waves, Proc. Russian Acad. Sc., FAO, **29**, **2**, 237–243, in Russian

Kravtsov Yu.A., 1968: On two new asymptotic methods in the theory of wave propagation in heterogeneous media, Acoustic Journal, **14**, 1, 1–11, in Russian

Kravtsov Yu.L., Yu.I. Orlov, 1980: *Geometrical Optics of Heterogeneous Media* (Nauka, Moscow), in Russian

Krivoshey M.I., 1976: Wind wave transformation on constant and pulsing current, Trans. State Hydrological Inst., **231**, 144–156, in Russian

Krylov Yu.M., 1948: Long wave propagation under ice field, Trans. State Oceanographic Inst., **8**(20), 107–111, in Russian

Krylov Yu.M., 1956: Statistical theory and wind sea calculation. Parts 1,2, Trans. State Oceanographic Inst., **33**, 5–79; **42**, 3–88, in Russian

Krylov Yu.M., 1966: *Spectral Methods of Wind Wave Research* (Gidrometeoizdat, Leningrad), in Russian

Krylov V.I., L.T. Shulgina, 1966: *Reference Book of Numerical Integration.* (Nauka, Moscow), in Russian

Krylov Yu.M., S.S. Strekalov, V.F. Tsyplukhin, 1976: *Wind Waves and their Force onto Constructions* (Gidrometeoizdat, Leningrad,), in Russian

Kudryavtsev V.N., S.A.Grodsky, V.A. Dulov, A.N. Bolshakov, 1995: Observation of wind waves in the Gulf Stream frontal zone. J. Geophys. Res. **100**, NC10, pp. 20,715–20,728

Labzovskiy N.A., 1973: Calculation of wave elements for flow surface, in *Hydrophysical Research of Lakes* (Nauka, Leningrad), in Russian

Lamb G., 1947: *Hydromechanics* (Gostekhizdat, Moscow), in Russian

Landau L.D., Ye.M. Lifshits, 1973: *Mechanics. Theoretical Physics.* Vol. 1 (Nauka, Moscow), in Russian

Landau L.D., Ye.M. Lifshits, 1974a: *Quantum Mechanics. Theoretical Physics.* Vol. 5 (Nauka, Moscow), in Russian

Landau L.D., Ye.M. Lifshits, 1974b: *Field Theory. Theoretical Physics.* Vol. 2. (Nauka, Moscow), in Russian

Landau L.D., Ye.M. Lifshits, 1979: *Physical Kinetics. Theoretical Physics.* Vol. 10. (Nauka, Moscow), in Russian

Lantsosh K., 1965: *Variational Principles of Mechanics* (Mir, Moscow), in Russian

Lavrenov I.V., 1980: On seismic presages of tsunami waves, Oceanology, **20**, 3, 373–380, in Russian

Lavrenov I.V., 1984: Reflection of long dispersive waves from submerged vertical obstacles, in *Probabilistic Analysis and Modelling of Oceanological Processes* (Gidrimeteoizdat, Leningrad) pp. 123–129, in Russian

Lavrenov I.V., 1985a: Meeting a "wave-killer", Marine Fleet, **12**, 28–30, in Russian

Lavrenov I.V., 1985b: Wave spectrum transformation in shallow water and on largescale currents, in *Wind Sea in the World Ocean* (Gidrometeoizdat, Leningrad) pp. 95–104, in Russian

Lavrenov I.V., 1986: On spectral behaviour of surface gravitational waves on horizontal non-uniform current, Proc. Acad. Sc. USSR, FAO, **22**, 5, 525–531, in Russian

Lavrenov I.V., 1987: On rip theory, Proc. Acad. Sc. USSR, FAO, **23**, **10**, 1060–1071, in Russian

Lavrenov I.V., 1988a: Calculation of wave elements along fetch on a current, Shipbuilding, **12**, 21–24, in Russian

Lavrenov I.V., 1988b: Wave transformation on non-uniform current and with changing a basin depth, in *Theoretical Principles and Methods for Wind Sea Calculation* (Gidrimeteoizdat, Leningrad) pp. 167–202, in Russian

Lavrenov I.V., 1989a: Significance of weak non-linear interactions in forming a wind sea on horizontal non-uniform current, in *Problems of Complex Automation of Hydrophysical Research*, Thes. rep. conf. (Marine Hydrophysical institute, Sevastopol), in Russian

Lavrenov I.V., 1989b: Wind sea development on a current, Meteorology and Hydrology, **4**, 78–87, in Russian

Lavrenov I.V., 1990a: Spectral model of swell propagation in oceans, in *Research on Oceanographic Processes in the Tropical Zone of the Pacific Ocean* (Gidrometeoizdat, Leningrad), pp. 35–41, in Russian

Lavrenov I.V., 1990b: The overfall spectral model, in *Advanced Experimental Techniques and Method in Ship Hydro- and Aerodynamics* 19th session scientific and methodological seminar on ship hydrodynamics, 1–6 October (Varna) pp. 31-1–31-8

Lavrenov I.V., 1991a: Weak non-linear evolution of wave spectrum in shallow water, Proc. Acad. Sc. USSR, FAO, **27**, **12**, p. 112–118, in Russian

Lavrenov I.V., 1991b: Weak non-linear interaction in rips spectrum, Proc. Acad. Sc. USSR, FAO, **27**, **4**, pp. 438–447, in Russian

Lavrenov I.V., 1992: *Mathematical Modeling of Wind Sea in the World Ocean under Conditions of its Spatial Non-uniformity.* Author's theses for phys. math. sc. doctor's degree (Arctic and Antarctic Res. Inst., St. Petersburg), in Russian

Lavrenov I.V., 1998a: *Wind Wave Mathematical Modeling in Non-uniforn Ocean*, (Gidrometeoizdat, St. Peterburg), in Russian

Lavrenov I.V. 1998b: Wave energy concentration in the Agulhas Current, South Africa Natural Hazards, **17**, pp. 117–127

Lavrenov I.V., 2001: Effect of wind wave parameter fluctuation on nonlinear spectrum evolution, J. Phys. Oceanogr. **31**, **4**, 861–873

Lavrenov I.V., A.V. Novakov, 2000: Numerical simulation of the interaction of gravity waves with elastic ice floes, J. Fluid Dynamics, **35**, **3**, 414–420, in Russian

Lavrenov I.V., F.J. Ocampo-Torres, 1999: Non-linear energy generation of waves opposite to the wind direction. The wind-driven air–sea interface. Proceedings of the Symposium on the Wind-Driven Air-Sea Interface. Sydney, Australia, 11–15 January 1999, 141–150

Lavrenov I.V., T.A. Pasechnik, 1989: Swell propagation calculation for ocean with registration of the Earth's surface sphericity, Meteorology and Hydrology, **6**, 73–81, in Russian

Lavrenov I.V., V.V. Ryvkin, 1986: Evolution of mean parameters of gravitational waves on the current, changing along its direction, Proc. Acad. Sc. USSR, FAO, **22**, **10**, 1089–1098, in Russian

Lavrenov I.V., V.V. Ryvkin, 1988: Interaction of wind sea and swell, in *Theoretical Principles and Methods for Wind Sea Calculation* (Gidrometeoizadat, Leningrad) pp. 167–182, in Russian

Lavrenov I.V., V.V. Ryvkin, 1989: Wind sea calculation in grid points on sphere, in *Problems of Complex Automation of Hydrophysical Research,* Rep. conf. (Marine Hydrophysical institute, Sevastopol) pp. 76–77, in Russian

Lavrenov I.V., V.V. Ryvkin, 1990: Wave transformation on non-uniform current, Water Resources, **4**, pp. 5–17, in Russian

Lavrenov I.V., T.V. Belonenko, Yu.D. Sharikov, 1992: Spatial non-uniformity of wind sea field in frontal oceanic zones, Proc. Acad. Sc. USSR, FAO, **28**, **2**, 185–195, in Russian

Lavrenov I.V., V.I. Dymov, T.A. Pasechnik, 1990: Numerical modeling of wind sea at the Bulgarian coast in the Black Sea, in *Problems of Coastal Protection of the Black Sea Near Bulgaria,* Theses of sc. techn. conference (Varna), pp. 58–59, in Russian

Le Blond P.H., A. Maysek, 1981: *Waves in Oceans. Parts 1,2.* (Mir, Moscow) pp. 360–480, in Russian

Leykin I.A., A.S. Monin, 1985: On rip spectra, Papers Acad. Sc. USSR, **284**, **3**, 1435–1439, in Russian

Lionello P., H. Gunther, P.A.E.M. Janssen, 1992: Assimilation of altimeter data in global third generation wave model, J. Geophys. Res., 14,453–14,474

Liu A.K., E. Mollo-Christensen, 1988: Wave propagation in a solid ice pack. J. Phys Oceanogr., **18**, 1702–1712

Liu A.K., B. Holt, P.W. Vachon, 1991: Wave propagations in the marginal ice zone: Model predictions and comparisons with buoy and synthetic aperture radar data, J. Geophys. Res., **96**, **C3**, 4605–4621

Long R.B., 1973: Scattering of surface waves by on irregular bottom, J. Phys. Oceanogr., **78**, 7861–7870

Longuet-Higgins M.S., 1957: On the transformation of a continues spectrum by refraction, Proc. Camb. Soc., **53**, 226–229

Longuet-Higgins M.S., 1962: Statistical analysis of a random moving surface, in *Wind Waves* (Inostrannaya literatura, Moscow) pp. 125–218, in Russian

Longuet-Higgins M.S., 1969: A nonlinear mechanism for the generation of sea waves, Proc. Roy. Soc., **A311**, 371–389

Longuet-Higgins M.S., 1976: On the nonlinear transfer of energy in the peak of a gravity-wave spectrum, Proc. Roy. Soc., **A374**, 311–328

Longuet-Higgins M.S., R.W. Stewart, 1960: Changes in the form of short gravity waves on long waves and tidal current, J. Fluid Mech., **8**, 565–583

Longuet Higgins M.S., R.W. Stewart, 1961: The changes in the amplitude of short gravity waves on steady non-uniform current, J. Fluid Mech., **10**, 529–549

Longuet-Higgins M.S., R.W. Stewart, 1962: Radiation stress and mass transport in gravity waves, with application to "surf beats", J. Fluid Mech., **13**, 481–504

Longuet-Higgins M.S., R.W. Stewart, 1964: Radiation stress in water waves, physical discussion with applications, Deep Sea Res., **11**, 529–562

Longuet-Higgins M.S., Cartwright, N.D. Smith, 1963: Observation of the directional spectrum of sea waves using the motion of a floating buoy, in Proc. Conf. Ocean Wave Spectra (Easton) pp. 111–132

Ludwig D., 1966: Uniform asymptotic expansions at a caustic, Com. Pure. Appl. Math., **19**, 215–250

Madsen P.A., O.R. Sorensen, 1993: Bound waves and trial interaction in shallow water, Ocean Engineering, **20**, **4**, 359–388

Makin V.K., 1987: Wave flows of impulse in a boundary layer above sea waves, Oceanology, **27**, 176–183, in Russian

Makin V.K., 1989: *Dynamics and Structure of Near-Water Layer Above a Sea.* Author's theses for phys. math. sc. doctor's degree (Inst. Oceanology, Acad. Sc. USSR, Moscow), in Russian

Makin V.K., V.D. Chalikov, 1986: Calculation of impulse and energy flows in developing waves, Proc. Acad. Sc. USSR, FAO, **23**, **12**, 1309–1316, in Russian

Makkaveyev V.M., 1937: On processes of increasing and decreasing small amplitude waves and about their dependence on distance in wind direction, Trans. State Hydrological Inst., **5**, 17–25, in Russian

Mallory J.K., 1974: Normal waves on the south-east of South Africa, Inst. Hydrog. Rev., **51**, 89–129

Marchenko A.V., 1988: On long waves in shallow water under ice cover, Applied Mathematics and Mechanics, **52**, **2**, 230–234

Marchenko A.V., 1997: Diffraction of flexural-gravity waves by linear inhomogeneities in an ice cover. Izv. Ros. Akad. Nauk. Mech. Zhidk. Gaza. **4**, 97–112, in Russian

Marchenko A.V., K. Volitak, 1997: Surface wave propagation in shallow water beneath an inhomogeneous ice cover. J. Phys Oceanogr., **27**, 1602–1613

Marchenko A.V., P. Purini, K. Voliak, 1995: Filtering surface waves by ice floes. In *Proc. 13th Intern. Conf. on Port and Ocean Eng. Under Arctic Cond. (POAC '95, Murmansk)*, **3** (St. Petersburg), pp. 134–142, in Russian

Marine Meteorology and Related Oceanographic Activities, 1986: Rep. 12, Supplement, pp. 1–44

Mase H., 1989: Goupiness factor and wave height distribution. J. Wat. Port Coast and Ocean Eng., **115**, **1**, 105–121

Maslov V.P., M.V. Fedoryuk, 1976: *Quasi-Classical Approximation for Quantum Mechanics Equations* (Nauka, Moscow), in Russian

Mass Ye.I., I.G. Kantargi et al., 1988: Calculation method for wind waves in big channels, Water Resources, **1**, 60–67, in Russian

Mass Ye.I., I.G. Kantargi, K. Marinov, 1990: Basic mathematical models and applied programs in marine coastal protection, in *Coastal Protection-89* (Sofia) pp. 41–45, in Russian

Massel S.R., 1996: Ocean surface waves: their physics and prediction, in *Advanced Series on Ocean Engineering* **11** (Singapore–New Jersey–London–Hong Kong)

Masson D., P.H. Le Blond, 1989: Spectral evolution of wind-generated surface gravity waves in a dispersed ice field, J. Fluid. Mech., **202**, 43–81

Masuda A., 1981: Nonlinear energy transfer between wind waves, J. Phys. Oceanogr., **10**, 2082–2093

Matushevskiy G.V., 1972: Calculation of wind wave mean period at a complicated shore line, Trans. State Oceanographic Inst., **112**, 65–71, in Russian

Matushevskiy G.V., 1975: Radiation tensions and mean wave level of irregular sea waves in coastal shallow water, Proc. Acad. Sc. USSR, FAO, **11**, **1**, 75–82, in Russian

Matushevskiy G.V., 1985: *Method of Determining Climatic Characteristics of Wind Sea and Estimation of their Validity* (VNIIGMI-MTsD, Obninsk), in Russian

Matushevskiy G.V., I.M. Kabatchenko, 1989: Parametric integral model of wind sea agreed to All-Union wave standards, Marine Hydrophysical Journal, **1**, 24–29, in Russian

Matushevskiy G.V., I.M. Kabatchenko, 1991: United parametric integral model of wind sea and its application, Meteorology and Hydrology, **5**, 45–50, in Russian

Matushevskiy G.V., G.V. Rzheplinskiy, L.N. Ikonnikova, 1979: Calculation of wind sea regime, in *Methodical Recommendations*, **42**, (State Oceanographic Inst., Moscow), in Russian

Mazova R.Kh., Ye.N. Pelinovskiy, 1982: Linear theory of tsunami wave running on a shore, Proc. Acad. Sc. USSR, FAO, **18**, **2**, 166–177, in Russian

Mc. Kee W.D., 1974: Waves on a shearing current: a uniformly valid asymptotic solution, Proc. Camb. Phil. Soc., **75**, 295–301

Methods of Examining Marine Currents from an Airplane, 1964, V.G. Zdanovich(Ed) (Acad.Sc.USSR, Moscow-Leningrad), in Russian

Meylan M.,V.A. Squire, 1994: The response of ice floe to ocean waves, J. Geophys. Res., **99**, C1, 891–900

Miles J.W., 1957: On the generation of surface waves by shear flow, J. Fluid Mech., **3**, 185–204

Miles J.W., 1960: On the generation of surface waves by turbulent shear flow, J. Fluid Mech., **7**, 469–478

Miller, H.C., C.L. Vincent, 1991: FRF spectrum: TMA with Kitaigorodskiis scaling. J. Water. Port Coastal Ocean Eng. **116**, 87–78

Mitsuyasu H. et al., 1980: Observation of the power spectrum of ocean waves using clover-leaf buoy, J. Phys. Oceanogr., **10**, 286–296

Mitsuyasu H., Rikiishi, 1975: On the growth of duration limited wave spectra, Rep. Res. Inst. Appl. Mech. Kyushu Univ., **23**, 31–60

Monin A.S., A.M. Yaglom, 1992: *Statistical Hydromechanics. Vol. 1* (Gidrometeoizdat, St. Petersburg), in Russian

Monin A.S., A.M. Yaglom, 1996: *Statistical Hydromechanics. Vol. 2* (Gidrometeoizdat, St. Petersburg), in Russian

Munk W.V., G.R. Miller, F.E. Snodgrass, N.F. Barber, 1963: Directional recording of swell from distant storms, Phil. Trans. Roy. Soc., Ser. A, **255**, 505–584

Non-Linear Waves. Propagation and Interaction 1981: A.V. Goponov-Grekhov (Ed.) (Nauka, Moscow), in Russian

Nikolayeva Yu.I., L.Sh. Tsimring, 1986: Kinetic model of wind wave generation by turbulent wind, Proc. Acad. Sc. USSR, FAO, **22**, **2**, 135–142, in Russian

Ocean Wave Modelling (SWAMP Group), 1985 (Plenum press, New York)

Ostrovskiy L.A., Ye.N. Pelinovskiy, 1975: Refraction of non-linear sea waves in a coastal zone, Proc. Acad. Sc. USSR, FAO, **11**, **1**, 75–82, in Russian

Pasechnik T.A., 1975: Numerical analytical algorithm of integrating the wave energy balance equation, in *Complex Research in the World Ocean* (Nauka, Moscow), pp. 88–91, in Russian

Pelinovskiy Ye.N., 1979: *Wave Propagation in a Statistical Non-Uniform Ocean. Non-Linear Waves* (Nauka, Moscow), in Russian

Peregrine D.H., 1976: Interaction of water waves and currents, Adv. Appl. Mechanics, **16**, 10–117

Peregrine D.H., R. Smith, 1975: Stationary gravity waves on non-uniform free streams: jet-like streams, Math. Proc. Camb. Phil. Soc., **77**, 415–438

Phillips O.M., 1957: On the generation of waves by turbulent wind, J. Fluid Mech., **2**, 417–445

Phillips O.M., 1958: The equilibrium rang in the spectrum of wind-generated waves, J. Fluid Mech., **4**, 426–434

Phillips O.M., 1960: On the dynamics of unsteady gravity waves of finite amplitude. Part 1, J. Fluid Mech., **9**, 193–217

Phillips O.M., 1980: *Dynamics of Upper Oceanic Layer* (Gidrometeoizdat, Leningrad), in Russian

Phillips O.M., 1985: Spectral and statistical properties of the equilibrium range in wind-generated gravity waves, J. Fluid. Mech., **156**, 505–531

Pierson W.J., L. Moskowitz, 1964: A proposed spectrum for fully developed wind seas based on the similarity theory of S.A.Kitaigorodskii, J. Geophys. Res., **69**, **24**, 5181–5190

Pierson W.J., G. Neuman, R.W. James, 1955: Practical method for observing and forecasting ocean waves by means of wave spectra and statistic, US Navy Hydrog. Office, **603**

Pierson W.J., L.J. Tick, L. Baer, 1966: Computer based procedures for preparing global wave forecasts and wand field analysis capable of using wave data obtained from a spacecraft, 6th Symp. on Naval Hydrodynamics, pp. 499–532

Plate E.J., M.J. Trawle, 1970: A note on the celerity of wind waves on a water current, J. Geophys. Res., **75**, 3537–3544

Pokazeyev K.V., A.D. Rosenberg, 1983: On observing the blockading effect of surface gravitational capillary waves by non-uniform current, Proc. Moscow State Univ., Series of Physics and Astronomy, **24**, **3**, 72–76, in Russian

Pokazeyev K.V., A.D. Rosenberg, M.V. Solntsev, 1986: Laboratory examination of weak wind sea on currents, Marine Hydrophysical Journal, **3**, 39–45, in Russian

Polnikov V.G., 1988: Analysis of non-linear transfer features of wave energy and its parametrization, VINITI on Oct. 18, 1988, N 7510-B88 (Sevastopol), in Russian

Polnikov V.G., 1989: Calculation of energy non-linear transfer in the surface gravitational wave spectrum, Proc. Acad. Sc. USSR, FAO, **25**, **11**, 1214–1225, in Russian

Polnikov V.G., 1990: Numerical solution of the kinetic equation for surface gravitational waves, Proc. Acad. Sc. USSR, FAO, **26**, **2**, 168–176, in Russian

Polnikov V.G., 1991: Spectral model of the third generation for wind waves, Proc. Acad. Sc. USSR, FAO, **27**, **8**, 867–878, in Russian

Polnikov V.G., 1993: Numerical formation of flux spectrum of surface gravity waves, Izv. Acad. Sci. USSR, ser. Phys. Atmosph. Ocean, **29**, N6, 1214–1225

Polnikov V.G., 1999: Numerical study of the equations of nonlinear swell in the directional approximation, Proc. Acad. Sc., FAO, **35**, **3**, 364–370, in Russian

Problems of Research and Mathematical Modeling of Wind Sea, 1995, I.N. Davidan (Ed.) (Gidrometeoizadat, St. Petersburg), in Russian

Rabinovich A.B., 1993: *Long Gravitational Waves in the Ocean: Capture, Resonance, Radiation* (Gidrometeoizdat, St. Petersburg), in Russian

Radder A.C., 1979: On the parabolic equation method for wave propagation, J. Fluid Mech., **95**, 159–176

Rayleigh Lord, 1876: On waves, Phil. Mag., **1**, 257–279

Resio D.T., 1981: The estimation of wind-wave generation in a discrete spectral model, J. Phys. Oceanogr., **11**, 510–525

Resio D.T., W. Perrie, 1991: A numerical study of nonlinear fluxes due to wave–wave interaction. Part 1. Methodology and basic results, J. Fluid. Mech., **223**, 603–629

Ris R.C., 1997: Spectral modelling of wind waves in coastal areas, Communications on Hydraulic and Geotechnical Engineering, June, TUDelft, **97-4**

Ris, R.C., N. Booij, L.H. Holthuijsen, 1999. A third-generation wave model for coastal regions. Part II, Verification. J. Geophys. Research, **104**, c4, 7667–7681

Rogers E.W., J.M. Kaihatu, N. Booij, L.H. Holthuijesen, 1999. Improving the numerics of a third-generation wave action model. Naval Research Laboratory, Stennis Space Center, MS 39529-5004

Romanova N.N., V.I. Shrira, 1988: Explosive generation of surface waves by wind, Proc. Acad. Sc. USSR, FAO, **24**, 7, 723–734, in Russian

Rosenberg A.D., 1986: Weak wind waves on currents in a laboratory trough, Oceanology, **XXVI**, 6, 126–132, in Russian

Rozhkov V.A., Yu.A. Trapeznikov, 1990: *Probabilistic Models of Oceanological Processes* (Gidrometeoizdat, Leningrad), in Russian

Russian State Building Standards and Rules, 1983: Loading on Hydrotechnical Structures (Moscow, Stroiizdat)

Ryabinin V.E., 1991a: Semi-Lagrangian algorithms for discrete spectral models of wind sea, Meteorology and Hydrology, **8**, 72–83, in Russian

Ryabinin V.E., 1991b: Calculation of wind sea energy propagation in discrete spectral models, Meteorology and Hydrology, **9**, 73–79, in Russian

Ryvkin V.V., 1990: Numerical realization of discrete-spectral model of wind sea. Author's theses for phys. math. sc. bachelor's degree, Arctic and Antarctic Res. Inst, Leningrad, in Russian

Rzhanitsyn N.A., V.S. Altunin, T.G. Voynich-Syanozhentskiy, Ye.I. Mass et al., 1988: *Recommendations of Hydraulic Calculation for Big Channels* (GKNT of the USSR, Soyuzgiprovodkhoz, Moscow), in Russian

Rzheplinskiy G.V., Yu.M. Krylov, G.V. Matushevskiy et al., 1969: New method for calculation and analysis of wind waves, Trans. State Oceanographic Inst., **93**, 5–52, in Russian

Sakai T., J. Iwagaki, 1983: Irregular wave refraction due to current, J. Hydraulic Eng., **109**, 1203–1215

Sand S., 1982: Wave grouping described by bundled long waves. Ocean Eng. (Gr. Brit.), **9**, 6, 567–579

Sanderson R.N., 1974: The unusual waves off South East Africa, Marine Observer, **44**, 180–183

Schuman E.H., 1976: High waves in the Agulhas Current, Mariners Weather, **20**, 1, 1–5

Shaw R.P., W. New, 1981: Long-wave trapping by oceanic ridges, J. Phys. Oceangr., Vol.**11**, 1334–1344

Shemdin O.H., K. Hasselmann, S.V. Hsiao, K. Herterich, 1978: Nonlinear and linear bottom interaction effects in shallow water, in *Turbulent Fluxes Through the Sea Surface* (Plenum Press, New York) pp. 347–372

Shuchman R.A., C.L. Rufenach, 1994: Extraction of marginal ice thickness using gravity imagery, J. Geophys. Res, **99**, C1, 901–918

Shuleykin V.V., 1959: Physical principles of wind sea prediction in the ocean, Proc. Acad. Sc. USSR, Geophys. Series, **5**, 710–724, in Russian

Skop R., 1987: Approximate dispersion relation for wave–current interactions, J. Waterway, Port, Coast and Eng., **113**, 2, 187–195

Smith R., 1976: Giant waves, J. Fluid. Mech., **77**, 417–431

Smirnov V.N., 1996: *Dynamical Processes in Sea Ice.* Hydrometeoizdat, St. Petersburg, 162p

Snodgrass F.E. et al., 1966: Propagation of ocean swell across the Pacific, Phil. Trans. Roy. Soc., **259(a)**, **1103**, 256–271

Snyder R.L., C.S. Cox, 1966: A field study of the generation of ocean waves, J. Mar. Res., **24**, 141–178

Snyder R.L., F.W. Dobson, J.A. Elliott, R.B. Long, 1981: Array measurements of atmospheric pressure fluctuation above surface gravity waves, J. Fluid Mech., **102**, 1–59

Snyder R.L., W.C. Thacker, K. Hasselmann, S. Hasselmann, G. Barzel, 1993: Implementation of an efficient scheme for calculating nonlinear transfer from wave–wave interaction, J. Geoph. Res., **98**, 14507–145245

Squire V.A, 1984: How waves break up inshore fast ice. J. Polar Record., **22**, **138**, 281–285

Sretenskiy L.N., 1977: *Theory of Wave Movements in a Liquid* (Nauka, Moscow), in Russian

Stevenson T., 1852: Observation on the force of waves, New Edin. Phil. J., **53**

Stewart R.H., J.W. Joy, 1974: HF Radio measurements of surface currents, Deep-Sea Res., **21**, 1039–1049

Strekalov S.S., 1986: Boundary layer of the atmosphere, in *Wind, Waves and Marine Harbours* (Gidrometeoizdat, Leningrad) pp. 6–62, in Russian

Sturm H., 1974: Giant waves, Ocean, **2**, **3**, 98–101

Sturova I.V., 1998: The oblique incidence of surface waves onto an elastic band. Proc. 2nd Intern. Conf. on Hydroelasticity in Marine Technology, Fukuoka, Japan, 1–3 Dec., Fukuoka Res. Inst. Appl. Mech. Kysushu Univ., pp. 239–245

Sturova I.V., 2000: Surface wave diffraction with non-uniform elastic plate. Appl. Mech. Tech. Phys. V**41**, N 4, 42–48

Sudolskiy A.S., V.I. Teplov, 1982: *Wave Generation on Flowing Water and Kinematic Parameters of a Flow*, in Papers of Conferences and Meetings in Hydrotechnics. Methods of Research and Calculation for Wave Forcing onto Hydrotechnical Constructions and Shores (Energoizdat, Leningrad) pp. 38–39, in Russian

Sverdrup G., W. Munk, 1961: Wind, wind sea and swell. Theoretical bases of prediction, in *Bases of Wind Wave, Swell and Surf Prediction* (Inostrannaya literatura, Moscow) pp. 15–87, in Russian

SWIM Group. A shallow water intercomparison of three numerical wave prediction models, 1985, Quart. J. Roy. Meteor. Soc., **111**, 1087–1112

Taylor, Sir Geoffrey, 1955: The action of a surface current used as a breakwater, Proc. Royal Soc., Ser. A, **231**, 466–478

Teplov V.I., 1980: Laboratory research of changing wind wave elements, depending on current velocity and wind speed, Trans. State Hydrological Inst., **263** 94–112, in Russian

Teplov V.I., (1989): Regularities of wind wave generation on flows and method for calculating wave transformation by currents. Author's theses for techn. sc. bachelor's degree, State Hydrol. Inst., Leningrad, in Russian

The Atlas of the Oceans. The Atlantic and Indian Oceans, 1977, (GUNiO MO, Leningrad) pp. 204–207, in Russian

Theoretical Bases and Methods for Wind Sea Calculation, 1988, I.N. Davidan (Ed.), (Gidrometeoizdat, Leningrad), in Russian

The WAM Model – A Third Generation Ocean Wave Prediction Model. WAMDI group, 1988, J. Phys. Ocean., **12**, 1775–1810

Thomas G.P., 1979: Water wave-current interaction, in Review Mech. Wave Induced Forces Cylinder. Symp. Bristol, 1978 (San-Francisco), pp. 179–204

Thomas G.P., 1981: Wave–current interactions: an experimental and numerical study. Part I. Linear waves, J. Fluid Mech., **110**, 457–474

Thornton E.B., R.T. Guza, 1983: Transformation of waveheight distribution, J. Geophys. Res., **88**, 5925–5938

Tolman H.L., 1991: A third-generation model for wind waves on slowly varying, unsteady and inhomogeneous depths and current, J. Phys. Ocean., **21**, **6**, 782–797

Tolman H.L., 1992: Effect of numerics on the physics in a third-generation windwave model, J. Phys. Ocean, **22**, 1095–1111

Tolman H.L., D. Chalikov, 1996: Source terms in a third-generation wind wave model, J. Phys. Ocean, **26**, **11**, 2497–2518

Tsimring L.Sh., 1989: Forming a narrow angular spectrum of wind sea with interaction of waves and wind, Proc. Acad. Sc. USSR, FAO, **25**, **4**, 411–420, in Russian

Tsuruya H., S. Nakano, S. Yanagishima, Y. Matsunobu, 1987: Development of wind waves generated on adverse currents, Rep. Port Harbor Research Inst., **26**, **4**, 35–56

Tubman M.W., J.N. Suhayda, 1976: Wave action and bottom movements in fine sediments, Proc. 15th Coastal Eng. Conf., **2**, 1168–1183

Uizem J., 1977: *Linear and Non-Linear Waves* (Mir, Moscow), in Russian

Uji T., 1975: Numerical estimation of sea waves in typhoon waves, Jap. Meteorol. Geophys., **26**, **4**, 199–217

Ursell F., 1956: Wave generation by wind, in *Surveys in Mechanics* (University Press, Cambridge), pp. 216–249

Van Vledder G.Ph., J.G. de Ronde, M.J.F. Stive, 1994: Performance of a spectral wind-wave model in shallow water, in *Proc. 24th Int. Conf. Coast. Eng. ASCE*, pp. 753–762

Vakhrameyeva L.A., L.M. Bugayevskiy, Z.A. Kazakova, 1986: *Mathematical Cartography* (Nedra, Moscow), in Russian

Vaynberg B.R., 1982: *Asymptotic Methods in Mathematical Physics Equations* (Moscow State University, Moscow), in Russian

Vilenskiy Ya.G., B.Kh. Glukhovskiy, 1955: Some regularities of wind wave development, Proc. State Oceanography Institute, **21**, N 41 (in Russian)

Vilenskiy Ya.G., B.Kh. Glukhovskiy, 1961: Wind sea in oceans. The results of researches and observations of waves and wind in the Northern Atlantic Ocean, Trans. State Oceanographic Inst., 62, in Russian

Voganov R.B., B.Z. Kantsenbaum, 1982: *Principles of Diffraction Theory* (Moscow, Nauka), in Russian

Voronovich A.G., 1976: Propagation of internal and surface waves in the geometrical optics approximation, Proc. Acad. Sc. USSR, FAO, **12**, **8**, 850–857, in Russian

Voronovich A.G., V.V. Goncharov, 1982: The influence of large-scale oceanic movement on internal wave propagation, Proc. Acad. Sc. USSR, FAO, **18**, **1**, 79–87, in Russian

Wadhams P., A. Squire, J.A. Ewing, R.W. Paskal, 1986: Effect of the marginal ice zone on the directional wave spectrum of the ocean, J. Phys. Oceanogr., **16**, C1, 901–918

Webb D.J., 1978: Nonlinear transfer between sea waves, Deep-Sea, **25**, 279–298

Weber S.L., 1991: Bottom friction for wind sea and swell in extreme depth-limited situations, J. Phys. Oceanogr., **21**, 149–172

Wells J.T., 1983: Dynamics of coastal fluid muds in low-, moderate- and high-tide-range environments, Can. J. Fish. Aquat. Science., **40**, 130–142

Whitham B.B., 1965: A general approach to linear dispersive waves using a Langangian, J. Fluid Mech., **22**, 273–283

Whitnam G.B., 1974: *Linear and Non-linear Waves* New York, Wiley

WMO Wave Programme. WMO(TD –N 35), 1991, Report for 1989 to 1990 on Wave Measuring Techniques, Numerical Wave Models and Intercomparisons, Marine Meteorology and Related Oceanographic Activities, **12**, Supplement N 3

Yefimov V.V., 1981: *Wave Process Dynamics in Boundary Layers of Atmosphere and Ocean* (Naukova Dumka, Kiev), in Russian

Yefimov V.V., V.G. Polnikov, 1985: Numerical experiments in simulating a wind sea, Oceanology, **25**, **5**, 725–732, in Russian

Yefimov V.V., V.G. Polnikov, 1991: *Numerical Simulation of Wind Sea* (Naukova Dumka, Kiev), in Russian

Yefimov V.V., Yu.P. Soloviev, 1979: Frequency-angular spectra and dispersion relation in wind waves, Proc. Acad. Sc. USSR, FAO, **15**, **11**, 1181–1196, in Russian

Yefimov V.V., Yu.P. Soloviev, 1984: Low frequency sea level oscillations and wind wave group structure. Izv. Acad. Sci. USSR. Atmos. and Oceanic Phys., **20**, **10**, 985–994, in Russian

Yefimov V.V., B.B. Krivitskiy, Yu.P. Solovyev, 1986: Research of sea wind wave energy in dependence on fetch, Meteorology and Hydrology, **11**, 68–75, in Russian

Yefimov V.V., V.G. Polnikov, Ye.N. Sychev, 1986: *Spectral Model and Numerical Realization of Experiment on its basis*(Marine Hydrophysical Institute, Sevastopol), in Russian

Young I., 1999: *Wind Generated Ocean Waves*, Elsevier

Young I.R., S. Hasselmann, K. Hasselmann, 1987: Computations of the response of wave spectrum to a sudden change in the wind direction, J. Phys. Oceanogr., **17**, 1317–1358

Yuen G., B. Lake, 1987: *Non-linear Dynamics of Gravitational Waves in Deep Water* (Mir, Moscow), in Russian

Zakharov V.Ye., 1968: Stability of finite amplitude periodic waves on deep liquid surface, Applied Mechanics and Technical Physics, **2**, 86–91, in Russian

Zakharov V.Ye., 1969: Stability of periodic waves of final amplitude on a deep liquid surface. PMTF, **2**, 86–94

Zakharov V.Ye., N.N. Filonenko, 1966: The energy spectrum for stochastic oscillation of a fluid surface, Doklady Akademii Nauk, **170**, 1292–1295, in Russian

Zakharov V.Ye., V.I. Shrira, 1990: On forming the angular spectrum of wind waves, ZhETF, 98, 6(12), 1941–1958,

Zakharov V.Ye., A.V. Smilga, 1981: On the quasi-one-dimensional spectra of weak turbulence, ZhETF, **4**, 1318–1326, in Russian

Zakharov V.Ye., M.M. Zaslavskii, 1982: Kinetic equation and Kolmogorov's spectra in week turbulent theory of wind waves, Proc. Acad. Sc. USSR, FAO, **18**, **9**, 970–980, in Russian

Zakharov V.Ye., M.M. Zaslavskii, 1983a: Form of energy spectral components, Proc. Acad. Sc. USSR, FAO, **19**, **3**, 282–290, in Russian

Zakharov V.Ye., M.M. Zaslavskii, 1983b: Wave parameter dependence on wind velocity, duration and fetch in week turbulent theory of wind waves, Proc. Acad. Sc. USSR, FAO, **19**, **4**, 406–415, in Russian

Zakharov V.Ye., A.N.Pushkarev, 1999: Diffusion model of interacting gravity waves on the surface of deep water fluid, Nonlinear Processes in Geophys., **6**, 1-10.

Zambresky L.F., 1989: A verification study of the global WAM model December 1987–November 1988, in *ECMWF Technical Report 63. Reading, ECMWF*

Zaslavskii M.M., 1989a: On long-wave cutting of wind wave spectrum, Proc. Acad. Sc. USSR, FAO, Vol.**25**, **11**, 1187–1194, in Russian

Zaslavskii M.M., 1989b: On the narrow-directed approximation of the kinetic equation for the wind wave spectrum, Proc. Acad. Sc. USSR, FAO, **25**, **4**, 402–410, in Russian

Zaslavskii M.M., 1997: Relation between narrow-directional, narrow-band and diffusive approximations of the nonlinear kinetic integral for surface gravity waves, Oceanology, **37**, **3**, 331–337, in Russian

Zaslavskii M.M., 1998: Quasi-kinetic approach for wind wave in shallow water, Proc. Acad. Sc., FAO, **34**, 3, 364–370 (in Russian)

Zaslavskii M.M., 2000: Nonlinear evolution of the spectrum of a swell, Proc. Acad. Sc., FAO, **36**, **2**, 275–283, in Russian

Zaslavskii M.M., V.P. Krasitskii, 1976: On wind wave generation in a sea with finite depth, Oceanology, **16**, **2**, 207–211, in Russian

Zaslavskii M.M., V.P. Krasitskii, 1993: About wave fluctuations of wind wave spectral parameters, Oceanology, **33**, **1**, 21–26, in Russian

Zaslavskii M.M., V.G. Polnikov, 1998: Three waves quasikinetic approximation of nonlinear spectrum evolution in shallow water, Proc. Acad. Sc., **34**, **5**, 677–685, in Russian

Zhevnovatiy V.G., 1971: Wind waves on currents, Nature and Economy of the North, **3**, 68–75, in Russian

Zubakin G.K., 1976: Calculation of wind wave parameters in a sea covered with broken ice, Trans. Hydrometorol. Centre, USSR, **164**, 11–19, in Russian

6th *International Workshop on Wave Hindcasting and Forecasting. Preprints.* Monterey, California, Nov. 6–10, 2000

Index

abnormal waves (*see also*, freak waves, killer waves, and rogue waves) 217, 225
action (*see also* wave action) 21, 25
adiabatic invariant 3, 20, 225
Adriatic Sea 140
Agulhas Current 217
air flow (over waves) 11–13, 247
Airy function 196, 198
angular distribution: *see* directional 130, 161, 179, 237, 255, 277, 296, 309
approximation to non-linear energy transfer
 diffusion 84
 discrete interaction 83
Arctic Sea 315
ARSLOE experiment 306
Atlantic Ocean 49, 141
atmospheric boundary layer 13, 317

Bering Sea 40
Bessel function 150, 259
Black Sea 39, 302
BOMEX experiment 104
bottom
 dissipation 28, 34
 effect 285
 friction 7, 34, 285, 301
 relief 254, 293
 roughness 292
 scattering 291
 stress 288
boundary
 condition 13, 18–20, 39, 60, 227, 241, 288, 315, 319
 layer 3, 4, 12, 25, 125, 138, 140, 285, 286

Boussinesq equations 260, 261
BSIE (Black Sea International Experiment) 302
Bubnov–Galerkin method 339

canonical (conjugate) variables 325, 331
cape roller (*see also* abnormal waves, freak waves, and rogue waves) 217
Cartesian system 12
caustic 153, 185, 190, 192, 214, 236, 260, 294
collision integral 3, 5, 7, 26, 83, 90, 105, 261
conservation law 11, 18, 20, 66
conservation of
 mass 11, 268
 momentum 11
 wave action 22
 wave density 15
coordinates 16, 343
 Cartesian 12, 22
 cyclic 26
 immovable 241, 244
 local 153, 254, 268
 polar 108
 rectangular 35, 38
 spherical 28
Coriolis force 11, 13, 18
Courant–Friedrich–Lewy (CFL) criterion 51, 54, 64, 79

deformation velocity tensor 12
density 12, 225, 297
 of ice 315
 of water 11
diffraction of waves 190

diffusion
  numerical   61
  operator approximation of non-linear transfer   84
directional distribution   255
discrete interaction approximation   7, 83
dispersion   130, 142, 267, 272
dispersion ratio   17, 19, 156, 233, 262, 269, 291
dissipation   2, 8, 28, 176, 289, 291
  due to bottom friction   258
  due to percolation   289
  due to whitecapping   128, 133
  of shallow-water waves   286
Doppler
  frequency   19, 155
  shift due to currents   155

Earth's
  radius   9, 11, 35, 43, 52, 295
  rotation   29, 35
  sphericity   9
eigen
  frequency   31
  functions/modes   20
energy, wave energy   58, 66, 86, 129, 153, 163, 179, 187, 224
  angular distribution   93, 99
  balance equation   2, 3, 7, 35, 36, 38, 39, 49–51, 54, 56, 59, 65, 74, 75, 78, 128, 137, 142, 152, 183, 294, 296, 317, 335
  diffusion   61
  dissipation   81, 127, 132, 174, 229, 287, 289, 294, 317
  flux   4, 28, 126, 183, 248, 255
  input   81, 138, 247
  non-linear transfer (between waves)   81, 90, 92, 96, 101, 108
  propagation of   79
  spectrum: see spectrum   200, 254, 294
  transfer (from wind to waves)   152, 176, 198, 263
equilibrium (range)   3, 69, 129, 130, 135, 163, 164, 166, 173, 176, 179, 185, 189, 267, 284, 308
Ermith function   216

Euler's method   70
exceptional waves (see also abnormal waves, freak waves, and rogue waves)   217

Feynman diagrams   28
finite difference method: see numerical scheme   51
floes   284, 335
four-wave interaction (see also non-linear energy transfer)   3, 267
Fourier
  integral   194, 292, 328
  transformation   24, 326
freak waves (see also abnormal waves, exceptional waves and rogue waves)   217
frequency   14, 51, 132, 292
  intrinsic   31
friction
  force   12
  velocity   12, 123, 139, 239

garden-sprinkler effect   50, 51, 54, 56, 62
Gaussian
  distribution   24, 139
  process   24
general problem formulation   11, 35, 253
geodesic line   36
geometric(al) optics approximation   5, 14, 18, 166, 193, 223, 260, 317
global distance   35
gravity
  acceleration   11, 336
  waves   335, 342
Green formula   258, 281
Green function   335
grid point   49, 51, 53, 59, 78, 263, 297, 300
group velocity   17, 18, 33, 52, 153, 158, 178, 188, 193, 214, 229, 241, 257, 258, 271, 296, 316, 320
gustiness effect   138

Hamilton–Jacobi equation   16, 17
Hamiltonian equations (of motion)   17, 25, 153, 154, 223, 317, 319

Hamiltonian function 16, 18, 26, 27, 29, 325, 326
Hasselmann equation 105
Hasselmann model 128
Heaviside function 40, 86, 162, 169, 203, 275, 321
Helmholtz equation 191, 194

ice
  cake 6, 316, 332
  field 6, 335
  floe 6, 335, 340, 342
  marginal ice zone (MIZ) 335
  scattering of waves 6
implicit
  form 241
  integration scheme 68
Indian Ocean 218, 220
inertia force 11
interaction
  coefficients 267
  integral 104
INTERPOL (interpolation-ray method) 59, 60, 62, 79

Jacobi weight functions 87
Jacobian 31, 154, 160, 195
JONSWAP 4, 90, 92, 94, 102, 126, 144, 151, 243, 289, 333

kinematic
  coefficient of viscosity 12
  condition 178, 211, 254, 274
  equation 15, 25, 33, 56, 138, 155, 185, 192, 237, 261, 266, 285
kinetic equation 54, 319, 331
Kolmogorov spectrum 123, 132, 176

Lagrange function 17
Laplace's equation 13, 315, 328, 335
Liouville's theorem (and conservation of wave action) 31, 154
local problem formulation 39
local sinusoidal waves 15
logarithmic boundary layer 12

mass acceleration 11
mathematical simulation 35
Miles' model 123, 132

Miles' theory (of wave generation) 3, 139, 140
MIZEX experiment 335
momentum 16, 126, 286
monochromatic wave 15, 160, 187, 196, 214, 268, 277, 286
Monte Carlo method 140
motion equation 31

Navier–Stokes equation 134
NEDWAM 39
non-linear (wave–wave) interactions: see non-linear energy transfer 5, 7, 261, 265, 266, 291
non-linear effect 254, 285
non-linear energy transfer (between waves) 3, 6, 7, 75, 132, 134, 144, 146, 149, 166, 173, 180, 261, 317, 333
non-uniform current 3, 7, 14, 18, 22, 23, 33, 153, 156, 161, 166, 173, 175, 183, 188, 193, 198, 209, 210, 212, 228, 236, 249, 268, 271, 277, 284, 293, 302, 317
non-uniform media 14, 199, 253
non-uniform ocean 11
non-uniformity 14, 56, 138, 144, 155, 209, 274, 293
North Atlantic 49
North Sea 52, 56, 60, 308
Norwegian Sea 60
numerical
  algorithm 58, 107
  diffusion 61
  grid 60
  implementation 49, 50, 58, 74, 294
  modelling 6
  simulation 25, 224, 299, 301, 335, 340, 343

Onega Bay 164
optimal algorithm non-linear energy transfer computation 85

Pacific Ocean 40, 43, 44
peakness 90, 94, 144, 149, 151
penetrability coefficient 290
phase space 26
phase velocity 256
Philips constant 276

Philips parameter   126, 135
Pierson–Moskowitz   151
   approximations   128
   spectrum   128, 166, 342
Poisson brackets   27
potential   12
predictor–corrector method   75, 146
probabilistic methods   4
propagation   6, 14, 18, 22, 35, 39, 43, 44, 47, 53, 156, 167, 175, 185, 202, 253, 277, 291, 317

quasi-oscillations   137

radiation stress   3
random
   distribution   128
   fluctuations   13
   functions   2
   process   2, 23, 24, 137, 139, 152
ray   15, 17, 156, 176, 192, 223, 294
Rayleigh equation   226
Rayleigh's wave propagation   35
reflection
   coefficient   340, 344
   of waves   171, 203
refraction (of waves)   34, 155, 253, 258, 285, 293, 298
remote sensing   209
resonance   341
Riemann metrics   37
rips   164, 174, 175, 183
rogue waves (*see also* abnormal waves, exceptional waves, freak waves, and cape roller)   217
root mean square error (RMSE)   71
rotation vector   11
roughness parameter   126
Runge–Kutta method   223, 296

scale
   global   33
   local   34
   regional   33, 34
   small   34
scattering   285
Schrödinger equation   216
seiche oscillations   261
self-similar asymptotic solution   120

semi-implicit
   method   107
   scheme   67
SMIWEH   141
soliton   166, 175
source function (source term)   28, 52, 65, 67, 68, 78, 81, 127, 132, 134, 142, 149, 153, 166, 241, 247, 254, 317
Southern Ocean   44, 47
spatial
   distribution function   52
   non-uniformity   9
spectral density   298
   function   128
spectral shape approximation
   Banner   94, 99
   Davidan   130
   Donelan   93, 99
   equilibrium (range) area   131
   JONSWAP   176
   theory   4
   TMA   307
spectrum   3, 52, 160, 208
   approximation   99, 242
   energy   168
   evolution of wave spectrum   139, 156, 171, 180, 262
   frequency   129, 130, 258
   frequency-(directional) angular   176, 254, 262, 303, 309, 320
   JONSWAP   106, 263, 307
   TMA   307
   wave   51, 60, 131, 153, 161
splitting method   73
statistical movements   24
storms   1
stress   286
surface gravity waves   6, 154
surface resistance coefficient   124
SWAMP   7, 8, 51
SWAN   8, 267, 313
swell   129, 136, 164, 193, 204, 215
   propagation (*see also* propagation)   39, 44, 50, 52, 59, 78, 223

three-wave interaction   267
transfer function   306, 323
transformation   34, 155, 177, 183, 237

transparency window 132
transport equation
  transfer equation 25
turbulence 12, 132

uneven
  bottom 291
  depth 268
  seabed 7

velocity
  current 153
  dynamic 131
  friction 123, 139
  group 56
  orbit 286
  phase 93
  potential 12, 341
  propagation 55
viscosity 12, 287
  coefficient 290, 318
von Karman constant 124
vorticity: see irrotational motion 12, 227

WAM model 6, 8, 50–53, 56, 62, 66, 79, 104, 126, 140
WAMDI 7, 8
water density 316
wave action 3, 30–32, 82, 86, 123, 132, 153, 154, 166, 176, 193, 195, 199, 218, 237, 266, 268, 294
  age 56
  amplitude 14, 193
  blocking 270, 282
  breaking 127, 135, 140, 161, 168, 179, 189, 206, 208, 254, 257, 277, 309
  crest 219
  crest collapse 3
  density 21, 25
  development 181, 245, 248
  diffraction 190
  dissipation/attenuation 144, 152, 166
  element 2, 5, 189, 210, 234, 243, 248, 277, 296, 301
  evolution 3, 5, 156, 174, 185, 188, 225, 258, 277

field 137, 166, 193, 206, 209, 214, 268, 288, 292
forecast 8
generation/growth 3, 11, 13, 225, 237, 247
group 136, 167, 238, 242
height 1, 42, 43, 52, 61, 62, 183, 186, 212, 219, 224, 234, 236, 245, 258, 266, 276, 302
interaction 3, 7, 49, 175, 218, 236, 237, 247, 261, 285, 343
length 14, 126, 184, 189, 215, 226, 245, 247, 276, 291, 302, 341
number 32, 33, 46, 185, 191, 199, 208, 214, 216, 228, 231, 236, 243, 253, 256, 266, 269, 285, 293, 296, 305, 307
packet 3, 16, 17, 30, 33, 36, 37, 39, 45, 51, 132, 153, 158, 177, 194, 198, 214, 232, 235, 253, 272
period 1, 58, 136, 187, 246, 247, 257, 266, 276, 300
prediction 2
profile 259
propagation 14, 56, 60, 62, 153, 158, 171, 175, 176, 191, 199, 227, 235, 240, 253, 260, 266, 270, 276, 279, 292, 300, 309, 319
reflection 210, 259
scattering 6, 291, 335
spectrum: see spectrum 4, 5, 152, 157, 166, 169, 171, 175, 178, 208, 224, 255, 266, 274, 284, 285, 298, 317, 335, 342
steepness 191
system 206, 208, 220
transformation 164, 189, 198, 225, 230, 247, 253, 261, 268, 281, 294, 299, 302
vector 14, 16–18, 21, 185, 199, 253, 292
wave number 316
wave packet 317
wave vector 320
wave–wave interaction (see also non-linear energy transfer) 3, 324
waves 217
WAVEWATCH 8

White Sea 164, 188, 250
wind
  fields 129, 138
  profile 127
  sea 126, 127, 131
  wave 1, 5, 23, 52, 124, 129, 136, 152, 179, 185, 206, 290
  wave calculation 2
  wave dynamics 133
  wave energy 2, 7, 124, 125, 127, 136
  wave evolution 2, 11, 136
  wave interaction 125

wave modelling 13, 35, 138, 153
wave models 4–6, 8, 11, 35, 50, 56, 133, 138, 145, 152, 166, 174, 267
wave spectrum 3, 25, 65, 81, 85, 131, 137, 145, 153, 176, 180, 199, 210, 234, 243, 302
WISE 8
WKB approximation 19

Young modulus 340

zero moment 40, 161, 242, 266, 305

Druck: Strauss Offsetdruck, Mörlenbach
Verarbeitung: Schäffer, Grünstadt